# Explainable Artificial Intelligence for Trustworthy Decisions in Smart Applications

Nicu Bizon · Bhargav Appasani
Editors

# Explainable Artificial Intelligence for Trustworthy Decisions in Smart Applications

Springer

*Editors*
Nicu Bizon
Piteşti University Centre
The National University of Science
and Technology Politehnica Bucharest
Piteşti, Romania

Bhargav Appasani
School of Electronics
Kalinga Institute of Industrial Technology
Bhubaneswar, Odisha, India

ISBN 978-3-031-97006-1 ISBN 978-3-031-97007-8 (eBook)
https://doi.org/10.1007/978-3-031-97007-8



This Springer imprint is published by the registered company Springer Nature Switzerland AG
The registered company address is: Gewerbestrasse 11, 6330 Cham, Switzerland

# Foreword

Today Artificial Intelligence (AI) is being used by a wide range of applications that include finance, urban infrastructure, education, and beyond. By harnessing the benefits of AI, these application areas are able to make decisions and recommendations with little human intervention. As a result, the demand for *transparency* in AI-driven systems has never been more pressing. The power of smart technologies stems not only from their operational efficiency and accurate predictions, but increasingly in their ability to *explain* their reasoning behind decisions and recommendations. This transparency is vital in their trustworthiness.

This edited volume, *Explainable Artificial Intelligence for Trustworthy Decisions in Smart Applications*, brings together a diverse and interdisciplinary group of researchers, practitioners, and visionary leaders who are pioneering efforts to make AI not only smarter but also more interpretable, ethical, and accountable. Across 13 comprehensive chapters, the contributors offer useful insights into both foundational principles and innovative applications of Explainable AI (XAI), paving the way toward responsible AI integration and deployment in smart systems.

The book starts with a thorough explanation of the relationship between trust and explainability in AI, followed by critical investigations of XAI models, interpretability challenges in financial technologies, and real-world implementations in smart cities. The rest of the book demonstrates how XAI enhances user understanding and confidence for application areas such as entertainment and education, energy distribution, and medical diagnostics.

This work presents extensive, rich, and interesting discussions on cutting-edge technologies such as Large Language Models (LLMs) and blockchain, and how explainability can bridge the gap between black-box AI and user-centric decision-making. Some of the chapters address the ethical issues of AI use, advocating for transparency, fairness, and regulatory oversight that are paramount in many sectors today.

This book promotes that XAI is not merely a technical enhancement but also plays a vital role in the AI revolution. The book makes a major contribution by not only addressing the "how" of explainability but also exploring the "why" aspects that include trust, accountability, and equity in the AI age.

I commend the editors and contributors for this timely and essential work. This book will surely serve as an important reference for educators, scholars, and practitioners striving to design and implement systems where smart decisions are strongly supported by explanations that are clear and easy to understand.

Sherali Zeadally, FBCS, FIET
University Research Professor
University of Kentucky Alumni Association Endowed Professor
College of Communication and Information
University of Kentucky
Lexington, USA

# Preface

Artificial Intelligence (AI) is no longer a distant promise—it is now an indispensable part of modern life. From diagnosing illnesses to managing city infrastructure, AI systems are transforming how decisions are made and services are delivered. Yet, as AI becomes more embedded in critical applications, so does the demand for *trustworthy* and *explainable* decision-making. This growing dependence on AI has brought the issue of transparency to the forefront, challenging researchers and developers to demystify the opaque nature of complex models that increasingly function as "black boxes."

This book, *Explainable Artificial Intelligence for Trustworthy Decisions in Smart Applications*, is born from the understanding that explainability is essential—not optional—for ensuring the responsible deployment of AI systems. The goal of this volume is to provide a comprehensive overview of emerging approaches and real-world applications of Explainable AI (XAI), offering a roadmap for making intelligent systems not only powerful, but also understandable and trustworthy. Through the contributions of experts across academia and industry, this collection spans theoretical frameworks, algorithmic innovations, and case-driven applications, painting a holistic picture of the XAI landscape.

Chapter 1 sets the foundation by exploring the connection between trust and explainability in smart applications. It examines the evolution of XAI and the challenges posed by black-box models, emphasizing that enhancing explainability is crucial for ethical AI adoption. Building on this, Chap. 2 introduces the concept of Verified and Explainable AI (VEAI), proposing a hybrid model called GEVAI that combines multiple levels of explainability—such as a priori, ad hoc, and ex post—in order to improve verification, efficiency, and interpretability within AI pipelines.

Chapter 3 transitions into the financial domain, where fairness and regulatory compliance are paramount. It applies SHAP and LIME techniques to credit card scoring, highlighting the impact of demographic variables and the role of XAI in mitigating bias and increasing transparency. Chapter 4 extends the conversation to smart cities, discussing the integration of XAI into urban systems for managing traffic, energy, and public services. Through real-world examples, it demonstrates how architectural models—centralized, decentralized, and layered—can make AI-driven decisions comprehensible to policymakers and citizens alike.

In Chap. 5, the focus shifts to entertainment, education, and environmental monitoring. Here, XAI is shown to play a pivotal role in content personalization, identifying biases in educational tools, and analyzing Internet of Things (IoT) data for environmental insights. Chapter 6 turns to the increasingly prominent use of Large Language Models (LLMs), especially in domains such as healthcare, finance, and education. It outlines how traditional XAI tools are being adapted—or challenged—by the complexity of these models, offering a forward-looking perspective on the evolution of explainability.

Ethical considerations take center stage in Chap. 7, which examines issues of fairness, privacy, and accountability in the implementation of XAI. It discusses the regulatory environment and underscores the importance of integrating explainability into ethical frameworks for AI. Chapter 8 introduces a unique geopolitical application of XAI, leveraging fuzzy logic and deep learning to analyze hegemonic trends in global power dynamics. By examining the developmental patterns of major nations, this chapter shows how XAI can inform strategic decision-making on an international scale.

In Chap. 9, the discussion turns to active distribution grids and the role of XAI in assisting decision-makers at energy utilities. As AI becomes instrumental in managing energy efficiency and prosumer integration, explainability is shown to be a vital asset for technical planning and public accountability. Chapter 10 continues in the domain of natural language processing, addressing the explainability of transformer-based abstractive summarization models. The chapter introduces the Fact Omission Explanation (FOE) framework, which reveals why certain facts are excluded from AI-generated summaries—an especially important advancement in mission-critical fields like law and medicine.

Chapter 11 offers insights into the design of photonic devices, such as surface plasmon resonance and tilted fiber Bragg grating sensors, using XAI techniques to understand how design parameters affect performance. This enhances both model transparency and alignment with physical principles. Chapter 12 presents a novel integration of AI and blockchain in a medical diagnosis system, offering a secure, transparent, and patient-centric platform for healthcare services that balances diagnostic accuracy with data privacy.

Finally, Chap. 13 closes the volume with a spatio-temporal platform for prioritizing high-demand resources during emergencies. Using novel clustering techniques, the framework transparently identifies vulnerable regions and explains the decision-making behind resource allocation, as exemplified in its application to COVID-19 caseload data in India.

Together, these chapters offer an expansive and deeply interdisciplinary view of how XAI is shaping the future of smart applications. From urban planning and financial regulation to national security and medical diagnostics, the importance of explainability is clear: it is not just a feature of responsible AI—it is a prerequisite. We hope this volume serves as a valuable reference for researchers, developers, and decision-makers committed to building AI systems that are not only intelligent, but also interpretable, ethical, and deserving of human trust.

We express our sincere gratitude to all contributing authors for their insights and dedication. We also thank the editorial and production team at Springer for their unwavering support throughout the development of this book.

Piteşti, Romania Nicu Bizon
Bhubaneswar, India Bhargav Appasani

# Contents

# Chapter 1
# Explainable Artificial Intelligence and Trust in Smart Applications: Definition, Evolution and Challenges

**Hatem Mosa and Qasem Abu Al-Haija**

**Abstract** This chapter addresses the critical link between Explainable Artificial Intelligence (XAI) and Trust in smart applications. As AI systems grow across multiple industries, including healthcare, banking, and transportation, the need for transparency and accountability becomes clearer. The chapter starts by describing essential topics such as smart applications, AI, and XAI, followed by a study of the emergence of XAI and its relevance in encouraging user confidence. It underlines the issues created by "black-box" models, which limit comprehension and lead to distrust among users. The chapter highlights the significance of giving explicit, interpretable explanations for AI-driven judgments through a review of several XAI strategies and their consequences for trust. Ultimately, it highlights that boosting explainability is vital for assuring AI technology's responsible and ethical deployment, driving increased acceptance and reliance on smart applications.

## 1.1 Introduction

Artificial Intelligence (AI) has emerged as a powerful force in smart applications, transforming industries and altering our everyday lives. AI systems powered by machine learning and deep learning algorithms are increasingly employed to automate processes, improve decision-making, and deliver tailored services across numerous areas. AI-driven smart apps have become vital in healthcare, banking, transportation, and personal support. However, as these systems develop in complexity and significance, the requirement for trust, openness, and accountability in their functioning has never been more vital [1].

H. Mosa
Department of Computer Science, Princess Sumaya University for Technology (PSUT), Amman, Jordan

Q. Abu Al-Haija (✉)
Department of Cybersecurity, Jordan University of Science and Technology (JUST), Irbid, Jordan
e-mail: qsabuhaija@just.edu.jo



N. Bizon and B. Appasani (eds.), *Explainable Artificial Intelligence for Trustworthy Decisions in Smart Applications*, https://doi.org/10.1007/978-3-031-97007-8_1

The opaque aspect of many AI models, commonly called "black-box" models, provides a fundamental obstacle. Although strong and accurate, these models lack the openness essential to explain their decision-making processes to consumers, stakeholders, and even developers. This opacity may lead to distrust, especially when AI systems are employed in sensitive domains such as healthcare and finance, where the stakes are high. Concerns regarding justice, prejudice, and accountability are regularly expressed, particularly when AI systems make judgments that directly touch people's lives, such as diagnosing medical problems, authorizing loans, or piloting autonomous cars [2]. In such circumstances, it is crucial that AI systems perform effectively and that their conclusions be intelligible and defensible to humans.

To address these problems, the notion of XAI has attracted substantial interest in recent years. XAI refers to a collection of strategies and techniques to make AI systems more interpretable and transparent, enabling users to understand how and why certain choices are made. The purpose of XAI is to bridge the gap between the capabilities of AI and human understanding, encouraging trust and allowing better-informed decision-making [3]. By offering explanations for AI-driven judgments, XAI allows users to verify, debate, or establish trust in the system's conclusions, thereby boosting the credibility and dependability of AI technology.

This chapter digs into the delicate link between XAI and trust in smart applications, analyzing the problems and possibilities that occur as AI continues to advance. We will begin by defining essential terms, including smart applications, AI, and XAI, before analyzing the emergence of XAI and its role in building trust. We will also explore the technical and ethical issues of deploying XAI in reality and give insights on how XAI might be incorporated into real-world smart applications.

### *1.1.1 Smart Applications*

Smart applications are clever software systems that employ AI and machine learning techniques to deliver users customized, adaptive, and context-aware services [4]. These programs are meant to learn from data, discover patterns, and make choices autonomously, generally with minimum human participation. Examples of smart applications include virtual assistants like Siri and Alexa, recommendation algorithms used by platforms like Netflix and Amazon, and autonomous systems such as self-driving automobiles.

The distinguishing characteristic of smart apps is their capacity to constantly improve their performance over time by learning from data and feedback. However, as these applications get more complex, their decision-making processes become challenging to explain, particularly when deep learning models with millions of parameters are involved. This lack of interpretability has led to worries about how smart apps arrive at some conclusions, especially when their effects are unexpected or undesirable.

For example, AI-powered diagnostic technologies are used in healthcare to analyze medical pictures and diagnose illnesses accurately. While these technologies offer immense promise, their usage raises questions about accountability—if an AI system mistakenly diagnoses a patient, how can physicians or patients trust the system without knowing its reasoning? Similarly, AI algorithms are employed in financial services to evaluate creditworthiness and identify fraud. Still, these systems might perpetuate biases or make unjust conclusions without transparency, leading to consumer mistrust [2].

### *1.1.2 Artificial Intelligence (AI)*

Artificial intellect (AI) is the more significant discipline of computer science that focuses on building computers capable of doing activities that traditionally require human intellect. These tasks include visual perception, voice recognition, decision-making, language translation, and more [5]. AI systems should replicate cognitive processes such as learning and problem-solving, allowing them to do complicated tasks with little human involvement.

Different forms of AI range from rule-based systems to sophisticated deep-learning models. Rule-based systems depend on established rules to make choices, while machine learning models learn from data to improve their performance over time. Deep learning, a form of machine learning, employs artificial neural networks to analyze massive volumes of data and detect patterns. While deep learning models have achieved impressive accuracy in different applications, they are sometimes criticized for their lack of interpretability, making it difficult to grasp how they arrive at particular results.

The emergence of AI has brought tremendous improvements in several areas. In healthcare, AI is utilized for medical imaging, medication discovery, and predictive analytics. In transportation, autonomous cars depend on AI to navigate and make real-time judgments. In finance, AI algorithms are used for fraud detection, risk assessment, and algorithmic trading. However, the growing dependence on AI has also prompted worries about the "black-box" nature of these models and their tendency to propagate prejudices, making the need for explainability all the more crucial [5].

### *1.1.3 Explainable Artificial Intelligence (XAI)*

XAI is an emerging paradigm in AI research that focuses on creating methodologies and strategies to make AI systems more interpretable, transparent, and intelligible to human users [6]. XAI's primary purpose is to explain the choices made by AI systems, allowing users to trust and verify their results. This is especially crucial in high-stakes applications, where the repercussions of AI choices might be considerable.

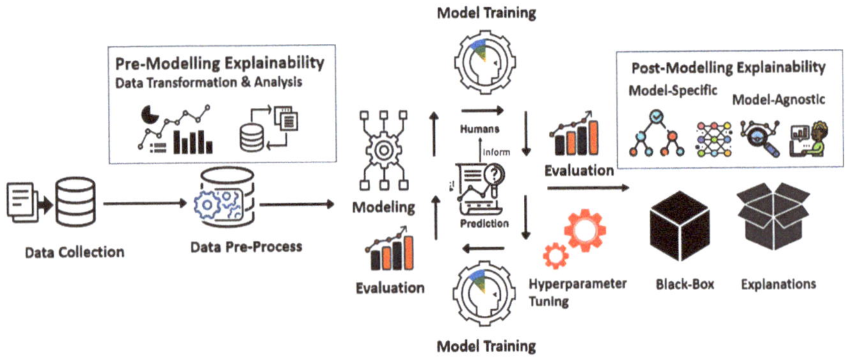


**Fig. 1.1** XAI for AI and ML engineers

XAI is not only about making AI models easier; it is about delivering relevant explanations that can be understood by diverse stakeholders, including developers, consumers, and regulators. For developers, XAI may assist in discovering mistakes or biases in the model, leading to more robust and dependable solutions. For users, XAI may boost confidence in the system's judgments, making them more inclined to trust and embrace AI technology. For regulators, XAI can offer the openness required to verify compliance with ethical and legal requirements, such as fairness and accountability [3].

Several strategies have been proposed to increase the explainability of AI systems. These include post-hoc explanation approaches like Local Interpretable Model-agnostic Explanations (LIME) and Shapley Additive exPlanations (SHAP), which give insights into the decision-making process of complicated models. Other model-specific approaches, such as decision trees and rule-based systems, are intrinsically interpretable. While these approaches have advanced the area of XAI, there is still considerable effort to ensure that explanations are accurate and intelligible to non-technical users [6] (Fig. 1.1).

### *1.1.4 The Importance of XAI in Building Trust*

Trust is vital in the acceptance and success of AI-driven smart apps. Users are more inclined to trust AI systems that offer clear, understandable reasons for their actions. This is especially essential when AI systems make judgments directly affecting people, such as in healthcare, finance, or criminal justice [3].

By enhancing the openness and interpretability of AI systems, XAI may assist in alleviating the trust gap that typically accompanies adopting "black-box" models. When people comprehend how an AI system concludes, they are more likely to trust its results, leading to better acceptance and adoption of AI technology. Moreover, by offering explanations, XAI allows people to contest or validate AI judgments, guaranteeing that the system is responsible for its actions [1].

In conclusion, XAI is vital in resolving the trust and transparency concerns connected with AI-driven smart applications. As AI advances and influences all facets of our lives, the necessity for interpretable and transparent AI systems will only rise. By offering meaningful explanations for AI judgments, XAI can create trust, increase accountability, and guarantee that AI technologies are utilized ethically and responsibly [6].

This chapter has discussed the critical link between XAI and trust, emphasizing the problems and possibilities that lie ahead in this quickly growing subject. As we proceed, it is critical to continue expanding XAI research and ensure that AI systems are trustworthy and transparent (Fig. 1.2).

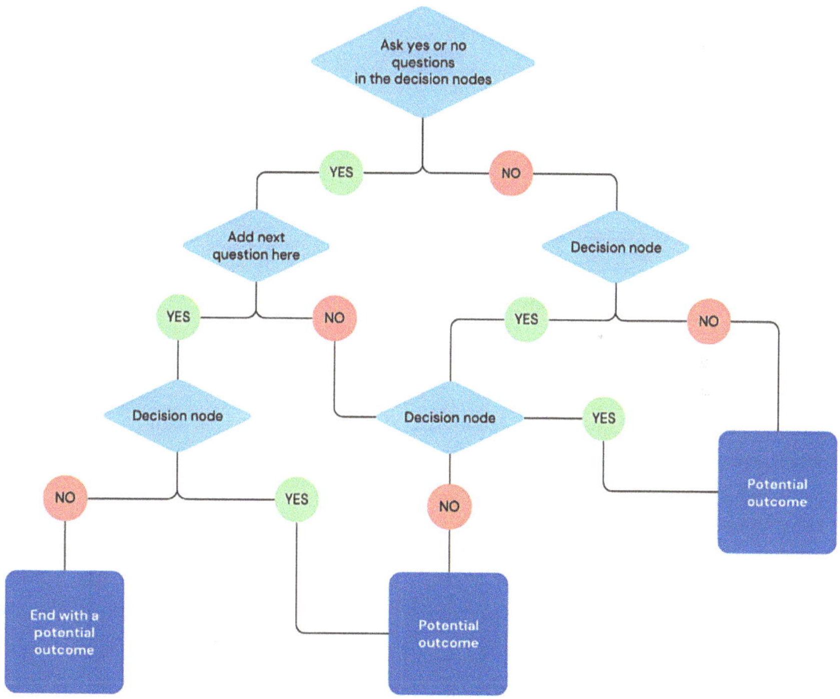


**Fig. 1.2** Example of decision tree

## 1.2 Explainable Artificial Intelligence (XAI)

### *1.2.1 Definition of XAI*

XAI is a multidisciplinary area that tries to construct AI systems whose behaviors and decision-making processes can be understood and explained by human users. The essential purpose of XAI is not just to guarantee that AI models give correct forecasts or judgments but also to provide clear, accessible explanations for these results. This openness is crucial for creating trust, facilitating successful human-AI cooperation, and assuring the appropriate adoption of AI technology across diverse domains [7].

XAI tackles one of the fundamental difficulties of current AI: the "black box" aspect of many machine learning and deep learning models. While these models may reach extraordinary performance levels, especially in tasks like image recognition, natural language processing (NLP), and autonomous systems, they generally do so without offering insight into how they arrived at their findings. This lack of transparency may lead to concerns about trust, especially in high-stakes industries, where the implications of AI choices can be considerable [8].

At its foundation, XAI includes many essential ideas that influence its development and application:

- **Transparency**: Transparency refers to the capacity to comprehend how an AI system arrives at its conclusions or forecasts. A transparent model gives transparency into the inner workings of its decision-making process, making it more straightforward for users to trace the logic behind a particular conclusion [8]. This notion is vital for allowing users to judge if an AI system is performing as planned and for discovering any possible flaws or biases in the model's logic.
- **Interpretability**: Interpretability is the degree to which a human can comprehend the cause of a decision made by an AI system [9]. Interpretability is a more technical element of explainability since it relates to the cognitive strain on users when they seek to grasp how the model works. Highly interpretable models enable users to comprehend the underlying variables leading to specific forecasts or suggestions, whereas low-interpretability models may need extra tools or strategies to help comprehension.
- **Explicability**: Explicability refers to the degree to which the underlying mechanics of a machine learning or deep learning system can be articulated in human words [10]. It highlights the necessity for technically correct and understandable explanations to non-experts, guaranteeing that varied audiences, including end-users, legislators, and regulators, can comprehend and trust the AI system's activities.

Different definitions of XAI have arisen in both academic and industry contexts, reflecting the varied viewpoints within the subject. For instance, in academics, Gunning and Aha describe XAI as "artificial intelligence systems that can explain their rationale, characterize their strengths and weaknesses, and convey an understanding of how they will behave in the future" [11]. This concept stresses the complete character of explainability, highlighting that AI systems should explain their present decision-making processes and assist users in predicting future behavior based on known strengths and shortcomings.

From an industrial viewpoint, XAI is generally characterized with a more practical emphasis. For example, IBM researchers characterize XAI as "the set of techniques and methods that allow human users to comprehend and trust the results and output created by machine learning algorithms" [12]. This concept highlights the relevance of XAI in gaining user trust since understandable AI systems are more likely to be adopted and relied upon in real-world applications.

Another significant definition comes from the Defense Advanced Research Projects Agency (DARPA), which began its XAI initiative in 2016. According to DARPA, XAI incorporates AI systems that can "explain their rationale, characterize their strengths and weaknesses, and convey an understanding of how they will behave in the future" [13]. This concept matches closely with the academic approach while highlighting the need to understand AI systems' limits and offer explanations that allow users to predict the system's future behavior.

### *1.2.2 History and Evolution of XAI*

The progress of XAI is intimately connected to the more extensive development of artificial intelligence systems. As AI technologies have become more powerful and prevalent, the requirement for explainability has become more evident, particularly as the complexity of these systems has expanded. The history of XAI may be followed through many critical eras, each reflecting a fundamental change in how AI systems are created, implemented, and understood.

1. Early Rule-Based Systems (1960s–1980s)

The original AI systems, built between the 1960s and 1980s, were generally based on clear rules and decision trees, making them more interpretable than many models today. These early algorithms, known as expert systems, depended on predetermined rules to replicate human decision-making processes. For example, MYCIN, a rule-based system created in the 1970s, was used to identify bacterial illnesses and prescribe therapies [14]. Because these systems followed clear logical principles, their decision-making processes were straightforward to explain and comprehend

During this time, AI researchers worked primarily on constructing systems that could reason using symbolic logic, enabling users to follow the system's reasoning route and understand why a particular conclusion was reached. This openness was a distinguishing aspect of early AI systems. Still, as AI technology advanced, the desire for better accuracy and performance led to the adoption of increasingly complicated, less interpretable models.

2. Rise of Machine Learning (1990s–2000s)

The emergence of machine learning in the 1990s signaled a dramatic change in AI research. Unlike rule-based systems, machine learning models learn from data and automatically modify their parameters to enhance performance. Algorithms like support vector machines (SVMs) and neural networks were more prevalent during this era because of their capacity to manage enormous datasets and enhance prediction accuracy [15]. However, these models compromised interpretability for efficiency, leading to the emergence of "black-box" computers whose decision-making processes were complex to explain.

As AI models got more data-driven, they also became more opaque. While they may generate outstanding outcomes, consumers and developers frequently found it challenging to grasp how these models made judgments. This trade-off between accuracy and explainability became a significant concern in AI research, especially when machine learning models started to be employed in high-stakes settings.

3. Deep Learning Revolution (2010s)

The introduction of deep learning in the 2010s changed the area of AI. Deep learning models, especially those based on neural networks, obtained state-of-the-art outcomes in tasks such as image recognition, natural language processing, and autonomous systems [16]. However, the intricacy of these models further aggravated the "black box" issue. Deep learning models typically consist of millions of parameters dispersed over numerous layers, making it practically hard for humans to grasp the inner workings of these systems.

The success of deep learning raised attention to the need for explainability, mainly as AI systems grew increasingly interwoven into daily life. The opacity of deep learning models raises questions about accountability, justice, and trustworthiness. The lack of transparency in these systems led to a growing demand for solutions that might explain how deep learning models arrive at their predictions.

4. Emergence of XAI (2010s–Present)

In reaction to the increased opacity of AI systems, academics focused on establishing explainable models and approaches. This decade witnessed the rise of XAI as a separate branch within AI research, aiming at tackling the interpretability issues faced by complicated machine learning models [17]. Several ways have been developed to increase the transparency and interpretability of AI systems, including post-hoc explanation techniques and intrinsically interpretable models.

One of the most notable advances in XAI research occurred in 2016 when DARPA unveiled its XAI initiative. This program considerably enhanced research efforts in the area and fostered the development of new ways for constructing AI systems that are both strong and intelligible [11]. Several significant achievements in XAI research include:

- **2016**: DARPA announced its XAI initiative, giving major money and assistance for research on explainable models [11].
- **2018**: The advent of the LIME (Local Interpretable Model-agnostic Explanations) algorithm by Ribeiro et al. provides a generally applicable mechanism for explaining individual predictions. LIME works by approximating the behavior of a complicated model with a smaller, more interpretable model that may give local explanations for particular predictions [18].
- **2019**: The invention of SHAP (Shapley Additive exPlanations) by Lundberg and Lee provides a consistent way to evaluate model predictions. SHAP gives significance ratings to each feature, offering insight into how each variable contributes to the model's overall prediction [19].
- **2020**: The European Union's proposal for AI rules underlined the significance of explainability in high-risk AI applications, demonstrating the rising acknowledgment of XAI as a vital part of AI ethics and governance [20].

In summary, the development and progress of XAI represent the continuous struggle between performance and interpretability in AI research. As AI systems have gotten more powerful, the requirement for openness and accountability has become more critical. XAI provides a crucial step in addressing these difficulties by providing methodologies and approaches that make AI systems more understandable and trustworthy. The invention of XAI has not only boosted the interpretability of AI models but also permitted their responsible deployment in high-stakes areas.

### *1.2.3 Types of Explainability*

Explainability in AI may be classed into numerous categories, each giving distinct insights into the operation of AI systems:

1. Post-hoc versus Intrinsic Explainability

Post-hoc explainability refers to strategies that explain a model's choices after the model has been trained [21]. These strategies try to understand the behavior of current "black box" models without affecting their underlying structure. Examples include LIME and SHAP.

Intrinsic explainability, on the other hand, entails building AI models that are naturally interpretable [22]. These models embed explainability into their design, making their decision-making more visible. Examples include decision trees and rule-based systems.

2. Local versus Global Explanations

Local explanations concentrate on understanding particular predictions or judgments an AI model produces [23]. They try to answer queries like "Why did the model make this specific prediction for this particular input?" LIME is an example of a local explanation approach.

Global explanations give insights into the general behavior of a model over its whole input space [24]. They try to expose the broad patterns and traits that the model employs to make judgments. Techniques like feature relevance rankings and partial dependency graphs give global interpretations.

3. Model-Specific versus Model-Agnostic Approaches

Model-specific techniques are geared to explain certain kinds of AI models. For instance, attention processes in neural networks reveal insights peculiar to that design [25].

Model-agnostic techniques may be used for any AI model independent of its underlying structure [26]. These techniques regard the model as a "black box" and concentrate on the link between inputs and outputs. LIME and SHAP are instances of model-agnostic approaches.

### *1.2.4 Techniques and Methods in XAI*

XAI includes various methodologies and procedures, each with its strengths and limits. Some of those significant methods are Rule-Based Systems, Decision Trees, Local Interpretable Model-agnostic Explanations (LIME), SHapley Additive exPlanations (SHAP), Attention Mechanisms, Counterfactual Explanations, Partial Dependence Plots (PDP).

The Rule-based systems employ a set of preset rules to make choices, making them naturally interpretable [27]. While they give straightforward explanations, they may struggle with complicated, high-dimensional data. On the other hand, Decision trees give a hierarchical representation of decision rules, enabling a visual and clear explanation of the model's logic [28].

LIME explains individual predictions by approximating the model locally using an interpretable model [18], and SHAP employs game theory principles to give priority levels to each feature for a specific prediction [19]. In deep learning models, especially for natural language processing tasks, attention mechanisms emphasize the bits of the most significant input to each output [29]. Counterfactual explanations demonstrate how the model's output might vary if the input were slightly altered, offering insights into the model's decision limits [30]. Finally, PDPs demonstrate the marginal influence of one or two characteristics on the projected result of a machine learning model [31]. Table 1.1 lists all the strengths and limitations of those methodologies.

As the area of XAI continues to advance, researchers are inventing new methods and improving old ones to meet the limits and constraints of present methodologies. The choice of the XAI approach frequently relies on the individual application, the kind of model being used, and the intended audience for the explanations.

**Table 1.1** Methodologies strengths and limitations

| Technique | Strengths | Limitations |
|---|---|---|
| Rule-based systems | – Highly interpretable<br>– Easy to verify and audit<br>– Can integrate domain expertise directly | – May not capture complicated patterns in data<br>– Labor-intensive to build and maintain<br>– Poor generalization |
| Decision trees | – Visually interpretable<br>– Handles both numerical and categorical data<br>– Captures feature interactions | – Struggles with high-dimensional data<br>– Prone to overfitting if not pruned<br>– Limited smooth decision boundaries |
| LIME | – Model-agnostic<br>– Intuitive local explanations<br>– Uncovers possible biases | – Inconsistent explanations across similar cases<br>– Sensitive to local neighborhood choice<br>– Struggles with high-dimensional data |
| SHAP | – Coherent, theory-based explanations<br>– Effective globally and locally<br>– Handles feature interactions | – Computationally expensive<br>– Difficult for non-experts<br>– Assumes feature independence |
| Attention mechanism | – Highlights model emphasis intuitively<br>– Improves model performance<br>– Provides clear insights | – May not fully represent decision processes<br>– Can mislead if misunderstood<br>– Limited to specific neural networks |
| Counterfactual explanations | – Offers actionable insights<br>– Explains model sensitivity<br>– Measures fairness and bias | – Computationally intensive<br>– Sensitive to distance metric<br>– May not cover all key model behaviors |
| Partial Dependence Plots (PDP) | – Provides global feature effects<br>– Exposes non-linear correlations<br>– Simple to understand | – Assumes feature independence<br>– Misses critical feature interactions<br>– Costly for large datasets |

In conclusion, XAI represents a key milestone in artificial intelligence, answering the rising demand for transparency and interpretability in AI systems. By making AI decision-making processes more intelligible to people, XAI plays a critical role in developing trust, assuring accountability, and allowing more effective human-AI cooperation.

## 1.3 Trust in Smart Applications

### *1.3.1 Defining Trust in AI Systems*

Trust is a vital part of human–computer interaction, especially regarding AI systems. In smart applications, trust may be described as the user's willingness to depend on an AI system's judgments, suggestions, or actions [32]. This idea of trust in AI is broad and involves many crucial dimensions:

- **Reliability**: The consistency and reliability of the AI system's performance over time and across numerous settings [33].
- **Transparency**: The degree to which the AI system's decision-making process is clear and transparent to users [34].
- **Fairness**: The lack of prejudice and discrimination in the AI system's outputs and decision-making processes [35].
- **Accountability**: The capacity to assign accountability for the AI system's actions and choices and to handle any negative outcomes [36].

These trust factors are interrelated and contribute to consumers' total trust in AI-driven smart apps. Lee and See present a model of trust in automation that stresses the role of these aspects in developing user trust [37]. They suggest that trust is dynamic and may alter depending on the user's experience with the system, the context of usage, and the perceived dangers and rewards.

### *1.3.2 The Role of Explainability in Building Trust*

Explainability is critical in creating and retaining confidence in AI-driven smart applications. By making AI systems more visible and interpretable, XAI approaches assist in addressing many essential issues of trust:

- **Understanding AI Decisions**: XAI helps users understand why an AI system made a certain choice, suggestion, or forecast. This information is critical for users to judge the system's dependability and suitability in diverse circumstances [2].
- **Verifying justice**: XAI may assist users and developers in uncovering possible biases in AI systems, supporting justice and non-discrimination [38].
- **Enabling Accountability**: By offering insights into the decision-making process, XAI simplifies the assignment of responsibility and allows the discovery and rectification of mistakes [39].
- **Enhancing User Trust**: When users can comprehend and verify AI judgments, they are more likely to gain trust in the system's capabilities and limits [40].

Several case studies and examples highlight how XAI has boosted confidence in smart applications:

- **Healthcare**: In research by Tjoa and Guan, employing XAI methods in a medical diagnostic system boosted doctors' confidence and acceptance of AI-generated diagnoses [41]. The system gave reasons for its predictions, enabling clinicians to validate the rationale and combine it with their clinical experience.
- **Financial Services**: Research by Bussmann et al. found that utilizing LIME and SHAP to explain AI-driven credit scoring algorithms enhanced clients' confidence in the loan approval process [42]. Applicants might comprehend the reasons impacting their credit ratings, leading to increased acceptance of the judgments.
- **Autonomous Vehicles**: Research by Koo et al. found that offering explanations for an autonomous vehicle's behaviors (e.g., reasons for route alterations or unexpected stops) greatly enhanced passengers' confidence and comfort levels [43].

### *1.3.3 The Impact of Lack of Trust*

When AI systems lack transparency and explainability, it may lead to substantial negative implications across numerous industries:

1. **Healthcare**: A lack of confidence in AI-driven diagnostic or treatment recommendation systems might result in healthcare practitioners dismissing potentially life-saving findings or patients refusing AI-assisted therapies [44].
2. **Autonomous Systems**: Insufficient confidence in autonomous cars might slow their adoption, thereby delaying the realization of their safety benefits and environmental advantages [45].
3. **Finance**: Opaque AI models in financial services might lead to unjust loan denials or biased investment recommendations, exacerbating economic disparities [46].
4. **Criminal Justice**: The employment of unexplained AI in recidivism prediction or sentence recommendations might lead to unfair results and destroy public faith in the justice system [47].

These situations show the practical ramifications of a lack of confidence in AI systems and pose serious ethical considerations. The opacity of AI decision-making processes might worsen existing social disparities, impinge on human rights, and erode democratic principles [48].

Moreover, the lack of confidence in AI systems might have larger social impacts:

- **Technical Resistance**: A broad suspicion in AI technology might lead to resistance to potentially beneficial breakthroughs, slowing down technical progress and economic growth [49].
- **Information Manipulation**: In the age of deep fakes and AI-generated content, a lack of confidence in AI systems might make it impossible for the public to discern between real and manipulated information, undermining social and political discourse [50].

- **Workforce Displacement**: As AI systems grow more ubiquitous in many areas, a lack of trust might intensify worries of job displacement, leading to societal instability and opposition to essential economic adjustments [51].

## 1.4 The Evolution of XAI and Trust in Smart Applications

### *1.4.1 Early AI Systems and Black-Box Concerns*

The early phases of AI research were characterized by a concentration on performance and accuracy, frequently at the sacrifice of explainability. Many early AI systems, especially in machine learning, were "black boxes"—systems whose internal workings were opaque and difficult to analyze [17].

***This lack of explainability led to various concerns:***

- **Little Understanding**: Users and even engineers have little understanding of how these algorithms arrive at their conclusions or predictions [8].
- **Difficulty in Debugging**: When mistakes occurred, it was tricky to identify and remedy the leading causes [47].
- **Potential for Bias**: Without transparency, it was impossible to discover and eliminate biases that could be present in the training data or the model itself [52].
- **Regulatory Challenges**: The opacity of AI systems makes it harder to guarantee compliance with current legislation and to design relevant new regulations [53].

These problems lead to a widespread cynicism and lack of confidence in AI systems, especially in high-stakes fields such as healthcare, finance, and criminal justice.

### *1.4.2 Shift Towards Transparency*

Acknowledging these concerns led to a gradual trend towards more transparent AI systems. Numerous factors triggered this transition:

1. **Research Advancements**: The development of new approaches for model interpretation and explanation, such as LIME and SHAP, gave tools for making AI systems more visible [18, 19].
2. **Ethical Considerations**: A growing understanding of the ethical consequences of opaque AI systems led to demands for more openness and accountability [54].
3. **Regulatory Pressure**: The advent of rules like the European Union's General Data Protection Regulation (GDPR) in 2018, which includes a "right to explanation" for automated decision-making, expedited the march towards XAI [55].

4. **Industry Initiatives**: Major tech corporations and industry associations started to produce and promote AI ethical standards that stressed the necessity of explainability [56].

This move towards openness has been mirrored in many AI ethical standards and legislative initiatives:

- The European Commission's Ethics Guidelines for Trustworthy AI, issued in 2019, explicitly lists openness as one of its fundamental needs [57].
- The IEEE Global Initiative on Ethics of Autonomous and Intelligent Systems has produced criteria for ethically aligned design, which include concepts of openness and explainability [58].
- The U.S. National Institute of Standards and Technology (NIST) has launched work to define standards for XAI [59].

### 1.4.3 Current Trends

The field of XAI is advancing rapidly, marked by several prominent trends. One significant trend is the integration of XAI in deep learning. Researchers are developing approaches to enhance the interpretability of complex deep learning models, utilizing techniques such as attention mechanisms in natural language processing [29] and saliency maps in computer vision [60].

Another notable focus is on causality in XAI. There is an increasing emphasis on causal explanations, which aim to provide insights into the cause-effect relationships identified by AI models [61]. This shift highlights the importance of understanding not just what decisions are made by AI, but why they are made.

Human-centered XAI is also gaining traction. Researchers are exploring the cognitive aspects of explanations, striving to create XAI systems that provide tailored explanations for diverse user groups and contexts [34]. This approach recognizes that different users may require different types of information to understand AI decisions fully.

In addition, there is a growing interest in XAI for reinforcement learning. As reinforcement learning becomes more significant in robotics and game playing applications, researchers are motivated to develop explanation methodologies specifically designed for these AI systems [62].

Lastly, adversarial XAI is emerging as a crucial area of study. Researchers are investigating how XAI approaches can be leveraged to enhance the resilience of AI systems against adversarial attacks [63]. This trend underscores the need for transparency and robustness in AI technologies.

The desire for openness in smart apps is also expanding across several sectors:

- **Healthcare**: There's an increasing focus on interpretable AI models for clinical decision support systems, driven by the requirement for physician confidence and regulatory compliance [44].

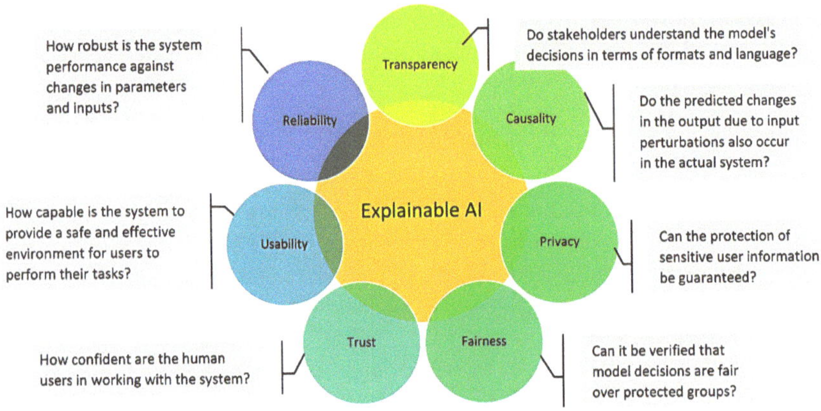


**Fig. 1.3** Goals of XAI

- **Finance**: The financial industry is experiencing significant use of XAI for credit rating, fraud detection, and algorithmic trading, partially driven by regulatory restrictions [9].
- **Autonomous Vehicles**: As self-driving vehicles get closer to broader deployment, there is a rising emphasis on building XAI systems for route planning, obstacle avoidance, and decision-making in complicated traffic situations [6].
- **Legal Tech**: The legal profession is researching XAI for jobs like contract analysis and legal research, where the capacity to explain AI-driven ideas is vital for generating confidence and assuring accuracy [64].
- **Education**: There is rising interest in employing XAI approaches to make AI-powered educational technologies more visible, helping students and instructors comprehend individualized learning suggestions [65].

In conclusion, the progress of XAI and trust in smart apps reflects a larger movement in the AI sector towards more responsible and human-centered AI research. The continuous research and development in XAI, along with increasing legislative frameworks and ethical principles, are forging a future where AI systems are robust but also transparent, fair, and responsible (Fig. 1.3).

## 1.5 Challenges in Implementing Explainable AI in Smart Applications

The deployment of XAI in smart applications involves a multidimensional collection of issues that span technical, user-centric, and regulatory realms. These issues typically interact and aggravate each other, making developing and implementing XAI systems a complicated task.

### 1.5.1 Technical Challenges

One of the key technological obstacles in adopting XAI is matching the complexity of AI models with the necessity for concise, accessible explanations [3]. Modern AI systems, especially deep learning models, may incorporate millions of parameters and complicated, non-linear interactions that are intrinsically difficult to analyze and describe in human-understandable terms. This difficulty emerges in numerous ways:

1. **Model Complexity versus Explanation Simplicity**: There is sometimes a conflict between the intricacy necessary for high-performance AI models and the simplicity needed for simple explanations. Highly sophisticated models with state-of-the-art performance may be rigid to explain without major simplification, jeopardizing the explanation's correctness [8].
2. **Local versus Global Explanations**: While local explanations (explanations for particular predictions) may be easier to construct, they may not give insights into the model's overall behavior. Global explanations, which characterize the model's behavior across all conceivable inputs, are frequently more complex to produce and express well [57].
3. **Feature significance versus Feature Interaction**: Many XAI strategies concentrate on feature significance, which might be deceptive when feature interactions play a vital role. Capturing and interpreting these interactions intelligibly remains a big issue [19].

Another significant technological difficulty is the trade-off between model performance and explainability [22]. In many circumstances, the most accurate models (e.g., deep neural networks) are also the least interpretable, whereas more interpretable models (e.g., decision trees) may not attain the same level of performance. This poses a problem for developers and consumers of AI systems, especially in high-stakes applications where accuracy and explainability are critical.

Researchers are currently working on tackling these difficulties via several approaches:

- Developing inherently interpretable models that preserve good performance [66].
- Creating post-hoc explanation approaches that can function with sophisticated black-box models [18].
- Exploring hybrid methods that use several AI techniques' characteristics to achieve high performance and explainability [67].

### 1.5.2 User-Centric Challenges

Different user groups' varying expectations and demands provide another key barrier in deploying XAI [34]. Users of AI systems may vary from AI professionals and developers to non-technical end-users, each with distinct degrees of AI literacy and varied needs for explanations.

Key user-centric problems in XAI include the need for tailored explanations. Different users may require varying degrees of information or distinct types of explanations. For instance, a data scientist might demand a comprehensive technical explanation, while a corporate executive prefers a high-level overview highlighting the significant elements influencing a decision [9].

Another critical issue is cognitive load. Explanations must be sufficiently detailed to provide meaningful insights without overwhelming the user. Achieving this balance is particularly challenging when dealing with advanced AI systems, where the complexity of the information can easily lead to confusion [68].

Confidence calibration is also an important consideration. Explanations should help users develop appropriate levels of trust in the AI system. If explanations are overly simplified, they may foster unwarranted confidence, whereas excessively complex explanations could result in skepticism or disengagement from the user [69].

Finally, there is a tension between interpretability and comprehensibility. A model may be technically interpretable, but not all users might easily understand this interpretation. Bridging this gap requires careful consideration of users' backgrounds and contexts, ensuring that explanations are accessible and relevant to a diverse audience [70].

Addressing these difficulties requires a user-centered design approach to XAI. This involves conducting user studies to understand the explanation requirements of various user groups [71]. Researchers can tailor explanations to better meet their needs by gathering insights directly from users.

Additionally, the development of adaptive explanation interfaces is crucial. These interfaces should be able to respond to user choices and competence levels, ensuring that the information presented is relevant and comprehensible for each user [72].

Moreover, integrating concepts from cognitive science and human–computer interaction can significantly enhance the effectiveness of explanations. By leveraging insights from these fields, researchers can create explanations that are not only informative but also align with how users process information [73]. This comprehensive approach is essential for fostering trust and understanding in AI systems.

### *1.5.3 Regulatory and Ethical Challenges*

The adoption of XAI also confronts substantial legal and ethical issues, particularly as AI systems become more prevalent in high-stakes decision-making processes. One of the primary regulatory challenges is compliance with existing regulations. For example, the European Union's General Data Protection Regulation (GDPR) includes provisions related to automated decision-making and the "right to explanation." Implementing XAI systems that adhere to these regulations can be difficult, especially given the lack of clear guidelines and a sufficient explanation [55].

Furthermore, the regulatory landscape is continually evolving. As AI technologies advance, it is expected that regulations will also shift. Therefore, XAI systems must be flexible enough to adapt to changing regulatory requirements [53]. This adaptability is essential for ensuring ongoing compliance and maintaining trust with users and stakeholders.

Additionally, there are sector-specific restrictions that complicate the implementation of XAI. Different industries, such as healthcare and finance, have unique regulations governing the use and explanation of AI systems. Navigating these sector-specific constraints adds another layer of complexity to adopting XAI, necessitating a tailored approach to meet diverse regulatory demands [74].

Ethical issues in providing explanations for AI decisions are significant and multifaceted. One major concern is fairness and bias. Explanations should help identify and minimize biases in AI systems; however, there is a risk that the explanations themselves could be biased or misused to justify unjust judgments [75]. This highlights the need for scrutiny in the development and deployment of XAI.

Another critical issue is related to privacy concerns. Detailed explanations may inadvertently reveal sensitive information about the training data or the model's internal structure. Striking a balance between the need for transparency and the imperative to protect privacy presents a significant challenge [76].

Responsibility is also a key ethical consideration. While XAI can promote accountability, it questions who is ultimately responsible for AI decisions and the explanations provided. Is it the developers, the users, or the organizations deploying the AI systems? This ambiguity complicates the ethical landscape surrounding XAI [77].

Lastly, the potential for manipulation and deception is an ethical issue that cannot be overlooked. There is a risk that explanations could be crafted to mislead users or to justify predetermined outcomes, raising serious questions about the trustworthiness of XAI systems [78]. This underscores the importance of maintaining integrity and transparency in the explanations provided by AI.

Addressing these difficulties demands a multidisciplinary strategy combining cooperation between AI researchers, ethicists, legal experts, and legislators. Initiatives such as formulating AI ethical standards and constructing regulatory frameworks for AI are steps toward resolving these concerns [56] (Fig. 1.4).

## 1.6 Use Cases of XAI for Smart Applications

The deployment of XAI in smart applications covers numerous areas, each with distinct needs and obstacles. Here, we investigate case studies and applications in healthcare, autonomous cars, finance, and human resources.

**Fig. 1.4** Challenges and downsides associated with AI

### *1.6.1 Healthcare*

XAI is vital in creating confidence in AI-driven diagnostic tools and treatment suggestions in healthcare. The enormous risks in medical choices make explainability especially crucial in this sector.

Case Study: XAI for Skin Cancer Detection

Researchers at the University of California, San Diego, created an explainable deep learning system for skin cancer diagnosis [79]. The system identifies skin lesions and offers visual explanations for its judgments using a Gradient-weighted Class Activation Mapping (Grad-CAM) method.

Key characteristics of the system designed for melanoma detection include high accuracy in identifying the condition, comparable to that of skilled dermatologists. Additionally, the system features visual heatmaps that highlight the sections of the image most significant to the diagnosis. Natural language explanations are also provided, summarizing the visual aspects contributing to the categorization.

The explainability properties of this system address three crucial demands in healthcare AI. First, it promotes transparency, allowing doctors to understand the reasons behind the AI's diagnosis. Second, it fosters trust-building, as patients can better comprehend and potentially trust the AI-assisted diagnosis. Lastly, the system can be an educational tool for medical students and residents, enhancing their learning experience.

However, some challenges and considerations need to be addressed. One significant challenge is balancing accuracy with explainability. While the system offers straightforward explanations, ensuring that these explanations accurately reflect the complex decision-making processes of the neural network remains difficult.

Additionally, the system must comply with strict healthcare standards, particularly those related to AI explainability and patient privacy. Meeting these regulatory requirements is essential for successfully implementing the system in clinical settings.

### 1.6.2 Autonomous Vehicles

In driverless cars, XAI plays a critical role in developing trust and assuring safety. The capacity to explain decision-making processes in real-time is highly significant in this situation.

Case Study: XAI for Autonomous Driving Decisions

MIT and Toyota Research Institute researchers created an XAI system for autonomous driving that delivers real-time explanations for the vehicle's behavior [80].

Key characteristics of the system designed for autonomous driving include the real-time generation of natural language justifications for driving choices. Additionally, it features visual attention maps that display which portions of the scene the AI is focusing on and counterfactual explanations that illustrate how different events might lead to varied behaviors.

The XAI features of this system address several essential demands in autonomous driving. First, they promote trust-building by allowing passengers to understand why the car makes specific decisions, enhancing their comfort and confidence in the system. Second, the explanations contribute to safety enhancement; by clarifying its judgments, the system empowers individuals to intervene if they observe any concerning behavior. Finally, the ability to explain decisions may help ensure compliance with anticipated future regulations that mandate transparency in autonomous systems.

However, there are notable challenges and considerations to address. Real-time constraints are significant challenges; explanations must be developed and presented quickly enough to remain relevant in rapidly changing driving conditions. Additionally, the system must carefully balance delivering sufficient information without overloading or distracting the human occupants. This equilibrium is critical for maintaining both safety and user engagement.

### 1.6.3 Finance

In the financial industry, XAI is vital for assuring openness and justice in algorithmic decision-making processes such as credit scoring, fraud detection, and risk assessment.

Case Study: XAI for Credit Scoring

A major financial institution has deployed an Explainable Artificial Intelligence (XAI) system for credit scoring to enhance transparency and comply with regulatory standards [81]. This system features several key characteristics, including feature relevance rankings demonstrating which elements most significantly affect credit scores. Additionally, it provides counterfactual explanations that offer concrete advice on how individuals can improve their credit ratings. Natural language explanations further clarify the credit determination process for users.

The XAI features of this system address several essential demands in financial services. First, they ensure regulatory compliance by helping to meet standards for transparent and explainable credit judgments. Second, customer satisfaction is enhanced as applicants better understand the reasons behind their credit ratings and how they might improve them. Finally, the system ensures fairness by recognizing and minimizing any biases in the credit rating process.

However, there are notable challenges and considerations. One significant challenge is model complexity; balancing the demand for highly precise credit risk projections with the necessity for interpretable models is crucial. Additionally, data privacy must be a priority, ensuring that the explanations provided do not inadvertently expose sensitive information about the model or other consumers. This careful attention to complexity and privacy is vital for maintaining trust and compliance in the financial sector.

### *1.6.4 Human Resources*

XAI is becoming more significant in human resources as AI technologies are employed in recruiting and recruitment procedures. Explainability is vital for guaranteeing fairness and preventing prejudice in these applications.

Case Study: XAI for Resume Screening

A large technology company has adopted an Explainable Artificial Intelligence (XAI) system for initial resume screening to enhance productivity and reduce bias in their recruiting process [82]. The system features several key characteristics, highlighting significant qualifications influencing the AI's decision to select or reject an applicant. Additionally, it explains how the candidate's credentials align with job criteria and includes bias detection alerts that identify potential unconscious biases in the screening process.

This system's XAI features meet several essential human resources (HR) demands. First, they promote fairness and non-discrimination by evaluating applicants based on relevant qualifications rather than irrelevant criteria. Second, the system enhances transparency, allowing HR professionals to understand and verify the AI's judgments. Lastly, the explanations aided in establishing compliance with equal employment opportunity regulations, reinforcing ethical hiring practice.

However, there are significant challenges and considerations to address. One major challenge is handling unstructured data; resumes often contain unstructured information, making it difficult to provide consistent and coherent explanations. Additionally, there is a need to balance automation with human judgment. Technology must offer explanations that empower human decision-makers rather than completely replace their judgment, ensuring that the final hiring decisions remain informed and thoughtful.

In conclusion, these case studies highlight the numerous uses of XAI across different industries. While XAI provides considerable advantages in openness, trust-building, and regulatory compliance, its implementation nevertheless comes with distinct hurdles in each discipline. As XAI technologies continue to improve, we should expect to see increasingly complex and customizable XAI solutions across many smart applications.

## 1.7 Future Directions for XAI and Trust in Smart Applications

As the area of XAI continues to advance, various potential paths are emerging that might considerably boost confidence in smart applications. These future advancements are anticipated to increase human-AI interaction, incorporate ethical issues more fully into XAI systems, and promote multidisciplinary cooperation.

### *1.7.1 Improving Human-AI Interaction*

The future of XAI depends on providing more natural and effective methods for humans to engage with AI systems. This requires enhancing the quality and relevance of explanations and designing better interfaces for conveying these explanations to users. Key areas of development include:

- **Personalized Explanations**: Future XAI systems may alter their explanations depending on the user's history, preferences, and situation. This customization might considerably boost the user's knowledge and confidence in the system [21].
- **Interactive Explanations**: Future systems could provide interactive interfaces that enable users to explore various facets of the AI's decision-making process rather than presenting static explanations. This might include features like configurable degrees of information or the option to ask follow-up inquiries [83].
- **Multi-modal Explanations**: Combining several modes of explanation (e.g., visual, written, and audio) might make explanations more accessible and complete for an enormous variety of consumers [84].

- **Continuous Learning from User Input**: Future XAI systems might continually use user input to modify and enhance their explanations. This feedback loop might assist in guaranteeing that explanations stay relevant and valuable over time [85].

The significance of user input in enhancing AI explanations is particularly crucial. Explainable Artificial Intelligence (XAI) systems can achieve several important outcomes by integrating user feedback and preferences. First, they can identify which explanations are most helpful for different user groups, ensuring the information provided is relevant and valuable. Second, these systems can detect and rectify errors or assumptions related to AI decision-making, improving overall accuracy and reliability. Finally, XAI systems can adapt the level of detail and complexity of explanations based on user needs and competencies, tailoring the experience to suit a diverse audience.

Research by Hoffman et al. demonstrates that this form of adaptive explanation system can significantly enhance consumer confidence and satisfaction with AI technologies [86]. By prioritizing user input, XAI systems become more effective and foster a greater sense of trust and engagement among users, ultimately leading to better acceptance and utilization of AI in various applications.

### *1.7.2 The Role of AI Ethics*

As AI systems become more prevalent in high-stakes decision-making, incorporating ethical issues into Explainable Artificial Intelligence (XAI) design is increasingly relevant. Future advancements in this area may include several important developments.

First, ethical transparency may become a key feature of XAI systems, allowing them to explicitly highlight the ethical concerns and value judgments that underpin their actions. This transparency can help users understand how a decision was made and the ethical context in which it was reached.

Second, future XAI systems could provide fairness-aware explanations that directly address issues of fairness and bias. This might involve identifying potential sources of bias in the data or model and detailing how these biases were mitigated, fostering a more equitable decision-making process.

Another potential advancement is the development of value-aligned AI, where explanations illustrate how the AI system's choices align with human values and societal standards. This alignment is critical for building trust in AI systems, especially in sensitive domains like healthcare or criminal justice.

Lastly, ongoing research aims to create explainable ethical AI systems capable of making ethical judgments and articulating the reasoning behind these decisions. This may involve embedding ethical reasoning frameworks directly into the AI's decision-making processes, enhancing the system's ability to navigate complex moral landscapes while providing clear explanations to users.

The incorporation of ethics into XAI systems involves various obstacles, including the necessity to manage opposing ethical ideals and the difficulty of converting abstract ethical notions into tangible algorithmic implementations. However, resolving these difficulties might considerably boost AI systems' credibility and public acceptability.

### *1.7.3 Interdisciplinary Research*

The future of XAI and confidence in smart applications will likely be defined by growing cooperation between AI researchers, psychologists, domain experts, and other key stakeholders. This multidisciplinary approach is vital for designing XAI systems that are technically sound, psychologically successful, and domain-appropriate.

Key areas for multidisciplinary cooperation include:

- **Cognitive Science and XAI**: Insights from cognitive science may guide the construction of explanations that correspond with human cognitive processes. This might lead to more intuitive and effective explanations [34].
- **Domain-specific XAI**: Collaboration with domain experts (e.g., physicians for healthcare AI, financial experts for fintech AI) helps guarantee that explanations are clear and relevant within certain settings [44].
- **Human–Computer Interaction (HCI) and XAI**: HCI researchers may help create more effective interfaces for delivering AI explanations, boosting usability and user engagement [87].
- **Ethics and XAI**: Collaboration between AI researchers and ethicists may assist in guaranteeing that XAI systems are created with ethical concerns at their heart [88].
- **Legal Experts and XAI**: As legislation surrounding AI transparency expand, cooperation with legal experts will be vital for designing XAI systems that fulfil legal criteria [89].

This multidisciplinary approach might lead to more comprehensive XAI systems that give technically correct explanations and address the larger sociological, ethical, and legal consequences of AI decision-making.

## 1.8 Conclusion and Remarks

### *1.8.1 Summary of Key Points*

XAI has become vital in promoting confidence in smart technologies. The need for transparency, interpretability, and accountability has substantially grown with the expanding presence of AI systems across sectors including healthcare, banking, and autonomous systems. Users and stakeholders need to understand what choices AI systems make and why and how those judgments are made.

Throughout this chapter, we have dug into several facets of XAI and its function in smart applications, focusing on the following major areas:

- **Definition and Evolution of XAI**: The beginnings of XAI may be traced back to the early rule-based systems in artificial intelligence, which were relatively straightforward to explain owing to their clear decision-making procedures. However, as AI advanced, especially with the emergence of machine learning and deep learning models, the decision-making processes grew more opaque. Modern AI systems typically behave as "black boxes," where even their engineers find it difficult to explain particular outputs. XAI intends to counter this tendency by making AI systems more accessible and interpretable.
- **Trust in AI Systems**: Trust in AI extends beyond technical performance. While precision and efficiency are vital, they are inadequate to generate confidence. Users and stakeholders must also be confident in AI systems' fairness, openness, and accountability. Trust is multidimensional and encompasses dependability, ethical concerns, the capacity to explain judgments, and the system's resilience in many real-world settings.
- **Technical Approaches to XAI**: XAI approaches fall into two basic categories: inherently interpretable models and post-hoc explanation procedures. Intrinsically interpretable models, such as decision trees and linear models, are meant to be visible from the beginning. In contrast, post-hoc approaches try to explain the behavior of complicated models, such as deep neural networks, after training them. Popular post-hoc approaches include LIME (Local Interpretable Model-agnostic Explanations) and SHAP (SHapley Additive exPlanations), which give means to understand particular predictions generated by opaque models.
- **Problems in Implementing XAI**: Several technological, user-centric, and regulatory problems limit the mainstream deployment of XAI. On the technological side, the issue is balancing model complexity with explainability. More sophisticated models tend to perform better in accuracy but are more complicated to comprehend. From a user standpoint, explanations must be customized to the user's individual demands and expertise levels. Legal and ethical issues are becoming increasingly significant as governments and organizations prioritize AI transparency and responsibility.
- **Case Studies**: We have analyzed case studies from diverse sectors, including healthcare, banking, and autonomous cars, where XAI has been deployed to

promote trust and accountability. In healthcare, for instance, XAI approaches have been used to explain diagnoses given by AI systems, allowing professionals to check and comprehend the logic behind the system's recommendations. Similarly, explainability is vital in banking for regulatory compliance, particularly in high-stakes sectors like credit scoring and fraud detection.

- **Future Directions**: As AI systems continue to advance, so will the XAI field. Future research is anticipated to make explanations more relevant and actionable, enhance human-AI interaction, and incorporate ethical issues directly into AI design. Moreover, multidisciplinary research encompassing AI, ethics, cognitive science, and domain-specific knowledge will play a significant role in furthering XAI.

The chapter also discussed important obstacles and prospects in the area of XAI. These difficulties include:

- **Model Complexity versus Explainability**: There is typically a trade-off between the complexity of a model and its interpretability. Complex models like deep neural networks may achieve more accuracy but are more complicated to explain than simpler models like decision trees.
- **User-Centric Explanations**: Explanations need to be adapted to the unique situation and the degree of competence of the end-user. A doctor, for example, may need a different explanation of a medical diagnosis than a patient would.
- **Accuracy versus Meaningfulness**: It is not enough for explanations to be technically correct; they must also be meaningful and actionable for the user. Explanations that are overly technical or abstract may fail to create confidence.
- **Ethical and Privacy Considerations**: The requirement for openness may sometimes collide with other concerns, such as preserving user privacy or intellectual property. Balancing these opposing agendas remains a problem for XAI researchers.
- **Legal and Legal Issues**: With the rising focus on AI transparency and accountability, legal frameworks are being developed that require AI systems to be explainable. Navigating this shifting legal framework will be a fundamental issue for firms that develop and use AI systems.

On the other side, XAI has various possibilities, including:

- **Building User Confidence**: Transparent AI systems that can explain their judgments are more likely to earn user confidence and adoption. This is particularly critical in high-stakes industries like healthcare, banking, and autonomous cars.
- **Improving Safety and Reliability**: XAI may help discover and fix flaws or biases in AI systems, leading to safer and more reliable decision-making processes.
- **Facilitating Human-AI cooperation**: XAI can promote cooperation between people and machines by making AI systems more intelligible. For example, XAI may help clinicians make better-informed judgments in areas like healthcare.
- **Advancing AI Research**: By making the internal workings of AI systems more visible, XAI may help academics better understand how these systems function and how they might be improved.

- **Ethical AI Development**: XAI may support more responsible AI development by ensuring that systems are transparent, fair, and aligned with ethical values.

### *1.8.2 Final Thoughts*

Looking to the future, it is apparent that XAI will continue to play a significant role in developing and implementing AI systems. The capacity to explain AI judgments is not simply a technological need but a social obligation. As AI systems grow increasingly embedded into our everyday lives and decision-making processes, the requirement for transparency and interpretability will only become more evident.

Several significant developments are expected to affect the future of XAI:

- **Holistic Explanations**: As AI systems grow more sophisticated and deeply interwoven into key elements of life, explanations must go beyond technical specifics to address ethical and social effects. This will demand a more holistic approach to XAI, where explanations are not only about how an AI system arrived at a given choice but also about its larger ramifications for society and people.
- **Personalization of Explanations**: Future AI systems are anticipated to deliver individualized explanations depending on the user's history, expertise, and unique demands. For example, a doctor may get a thorough explanation of a diagnosis, while a patient would receive a simplified version matched to their level of medical understanding.
- **XAI as a basic Design Principle**: Rather than considering explainability as an afterthought, AI engineers will progressively incorporate XAI concepts into the basic design of their systems. This will guarantee that AI models are robust, efficient, transparent, and interpretable.
- **Robust Regulatory Frameworks**: As governments and regulatory organizations globally adopt AI transparency and accountability regulations, AI developers will need XAI. Regulatory regimes will likely encourage the development of new XAI methodologies and standards.
- **Interdisciplinary cooperation**: The future of XAI will involve deeper cooperation between AI researchers, ethicists, domain specialists, and policymakers. By working together, these various groups may design AI systems that are not only explainable but also fair, ethical, and aligned with human values.

Recommendations for Future Research:

- **Multidisciplinary Research**: XAI research should be multidisciplinary, drawing ideas from subjects like cognitive science, ethics, and domain-specific domains such as healthcare and law. This will guarantee that XAI approaches apply to the real-world circumstances in which they are deployed.
- **Uniform Evaluation Measures**: Uniform measures are needed to assess XAI systems. These measurements should go beyond technical performance to examine variables such as user understanding, trust, and the meaningfulness of explanations.

- **User Feedback Methods**: XAI systems should constantly use user feedback methods to enhance the quality of explanations. By receiving user feedback, developers may fine-tune explanations to ensure they match the requirements of diverse users.
- **Focus on High-Stake Areas**: Future research should emphasize the development of XAI approaches for high-stakes areas where transparency and accountability are crucial, such as healthcare, banking, and autonomous systems.
- **Culture of openness**: Finally, developing a culture of openness and explainability within AI development teams is vital. AI developers should be encouraged to address explainability from the earliest phases of system design, ensuring that openness is a key element of the AI development process.

In conclusion, although XAI has considerable problems, it also provides excellent promise for boosting AI systems' trustworthiness, safety, and ethical standards. By making AI decision-making processes more visible, XAI can transform how people engage with smart apps and support more ethical and successful AI development in the future. The objective should always be to design AI systems that are accurate, efficient, transparent, fair, and aligned with human values.

## References

1. Gunning D, Aha D (2019) DARPA's explainable artificial intelligence (XAI) program. AI Mag 40(2):44–58
2. Adadi A, Berrada M (2018) Peeking inside the black-box: a survey on explainable artificial intelligence (XAI). IEEE Access 6:52138–52160
3. Guidotti R, Monreale A, Ruggieri S, Turini F, Giannotti F, Pedreschi D (2018) A survey of methods for explaining black box models. ACM Comput Surv 51(5):1–42
4. Weiser M (1991) The computer for the 21st century. Sci Am 265(3):94–104
5. Russell S, Norvig P (2020) Artificial intelligence: a modern approach, 4th edn. Pearson
6. Samek W, Montavon G, Vedaldi A, Hansen LK, Müller KR (eds) Explainable AI: interpreting, explaining and visualizing deep learning. Springer
7. Došilović FK, Brčić M, Hlupić N (2018) Explainable artificial intelligence: a survey. In: 41st International convention on information and communication technology, electronics and microelectronics (MIPRO), pp 0210–0215
8. Lipton ZC (2018) The mythos of model interpretability. Queue 16(3):31–57
9. Doshi-Velez F, Kim B (2017) Towards a rigorous science of interpretable machine learning. arXiv:1702.08608
10. Arrieta AB et al (2020) Explainable artificial intelligence (XAI): concepts, taxonomies, opportunities and challenges toward responsible AI. Inf Fusion 58:82–115
11. Gunning D (2017) Explainable artificial intelligence (XAI). In: Defense advanced research projects agency (DARPA), nd Web, vol 2
12. Arya V et al (2019) One explanation does not fit all: a toolkit and taxonomy of AI explainability techniques. arXiv:1909.03012
13. Gunning D et al (2019) XAI—explainable artificial intelligence. Sci Robo 4(37)
14. Shortliffe EH, Buchanan BG (1975) A model of inexact reasoning in medicine. Math Biosci 23(3–4):351–379
15. Holzinger A (2018) From machine learning to explainable AI. In: 2018 world symposium on digital intelligence for systems and machines (DISA), pp 55–66

16. LeCun Y, Bengio Y, Hinton G (2015) Deep learning. Nature 521(7553):436–444
17. Gilpin LH et al (2018) Explaining explanations: an overview of interpretability of machine learning. In: 2018 IEEE 5th international conference on data science and advanced analytics (DSAA), pp 80–89
18. Ribeiro MT, Singh S, Guestrin C (2016) Why should i trust you?": Explaining the predictions of any classifier. In: Proceedings of the 22nd ACM SIGKDD international conference on knowledge discovery and data mining, pp 1135–1144
19. Lundberg SM, Lee SI (2017) A unified approach to interpreting model predictions. Adv Neural Inf Process Syst 30:4765–4774
20. European Commission (2021) Proposal for a regulation laying down harmonised rules on artificial intelligence
21. Hoffman RR, Mueller ST, Klein G, Litman J (2018) Metrics for explainable AI: challenges and prospects. arXiv:1812.04608
22. Rudin C (2019) Stop explaining black box machine learning models for high stakes decisions and use interpretable models instead. Nat Mach Intell 1(5):206–215
23. Wachter S, Mittelstadt B, Russell C (2017) Counterfactual explanations without opening the black box: automated decisions and the GDPR. Harv. JL & Tech. 31:841
24. Ibid, pp 841–887
25. Du M, Liu N, Hu X (2019) Techniques for interpretable machine learning. Commun ACM 63(1):68–77
26. Molnar C (2019) Interpretable machine learning: a guide for making black box models explainable. Lulu.com
27. Clancey WJ (1983) The epistemology of a rule-based expert system—a framework for explanation. Artif Intell 20(3):215–251
28. Breiman L, Friedman J, Stone CJ, Olshen RA (1984) Classification and regression trees. CRC Press
29. Vaswani A et al (2017) Attention is all you need. Adv Neural Inf Process Syst, pp 5998–6008
30. Wachter S, Mittelstadt B, Russell C (2018) Counterfactual explanations without opening the black box: automated decisions and the GDPR. Harv J Law Technol 31(2):841–887
31. Friedman JH (2001) Greedy function approximation: a gradient boosting machine. Ann Stat 29(5):1189–1232
32. McKnight DH, Carter M, Thatcher JB, Clay PF (2011) Trust in a specific technology: an investigation of its components and measures. ACM Trans Manag Inf Syst 2(2):1–25
33. Krafft PM, Young M, Katell M, Huang K, Bugingo G (2020) Defining AI in policy versus practice. In: Proceedings of the AAAI/ACM conference on AI, ethics, and society, pp 72–78
34. Miller T (2019) Explanation in artificial intelligence: Insights from the social sciences. Artif Intell 267:1–38
35. Barocas S, Hardt M, Narayanan A (2019) Fairness and machine learning. fairmlbook.org
36. Doran D, Schulz S, Besold TR (2017) What does explainable AI really mean? A new conceptualization of perspectives. arXiv:1710.00794
37. Lee JD, See KA (2004) Trust in automation: designing for appropriate reliance. Hum Factors 46(1):50–80
38. Bellamy RKE et al (2019) AI fairness 360: an extensible toolkit for detecting and mitigating algorithmic bias. IBM J Res Dev 63(4/5)4:1–4:15
39. Mittelstadt BD, Allo P, Taddeo M, Wachter S, Floridi L (2016) The ethics of algorithms: mapping the debate. Big Data Soc 3(2):1–21
40. Schmidt P, Biessmann F (2019) Quantifying interpretability and trust in machine learning systems. arXiv:1901.08558
41. Tjoa E, Guan C (2021) A survey on explainable artificial intelligence (XAI): toward medical XAI. IEEE Trans Neural Netw Learn Syst 32(11):4793–4813
42. Bussmann N, Giudici P, Marinelli D, Papenbrock J (2021) Explainable AI in credit risk management. Comput Econ 57:203–216
43. Koo J et al (2015) Why did my car just do that? Explaining semi-autonomous driving actions to improve driver understanding, trust, and performance. Int J Interact Des Manuf 9:269–275

44. Holzinger A, Langs G, Denk H, Zatloukal K, Müller H (2019) Causability and explainability of artificial intelligence in medicine. Wiley Interdiscip Rev Data Min Knowl Discov 9(4):e1312
45. Gasser TM, Westhoff D (2012) BASt-study: definitions of automation and legal issues in Germany. In: Proceedings of the 2012 road vehicle automation workshop
46. Lapuschkin S et al (2019) Unmasking Clever Hans predictors and assessing what machines really learn. Nat Commun 10(1):1096
47. Angwin J, Larson J, Mattu S, Kirchner L (2016) Machine bias. ProPublica
48. Burrell J (2016) How the machine 'thinks': understanding opacity in machine learning algorithms. Big Data Soc 3(1):1–12
49. Brynjolfsson E, McAfee A (2014) The second machine age: work, progress, and prosperity in a time of brilliant technologies. W. W. Norton & Company
50. Chesney R, Citron DK (2019) Deep fakes: a looming challenge for privacy, democracy, and national security. Calif Law Rev 107:1753–1820
51. Frey CB, Osborne MA (2017) The future of employment: how susceptible are jobs to computerisation? Technol Forecast Soc Chang 114:254–280
52. Zerilli J, Knott A, Maclaurin J, Gavaghan C (2019) Transparency in algorithmic and human decision-making: is there a double standard? Philos Technol 32:661–683
53. Goodman B, Flaxman S (2017) European union regulations on algorithmic decision-making and a "right to explanation." AI Mag 38(3):50–57
54. Floridi L et al (2018) AI4People—an ethical framework for a good AI society: opportunities, risks, principles, and recommendations. Mind Mach 28:689–707
55. Wachter S, Mittelstadt B, Floridi L (2017) Why a right to explanation of automated decision-making does not exist in the general data protection regulation. Int Data Priv Law 7(2):76–99
56. Jobin A, Ienca M, Vayena E (2019) The global landscape of AI ethics guidelines. Nat Mach Intell 1:389–399
57. High-Level Expert Group on Artificial Intelligence (2019) Ethics guidelines for trustworthy AI. European Commission
58. The IEEE Global Initiative on Ethics of Autonomous and Intelligent Systems (2019) Ethically aligned design: a vision for prioritizing human well-being with autonomous and intelligent systems, 1st edn. IEEE
59. Hacker P, Krestel R, Grundmann S, Naumann F (2020) Explainable AI under contract and tort law: legal incentives and technical challenges. Artif Intell Law 28:415–439
60. Selvaraju RR et al (2017) Grad-CAM: visual explanations from deep networks via gradient-based localization. In: Proceedings of the IEEE international conference on computer vision, pp 618–626
61. Pearl J, Mackenzie D (2018) The book of why: the new science of cause and effect. Basic Books
62. Puiutta A, Veith E (2020) Explainable reinforcement learning: a survey. Mach Learn Knowl Extr 77–95
63. Slack D, Hilgard S, Jia E, Singh S, Lakkaraju H (2020) Fooling LIME and SHAP: adversarial attacks on post hoc explanation methods. In: Proceedings of the AAAI/ACM conference on AI, ethics, and society, pp 180–186
64. Aletras N et al (2016) Predicting judicial decisions of the European court of human rights: a natural language processing perspective. PeerJ Comput Sci 2:e93
65. Baker R, Siemens G (2014) Educational data mining and learning analytics. In: Sawyer RK (ed) The Cambridge handbook of the learning sciences, 2nd edn. Cambridge University Press, pp 253–272
66. Bastani O, Kim C, Bastani H (2017) Interpreting blackbox models via model extraction. arXiv: 1705.08504
67. Kim B, Wattenberg M, Gilmer J, Cai C, Wexler J, Viegas F, Sayres R (2018) Interpretability beyond feature attribution: quantitative testing with concept activation vectors (TCAV). In: Proceedings of the 35th international conference on machine learning, pp 2668–2677
68. Mueller ST, Hoffman RR, Clancey W, Emrey A, Klein G (2019) Explanation in human-AI systems: a literature meta-review, synopsis of key ideas and publications, and bibliography for explainable AI. arXiv:1902.01876

69. Pu P, Chen L (2006) Trust building with explanation interfaces. In: Proceedings of the 11th international conference on intelligent user interfaces, pp 93–100
70. Weller A (2019) Transparency: motivations and challenges. In: explainable AI: interpreting, explaining and visualizing deep learning, Springer, pp 23–40
71. Kizilcec RF (2016) How much information? Effects of transparency on trust in an algorithmic interface. In: Proceedings of the 2016 CHI conference on human factors in computing systems, pp 2390–2395
72. Eiband M, Buschek D, Kremer A, Hussmann H (2019) The impact of placebic explanations on trust in intelligent systems. In: Extended abstracts of the 2019 CHI conference on human factors in computing systems, pp 1–6
73. Kulesza T, Stumpf S, Burnett M, Yang S, Kwan I, Wong WK (2013) Too much, too little, or just right? Ways explanations impact end users' mental models. In: 2013 IEEE symposium on visual languages and human centric computing, pp 3–10
74. Selbst AD, Powles J (2017) Meaningful information and the right to explanation. Int Data Priv Law 7(4):233–242
75. Barocas S, Selbst AD (2016) Big data's disparate impact. Calif Law Rev 104:671–732
76. Veale M, Binns R, Edwards L (2018) Algorithms that remember: model inversion attacks and data protection law. Philos Trans R Soc A Math Phys Eng Sci 376(2133):20180083
77. Citron DK, Pasquale F (2014) The scored society: due process for automated predictions. Wash Law Rev 89:1–33
78. Edwards L, Veale M (2017) Slave to the algorithm? Why a 'right to an explanation' is probably not the remedy you are looking for. Duke Law Technol Rev 16(1):18–84
79. Ibid (2020) 38(4):523–531
80. Kim J, Rohrbach A, Darrell T, Canny J, Akata Z (2018) Textual explanations for self-driving vehicles. In: Proceedings of the European conference on computer vision (ECCV), pp 563–578
81. Doshi-Velez F et al (2017) Accountability of AI under the law: the role of explanation. arXiv: 1711.01134
82. Raghavan M, Barocas S, Kleinberg J, Levy K (2020) Mitigating bias in algorithmic hiring: evaluating claims and practices. In: Proceedings of the 2020 conference on fairness, accountability, and transparency, pp 469–481
83. Eiband M, Schneider H, Bilandzic M, Fazekas-Con J, Haug M, Hussmann H (2018) Bringing transparency design into practice. In: 23rd international conference on intelligent user interfaces, pp 211–223
84. Mohseni S, Zarei N, Ragan ED (2021) A Multidisciplinary survey and framework for design and evaluation of explainable AI systems. ACM Trans Interact Intell Sys 11(3–4):1–45
85. Samek, Wojciech, Müller K-R (2019) Towards explainable artificial intelligence. In: Explainable AI: interpreting, explaining and visualizing deep learning, pp 5–22
86. Hoffman RR, Mueller ST, Klein G (2017) Explaining explanation, part 2: empirical foundations. IEEE Intell Syst 32(4):78–86
87. Amershi S, Weld D, Vorvoreanu M, Fourney A, Nushi B, Collisson P, Suh J, Iqbal S, Bennett PN, Inkpen K, Teevan J (2019) Guidelines for human-AI interaction. In: Proceedings of the 2019 CHI conference on human factors in computing systems, pp 1–13
88. Floridi L, Cowls J (2019) A unified framework of five principles for AI in society. Harv Data Sci Rev 1(1)
89. Selbst AD, Barocas S (2018) The intuitive appeal of explainable machines. Fordham Law Rev 87(3):1085–1139ss

# Chapter 2
# Extracting Specifications Through Verified and Explainable AI: Interpretability, Interoperability, and Trade-Offs

**Giacomo Bergami, Oliver Robert Fox, and Graham Morgan**

**Abstract** This chapter introduces new Verified and Explainable AI terminology to better describe such algorithms by pigeonholing them. The combined provision of three types of explainability (a priori, ad hoc, and ex post) as advocated by *hybrid explainability* helps with verification while improving efficiency results by better structuring the specification space and accuracy ones, by enabling both machine and human explainability of the data while providing structured and more machine-readable representation. These can then be contextualised in a general framework named General Explainable and Verifiable Artificial Intelligence (GEVAI) showcasing the possibility of combining already-existing algorithms in a full data-driven pipeline. This observation is backed up by the latest results on hybrid explainability, which outperform the current state-of-the-art results. Last, we pose some final challenges for extending GEVAI into a data science pipeline while drawing similarities with evolutionary data-aware microservice orchestration.

## 2.1 Introduction

This chapter investigates the interpretability of several AI approaches by controlling the correctness of the training phase and explaining the classification outcome. This makes the peculiarities of the algorithms' inference phase stand out while discussing the different possible data representations, as these will naturally lead to different specifications, leading to varying amounts of specification declarativity. The chapter ponders the methodologies of extracting declarative specifications from raw data that

G. Bergami (✉) · O. R. Fox · G. Morgan
Newcastle University, Newcastle Upon Tyne, UK
e-mail: giacomo.bergami@newcastle.ac.uk

O. R. Fox
e-mail: o.fox3@newcastle.ac.uk

G. Morgan
e-mail: graham.morgan@newcastle.ac.uk



N. Bizon and B. Appasani (eds.), *Explainable Artificial Intelligence for Trustworthy Decisions in Smart Applications*, https://doi.org/10.1007/978-3-031-97007-8_2

still require a preliminary pre-processing phase and discusses how various levels of explainability might improve the training phase.

This chapter will bridge Verified with Explainable Artificial Intelligence through a cohesive framework as outlined in Fig. 2.1: this not only provides an index to the main sections and approaches being discussed in this chapter but also provides a generalisation over the main mining and learning algorithms (Sect. 2.4).

Unlike former surveys [32], this chapter does not differentiate such algorithms by the type of classification tasks rather than by the different types of explanations used and their function in the training or testing phase. The former analysis does not go at the essence of several types of explanations, mainly considering a *static ex post* explainability, which is just a minor part of this chapter (Sect. 2.3.3). Last, this chapter will support the need of pipelining different algorithms to actually achieve robust explainable and effective AI providing logical representation of the data, while showcasing the limitations of loss minimisation driven algorithms, thus remarking on the limitations derivable from mainly using LLMs [6, 47] for Artificial General Intelligence (AGI) as presented in [31] over biased and contradictory data.

**Attention**
This chapter works on some of the criticisms raised on the Penrose-Lucas argument by assuming one of the two possible alternatives: general intelligence can be achieved through Turing Machine reasoning using some sort of paraconsistent logic [54, 60] (Sect. 2.3.2.2). The other alternative entails that intelligence cannot be framed within a logical/computable system.

## 2.2 Verified Artificial Intelligence

The ultimate goal of AI is to adopt verification mechanisms through data interpretability and correctness checking, also referred to as formal verification, thus supporting the notion of making such explanations machine-readable and promptly verifiable while streamlining human- and automating machine-driven verification. This set of algorithms leads to the intuition that explainability is not necessarily human-centred, as machines can benefit from such explanations by modelling the environment through specifications. Any Verified Artificial Intelligence algorithm [72] relies on the notion of a system $S$, being the object of our study, and an environment $E$, in which such a system interacts. Specifications $\varphi$ can either check whether $S$ abides by it within the given environment (*formal verification*: $S, E \models \varphi$) or summarise system's properties (*specification mining*). The majority of the works within

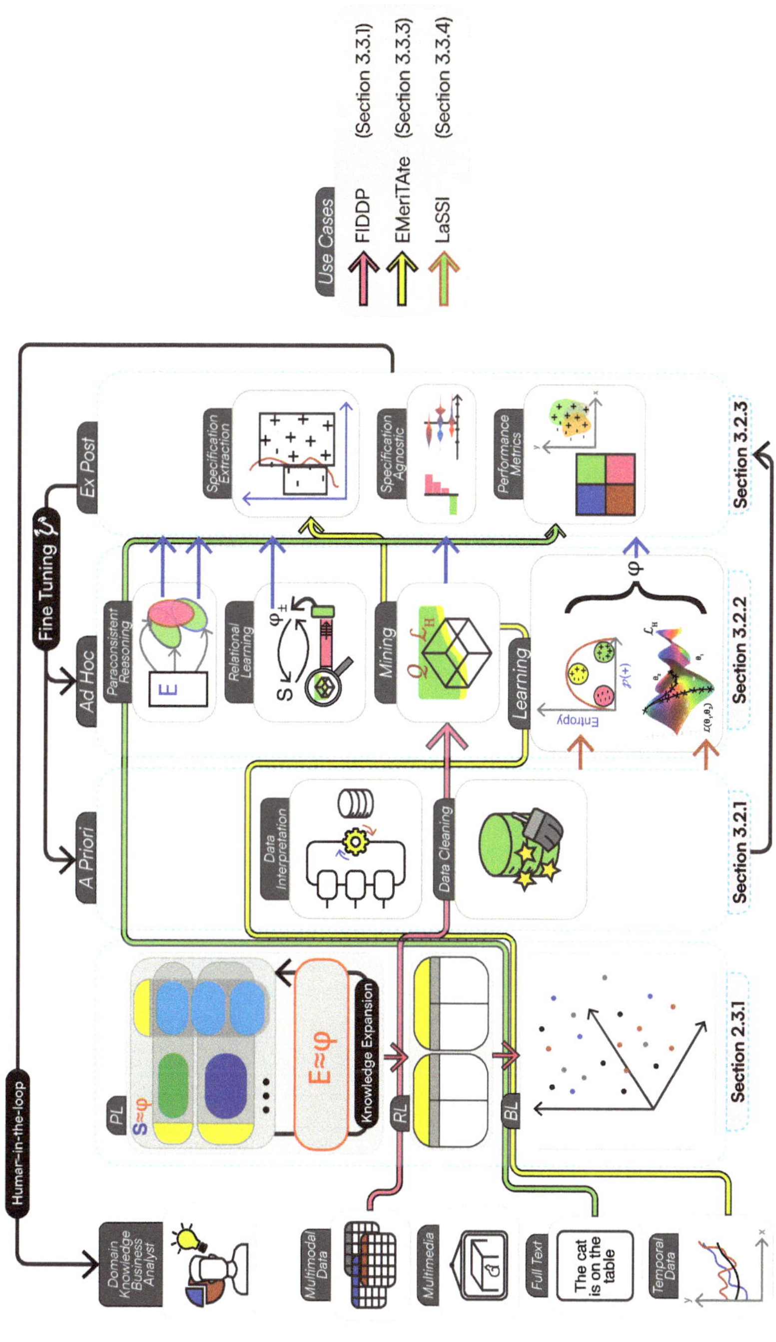


**Fig. 2.1** Our proposed General Explainable and Verifiable Artificial Intelligence (GEVAI) framework

this field freely assume that specifications are explainable as they are expressible in logic form, as the definition of specifications leads to the generation of a system satisfying all the requirements given (*formal synthesis*).

This formalisation, adopted initially from [72], has been recently extended also to support the definition of a dynamic system $S$ changing state in time, thus considering a set of snapshots $S = \{S_1, \dots, S_n\}$ [14]. This establishes an association between each possible state $c \in \mathcal{Y}$ the system is at a time $S_t$, and a specification $\varphi_c$ satisfying the formal verification requirement, thus modelling its behaviour as $\mathcal{M}(S_t) = \varphi_c$. This extension does not consider the Verified Artificial Intelligence algorithm as a real agent interacting with such a system: the agent should also act on the environment as a reaction to the perceived state $c$ which, consequently, can respond with a possible internal state change. This requires extending the former to capture a reinforcement learning problem [75] where the agent implicitly learns an action function $\alpha \colon \mathcal{Y} \to \mathrm{Act}$ as a response to which the system will potentially react by evolving into a next state using a transition function $\delta : \mathrm{Act} \times S \to S$. This describes the temporal evolution of a binary system comprising of an agent, being the Verified Artificial Intelligence algorithm, and the environment as the fixpoint of this function:

$$\mathrm{Evolution}\,(S_t) = \delta\,(\alpha\,(\mathcal{M}\,(S_t))\,,\,S_t)$$

The agent can break the evaluation of the former at any point to allow itself to learn from the novel experience if considering external stimuli from $S_{t+1}$ (Sect. 2.4), thus naturally leading to continuous learning environments [59]. This can be further generalised by considering the agent as an orchestration of different services rather than a single monolith, thus leading to data-science-driven evolutionary architectures [9].

**Background Information**

At the time of writing, much literature abides by the terminology above and, therefore, can be easily categorised [72]. Some domain-specific literature is divided into the terminology to be adopted, thus giving the readers the wrong impression that distinct terms refer to unrelated concepts. For example, within the field of Business Process Management [77], formal verification is referred to as *conformance checking*, specification mining is referred to as *process mining*, and formal synthesis is referred to as *replay*. In this field, the notion of an environment is ignored, while the system is considered as a collection of traces describing the running of multiple distinct system instances (*log*). Specifications can be either procedural, thus being described as either Petri Nets [13] or State Machines [12], or declarative, thus being described in a specific temporal language of choice, such as either $\mathrm{LTL}_f$ [18] or DECLARE*d* [8]. Furthermore, temporal specification learning is a broader definition from Business Process Management's *deviance mining*, for which current literature only considers having as many system snapshots as classes of interest.

Last, concerning mainstream AI terminology, specifications are referred to as hypotheses [28], models [77], or functions. *Specification* here refers to all these concepts in a unified way, while the (background) knowledge base [28] is assimilated to environment, as $S, E \models \varphi$ in relational learning is often used for checking whether **PL** data $S$ is covered by the previously mined specifications $\varphi$ while satisfying the rules in $E$. Data and system are used interchangeably and denoted as $S$.

## 2.3 Explainable AI

Figure 2.1 summarises most of current literature on Explainable AI through the lenses of a General Explainable and Verifiable Artificial Intelligence (GEVAI)[1] framework. At the time of the writing, explainable approaches can fit into two macro-categories, the *static* approaches (Sect. 2.3.2), considering only one possible kind of Explainable AI methodology where each algorithm might be considered as one single instantiation of a specific subcategory, and the *hybrid* ones (Sect. 2.3.3), considering a combination of multiple possible AI explanation technologies. The chapter identifies three kinds of static explainability in the form: a priori (Sect. 2.3.2.1), ad hoc (Sect. 2.3.2.2), and ex post (Sect. 2.3.2.3). Discussing how different data representations help with subsequent explainability will substantiate the essence of both approaches.

### 2.3.1 *Structured Data Representations*

Talking about AI explainability requires first dealing with machine-level expressiveness by adding more contextual information to our data by increasing the amount of structure. Current literature recognises the current hierarchy of structured data representations [28] as shown in Fig. 2.2:

Boolean (BL): each data entry is expressed as a single atom or 0-ary predicate associated with a truth value over a possible world. As specific data values can also be considered atomic predicates [68] that can be tracked down while processing data [26], simple data values (numbers, strings) can be represented as a single atom. This notion can be further extended with uncertainty scores or probabilistic values by associating provenance information [43].

Attribute-Value (AV): each data entry is expressed as a single record with no external dependencies, where values can be either nominal (or categorical), ordinal (i.e., totally ordered), or continuous (in $\mathbb{R}$ or any floating-point truncation). It is often

[1] https://github.com/lo-ds/gevai.

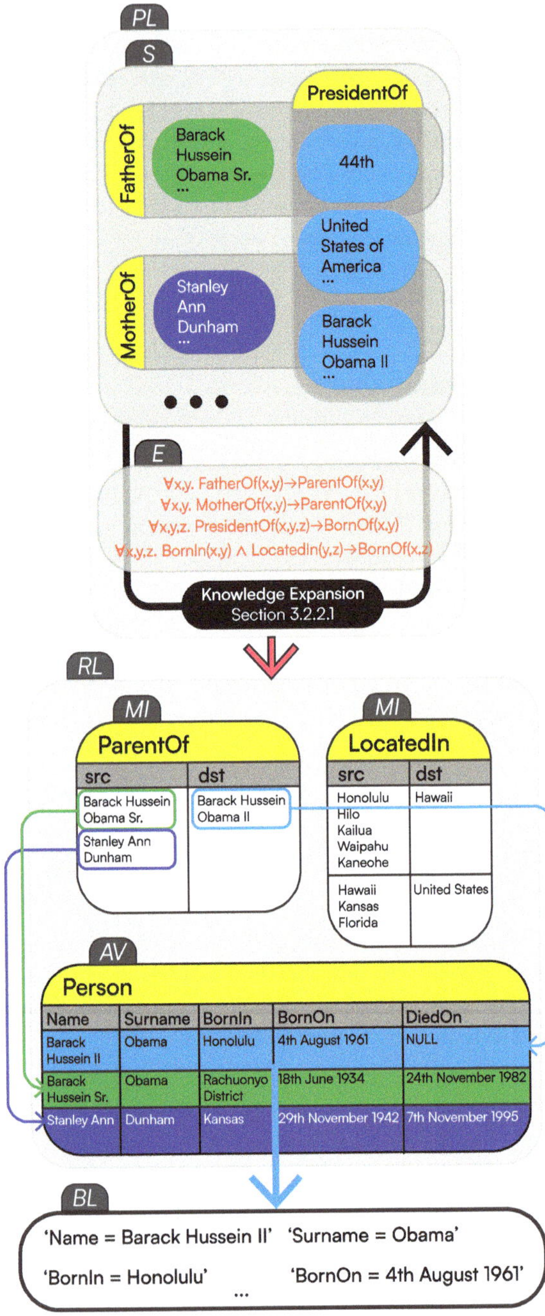


**Fig. 2.2** Expressing disparate structured data representation formats with structural degradation

assumed that each record shall represent at most one fact, which shall also be associated with a probabilistic [41] or uncertainty [3] score.

Multi-instance (MI): each record, as the former, can be further extended by associating nested content, usually expressing is-a or part-of hierarchical entailment. This data can be represented within all the representations violating the First Normal Form (1NF) [35], thus including nested relational or semistructured data. This can be accommodated into object-oriented data models with uncertainty scores [11].

Relational (RL): **RL** represents multiple records that might belong to different separate relationships. Given that this level is designed to be a generalisation of the **MI** model, **RL** can be extended to fit both object-oriented [11] and semistructured data models as a further generalisation of **MI** representations.

(Logic) Programs (LP): while **MI** implicitly provided the notion of entailment through is-a or part-of relationships expressed through the notion of containment and **RL** considers multiple types of relationships, **LP** extend the former by explicitly providing logical connectives for data ($S$), thus including the notion of entailment when also describing rules ($E$), while supporting the definition of recursive predicates. As any logical framework being as expressive or more than First-Order Logic (FOL) is at least semi-decidable, this greatly undermines the ability to create converging inference algorithms. Thus, data science often considers fragments of FOL, such as AE, where all the universal quantifiers (A) precede the existentials (E) [45], and the Horn clauses commonly used in Prolog [25]. Algorithms can deal with an higher level of content by considering programs as data. This is considered the representation goal within Verified Artificial Intelligence domains.

Given that these representations are already in a hierarchy of expressiveness [28], these lead to the following statements:

Interpretability: while the logical programs provide a more accessible level of human interpretability, AI algorithms dealing with lower representation within the hierarchical level can be easier to implement and interpret (e.g., debug or test). Nevertheless, due to their higher contextual nature, the higher level of data representation provides better chances for verifiability and interpretability.

Interoperability: a higher representation form might be structurally simplified into a simpler one, while the converse approach, providing more structure to the data, would require *specification mining* (see "Climbing up the Data Representation Ladder (Specification Mining)") considering more contextual information. This helps with testing the training abilities of different algorithms assuming different data representation inputs.

Trade-Off: as most of the traditional AI algorithms consider less expressive forms of data representation (e.g., are not considering data as logical programs), any rewriting to a lesser form of expressiveness allows feeding the data to off-the-shelf algorithms at the cost of an exponential blowup on the final data representation size. Easier data formats lead to easier algorithms to write and maintain.

## 2.3.2 *Static Explainability*

### 2.3.2.1 A Priori

The characterisation in Sect. 2.3.1 does not fully capture the information content for non-structured data. When considering images, the machine cannot immediately derive complex relational information by just acquiring its raw representation, thus requiring a preliminary explanation step for representing the content of an image. After doing this, the AI algorithm can explain the data it is referring to and potentially extract information concerning the entities within the picture and their spatio-temporal relationship and state [81]. Similarly, when considering raw full text information, a machine cannot directly understand its sentence's logical representation [40] nor complex argumentation structure [26]. For this, algorithms have to interpret the raw data information while representing it in one of the forms compatible with the representations above. Similar considerations can also be carried out for temporal data, where convenient representations include data trends [49], mainly considering different growth and decrease patterns within the data and declarative activity correlations [8], which can compose the former trends to establish temporal correlations across different temporal dimensions (Sect. 2.3.3.3).

**Important**
Finally, an a priori explanation algorithm works like a transformation function of input data towards a novel representation expressing the information needed for the learning algorithm. The a priori categorisation does not consider the explainability of the algorithm itself, rather than the explanation that can be added to the data.

Any a priori explanation algorithm also providing a logical program describing the data is then considered a *specification mining* algorithm (see "Mining by Quality-Driven Specification-Space Navigation").

This Section discusses two classes of a priori explainability algorithms: *data cleaning and integration*, and *specification mining*. These provide different degrees of **interpretability**: while the former allows the machine to correctly interpret the data by rectifying the information contained in it but not increase the amount of data information by extending it with further contextual information, the latter also enriches the data representation by also climbing up its hierarchy through contextual information. The latter type allows for a greater form of algorithmic **interoperability** by representing disparate data formats into a common and machine-readable representation, thus allowing the use of the resulting information in other domains of interest. The following paragraphs discuss the **trade-offs** per specific class of algorithms.

Data Cleaning and Integration

Data cleaning helps with data **interpretability** in two ways: first, the correctness of the data ensures the correctness of downstream algorithms, such as data deduplication (Sect. 2.3.3.1) or, generally speaking, mining or learning (Sect. 2.3.2.2).

A simple data transformation function concerns the data cleaning process for single data fields. These techniques mainly deal with **AV** data. This cleaning might adopt statistical techniques [85], such as simple statistical rules learned from the data, or derived from hardcoded functional dependencies [64]. After considering each attribute as a random variable, the data cleaning process can be seen as the way to associate the goal variable with the most frequent value given the same values from the target variables. The embedding of variables in a probabilistic model (e.g. factor graphs [52]) enables the refinement of the former simplistic approach by changing the former consideration across multiple functional dependency rules, thus boiling down the data cleaning to a probabilistic inference and data imputation [68]. A specific class of data cleaning algorithms considers the extension of data records by replacing missing values with ones being implicitly inferred from the record of interest; these are referred to as data imputation methods [50].

A more advanced set of data cleaning tools entails rewriting rules and detecting potential violations and fixes. As the detection of violations can be easily representable as database-wise integrity constraints, these algorithms often assume **RL** data representations, thus considering data cleaning problems across different relationships with the same database. Among all the possible database repairs, the one that minimises the number of changes are usually preferred [12] while minimising the number of violations on the target database [27].

Last, data integration problems are a higher form of data cleaning problem, ensuring homogeneous data representation across databases. As these approaches often entail the usage of algorithms exploiting the notion of description logic, these approaches often assume data to be represented in **LP**, where entities are associated with rich data property information. A data integration task aims at reconciling entities [61], generalisable as concepts, while considering a data transformation function and an uncertainty score for the alignment [9]. To ensure a uniform data representation across databases, it is often preferred to generate a uniform schema representation so to represent all the possible entities across databases [36], for then transforming the original databases to this ultimate data format via the alignment information [44]. Given such approaches' mechanic and automated nature, these can be easily representable as fully interpretable ad hoc approaches.

Concerning the **trade-offs** for these algorithms, each algorithm working at a higher level of data abstraction cannot necessarily capture the minute operations expressed at a lower level, and vice versa: data integration algorithms often assume the data to be already cleaned and thus are not necessarily ensuring the cleanness of the original data sources before conducting the main data integration task. On the other hand, algorithms for data cleaning and imputation consider the repair of records at a lower representation level and do not consider data rewriting operations, allowing the transformation of a record to comply with a target schema. This suggests that a full data science pipeline should support all of these algorithms to be applied in an increasing level of abstraction. This is also confirmed by standard practices in data deduplication pipelines, which first clean the data, repair the record at the value level, and then remove duplicate data after assumedly clean data [3]. Such features are currently missing from general-purpose data science pipelines [9].

Last, any *knowledge (base) expansion* or *grounding* process is a way to "clean" and explain the data by generating further explanations derivable from the initially available information. Rule-based systems such as the early MYCIN [86] or description logic reasoners [5] rewriting factoid information represented in **RL** (e.g., ABox) using explainable rules in **LP** (e.g., TBox) are two possible examples: each rule works by matching several factoid elements for then generating new knowledge from its application. By tracking each piece of generated data, the original data that was used to derive it, as well as the rule generating such data [7] through the usage of provenance techniques (Sect. 2.3.3.4), the derived information is explained in terms of the initially available data and the rules being applied. This approach can also be refined to consider probabilistic graph models for dealing with uncertainties [23] or conflicting information [83]. Still, these are not well suited to capture conflicting and contradictory information within the data [53] as "*rational acceptance based on a probability threshold cannot be closed under logical consequence*" and, most importantly, "*it will not be closed under conjunction*" as this will always generate sentences with lesser probability and potentially draw arbitrary conclusions from any inconsistency [17]. This then motivates an orthogonal approach by reasoning in a purely logical way and paraconsistently by ignoring the uncertainty scores, to then use later on within a subset of consistent information (Paraconsistent Reasoning).

#### Climbing up the Data Representation Ladder (Specification Mining)

Data transformation algorithms extend raw or low-level data representations by gradually increasing the amount of structured information. These techniques leverage, to some extent, contextual information to infer additional structure information. Algorithms for (multi-)word named entity recognition [39] aggregate distinct near words into one single entity while associating its type information (e.g., geographical location, GeoPolitical Entity (GPE), organisation, person). By considering each word as part of **BL**, this representation transforms data to the **RL** level. Such algorithms not only extend each single word with grammatical and lexical information, but also provide multiple levels of **MI** information: first, by representing each word as being part of a possible multi-word entity, and then by associating each entity (be that single or multi-word) to type information. In addition to the relationship among words, entities, and type, such information extraction can further support the extraction of relationships among entities as per the original full text; in their simplest form, such relationships can be immediately inferred from verbs expressing actions, properties, or states [40]. This interpretation of entities and verbs occurring in the text can be leveraged to extract further domain-dependant relationships, such as *location_of* and *employee_of* [30], providing greater human-level explainability than the former through information reduction. This stage reaches full **RL** data representation, where relationships contain metadata information and express relationships across previously mined entities. This intermediate representation makes it relatively straightforward to represent each sentence in **LP**, and then transform the logical connectives occurring as relationships in **RL** into proper logical connectives [40].

In image data analysis, the named entity recognition task is often called *object detection* [89]. These algorithms work by identifying a bounded (often rectangular) region of space as containing an object, and then classifying such object to a specific type. The algorithm can also identify several different entities within the same image, thus allowing the recognition of multiple entities. Very recently, this stream of research also showed the possibility of not only extracting objects/entities but also extracting multiple relationships from text (e.g., relative position of objects in space) [81], thus achieving full **RL** data representation, which can be then easily represented in **LP** as single prepositions. Similar considerations can be applied to time series analysis. As the current state-of-the-art algorithm requires combining various explainability steps, it is deferred when discussing hybrid explainability (Sect. 2.3.3.2).

Last, specification mining algorithms often achieve increasing data generality by stacking together different representation aims (e.g., extracting the entities, then the relationships, for finally deriving a logical representation). By pipelining the algoirthms in increasing level of structural data complexity, ensures both increases in **interpretability** as well as of pipeline **interoperability**. Consequently, the major **trade-offs** these specification mining algorithms pose are related to the amount of achievable expressiveness, thus generally preferring algorithms returning **LP** representations to **BL** or **AV** ones to the cost of an increased running time. The next Section will further solidify these claims by contextualising how data representations affect the explainability of the inference process itself.

#### 2.3.2.2 Ad Hoc

**Important**
An algorithm supports **ad hoc** explainability if it can motivate the choices for a specific decision by associating a descriptive value to the decision being taken *during* the learning process.

This substantially differs from common *ex post* explainability approaches (Sect. 2.3.2.3), which are usually taken to extract the knowledge from a learned or mined specification after the ad hoc phase. The definition of all the forthcoming algorithms is completely independent of the specific structured data representation of choice, as the categorisation of such algorithms depend on how the act of generating new specifications leverages the data properties in an explainable way, thus helping with human debugging. These are summarised in Fig. 2.3.

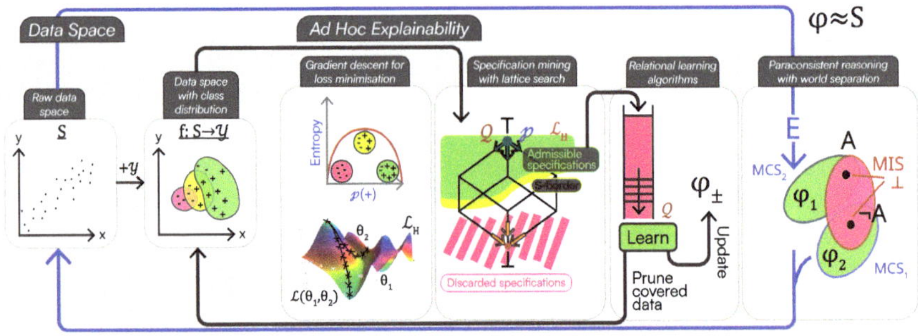


**Fig. 2.3** Showcasing different ad hoc explainable approaches as well as their interconnections

The forthcoming AI algorithms are introduced in terms of increasing explainability achieved *during* the learning process by better structuring the specifications' search space, while providing suitable insight for the derivation of a specification. These show that stronger assumptions on the data representation improve the decision's interpretability while imposing trade-offs in task learning's generality.

Learning as a Loss Minimisation Problem

Given an annotated dataset described as a mapping $f : S \to \mathcal{Y}$ between the data examples $S \subseteq \mathcal{L}_E$ and some classes in $\mathcal{Y}$, the goal of a **(classical) machine learning** algorithm is to learn a function $M \simeq \varphi$ called *specification* expressed in a specific target language $\mathcal{L}_h$ while minimising the number of classification errors over $f$ [28]. Thus, given a dataset in which data points are associated with a positive and a negative class, $\mathcal{Y} = \{\mathbf{1}, \mathbf{0}\}$ , $f^{-1}(\mathbf{1})$ denotes the *positive* dataset, and $f^{-1}(\mathbf{0})$ the *negative* dataset.

Such $\varphi$ can be refined via a *loss function* minimisation problem [46]: given a set of classes $\mathcal{Y}$ and a distance function between classes $\delta : \mathcal{Y}^2 \to \mathbb{R}$, a loss function $L_\delta(f, \varphi)$ over $\delta$ is defined as an arbitrary and domain-dependant aggregation measure over each classification error $\delta(f(\mathbf{x}), \varphi(\mathbf{x}))$ for each $\mathbf{x} \in \mathrm{dom}(f)$. This does not consider the specificity of the learning algorithm itself, which might define its own way of navigating the search space. Still, the learning task can be seen as seeking $\varphi$:

**Definition 1** (*Classical Machine Learning*) The goal of learning a specification within a target language $\mathcal{L}_h$ minimising the classification error from an annotated dataset $f$ is defined as [28]:

$$\texttt{Learn}(f, \mathcal{L}_h) = \arg\min_{\varphi \in \mathcal{L}_h} L_\delta(f, \varphi)$$

A specific learning algorithm considers a target specification language for fitting a specification: e.g., a linear regression problem has $\mathcal{L}_h$ as the set of all the possible hyperplanes, a Deep Neural Network (DNN) training algorithm will have $\mathcal{L}_h$ as all the network configurations (e.g. dropout) obtainable from one or more initial seed configurations as well as its neuron weights. Thus, the learning algorithm looks for the specification that best fits the data by exploiting an algorithm ensuring that the specification space is visited to minimise the loss and find a local minimum. While doing so, these algorithms usually start from an initial specification and refine it while performing the above search. Different loss functions and minimisation techniques depend on the specification language of choice: gradient descent algorithms relates to specifications that can be expressed as differential multivariate equations [16], while maximum likelihood estimation is used to fit most statistical distributions [34]. Other approaches, such as decision trees [82], exploit impurity functions such as entropy measures or diversity indices [87].

Please observe that the problems above have different expressive powers depending on the specification language of choice. While Graph Neural Networks (GNNs) can solve approximations of NP-complete problems [66], transformers are likely not to capture simple closed Boolean formulae [74]. Furthermore, these approaches only return one possible fitted specification by design, albeit a probabilistic interpretation of the data might approximate this, rather than returning a set of all possible specifications compliant with the data. Conversely, this appears to be crucial when considering multiple possible alternatives over a data representation containing conflicting and contradictory information. This issue is not entirely solved by the adoption of Logical Neural Networks [55], as classical logic is still hampered by *reasoning by contradiction*, thus requiring orthogonal approaches. This, along with the aforementioned impossibility of reasoning with conflicting information in a probabilistic way (Paraconsistent Reasoning), further motivates the introduction of paraconsistent reasoning.

The major **trade-off** is posed by the possibility of solving many problems, albeit approximately, while not providing an immediate understanding of the thought process while deriving the specification, thus including its correctness or quality. This is a direct consequence of how the specification search is structured: these are often guided by (hyper)parameters restricting the number of possible specifications that can be obtained, thus often requiring heuristics to fasten the search space. On top of that, Neural Networks have initialisation parameters, such as initial weights and biases, the number of layers, and their activation functions. Decision trees have the maximum tree depth or an acceptable purity level. Thus, these restrict the specifications to return, thus forcing the algorithm to perform (hyper)parameter tuning for searching across a larger subset of all possible specifications. A possible solution to overcome this limitation is to exploit hyperparameter optimisation techniques [38]. Due to the minimisation problem not establishing an explicit explainable correlation between specification $\varphi$, quality measure (loss function) $L_{\mathcal{S}}$, and data space $(S, f)$, these methodologies cannot leverage a better specification search, as this often boils down to data space and parameters navigation. Consequently, the **interpretability** of the decision process is limited to loss minimisation, with no guarantee to reach the global minimum given the hyperparameter assumptions and no direct control over the quality of the specification.

Mining by Quality-Driven Specification-Space Navigation

*Mining algorithms* [28] address some of the limitations of the learning problems. By structuring the search space as a lattice, they start the search from the most general (or most specific) specification and then descend (or ascend) through a refinement $\rho$ (or most general unifier) operation defined over a general-to-specific relationship $\preccurlyeq$. Efficiency-wise, general-to-specific (i.e. top-down) are generally preferred to the specific-to-general ones (i.e. bottom-up) [63]. Thus, different from the previous approaches, mining algorithms navigate the specifications space rather than the data or parameter space, thus giving us direct control on the quality of the specifications. This can be selected through monotone quality measures $Q$ (e.g., support) stopping the generation of specifications as soon as the set of possible specifications starts to deteriorate. Such metrics can exploit the notion of the *coverage* $\mathbf{co}(\varphi, f)$ of a specification $\varphi$ over a subset of data $S' \subseteq \mathrm{dom}(f)$, thus referring to the number of possible examples captured by our specification of choice.

**Definition 2** (*Data Mining*) The goal of mining a set of specifications $\varphi$ within a target language $\mathcal{L}_h$ and satisfying a quality constraint $Q$ over the annotated dataset $f$ is defined as [28]:

$$\texttt{Mining}_Q(f, \mathcal{L}_h) = \{\varphi \in \mathcal{L}_h | Q(\varphi, f) \quad holds\}$$

Please observe that the specification space $\mathcal{L}_h$ should not necessarily coincide with the data space $S$: while these coincide for *frequent itemset* mining [1] by considering the set of all the possible itemsets [28], *association rules* $X \mapsto Y$ [76] slightly refine over this search space by structuring specifications with a head (or consequence) $Y$ and a tail (or premise) $X$. With respect to temporal declarative languages such as DECLARE*d*, the notion of logical entailment across different declarative clauses structures the search space and guarantees an efficient monotonic search [10]. When considering **RL** data, this algorithm can also be used to derive integrity constraints from clean data [29].

The way the specification space is structured increases the **interpretability** of the learning approaches by ranking the specifications according to a quality metric embedded in the predicate $Q$, and $\preccurlyeq$-generality. Notwithstanding the former, this quality control might skew the specification generation when considering biased information. As coverage-based quality predicates $Q$ might favour over-represented misinformation data than plausible ones supported by trustworthy domain knowledge, this might bias the mining of the available data when considering misinformation scenarios. Biased content is relatively abundant as it is known to travel faster than reliable information [80]. This then poses a major **trade-off** with respect to the specification's quality: despite the possibility of exploiting these techniques to navigate the specification space efficiently, thus minimally explaining the inference process in terms of generality and coverage [10], mining algorithms might produce biased specifications, which are then heavily reliant on the cleanness and fairness of the underlying data. This problem, which also affects learning algorithms, can be partially mitigated in these ones by regularising the loss function to take into account on the skewed data distribution [51]. Last, the only way to increase the interpretability of the decision process is to make stricter assumptions on the specification and data representation of choice to validate the entire process with environment $E$ acting as background knowledge. To achieve this, the next Section will consider specifications where $\mathcal{L}_h$ is the set of all the possible **LP** for a decidable fragment of choice, in which we also express $\varphi$, $E$, and $S$.

Relational Learning Algorithms

Relational Learning algorithms work with specifications and data expressed in **LP**, usually as Horn clauses, and by generating multiple possible specifications as per mining algorithms, rather than one single specification as per the learning problems. These approaches easily overcome limitations posed by data biases as per the former approaches, as Relational Learning methods do not require statistically independent and uniformly distributed examples [64]. Similarly to the former, these approaches follow a general-to-specific approach [28] with some important exceptions when the search is conducted in a specific-to-general approach to ensure good generality properties [64]. Relational Learning algorithms also assume to derive novel specifications from an annotated dataset $f$, distinguishing between positive and negative examples, while providing a set of knowledge base rules (TBox) acting as our environment of interest $E$ to strengthen the quality of the derived specification.

From the following description, these algorithms are a specialisation of the mining ones: when assuming clean data being perfectly annotated in $f$ for the positive and negative dataset, the classical FOIL algorithm as generalised in [28] generates new hypotheses only covering the positive examples by firstly $\rho$-refining the most generic specification $\varphi$ until it stops maximising the information gain [65]: after retaining a new clause in $\varphi$ (partially) covering the positive examples in the final set of the specifications to be returned, the algorithm removes from the positive examples the ones verifying all the previously mined clauses as well as $E$. Notwithstanding the former, FOIL still learn clauses over one single **RL** instance without the possibility of learning clauses over **PL** data considering different databases. Thus, the **LP** generated by the former cannot be generalisable across different database schemas and cannot be used to learn over **PL** data.

Recent research increased the **interoperability** of such algorithms over different data representations by increasing their adaptability to more realistic and broader use cases. Such research showed the possibility of achieving this goal when considering $E$ including multivalued functional dependencies while considering databases with different schemas but equivalent *fourth normal form* [35]: a *multivalued functional dependency* $X \twoheadrightarrow Y$ for a relationship $\mathfrak{R}$ is such that $\forall r, r' \in \mathfrak{R}.r[X] = r'[X] \Rightarrow r[Y] = r'[Y]$, while a fourth normal form decomposition $\mathbf{R} = \{R_1, \ldots, R_n\}$ of a relationship $\mathfrak{R}$ is such that $\mathfrak{R} = \Join_{R_i \in \mathbf{R}} \pi_{R_i}(\mathfrak{R})$ while considering one table per multivalued functional dependency occurring in $\mathfrak{R}$. These assumptions entail that clause learning algorithms visiting the hypothesis space specific-to-general are schema-invariant and thus generalisable to different databases [65]. Subsequent research refined the same algorithm also to consider learning from inconsistent databases, where such inconsistencies were determined as violations of the aforementioned (multivalued) functional dependencies. The resulting algorithm then guarantees that the learned rules hold true over each possible instance of a repaired database in compliance of such rules [64].

Compared to the former approaches, these algorithms maximise the **interpretability** results by adopting an expressive specification language while ensuring that they learn from **LP** data. Still, their specific choice poses a major **trade-off** in their ability to deal with contradictory data: to avoid generating inconsistent hypothesis formulations from the data, these algorithms are not considering clauses with neither negations nor the universal falsehood as their heads. This also entails that all the instances of the **LP** data or $S$ abolishes logical connectors. This entails that these approaches cannot even be used to detect data inconsistencies also considering hierarchical data information. Furthermore, the same strong assumption over the rule formulation limits the applicability of the learning algorithms, as they cannot be considered in other more realistic scenarios requiring full data transformation operations. These will still be considered inconsistencies due to the missing of specific data rewriting functions. Notwithstanding the possibility of inducing specifications from databases containing inconsistencies, the last presented algorithm assumes the possibility of correctly deriving a database repair over which then deriving accurate specifications while assuming to have data explicitly annotated in $f$. On the other hand, real data does not come with such an explicit annotation, thus requiring to entirely trust the rules in $E$ for separating conflicting information.

Paraconsistent Reasoning

The only way to extend Relational Learning algorithms' expressiveness is by extending the specification language to include negations and universal falsehood to represent conflicts occurring in the data. This requires avoiding the explosion problem, due to which arbitrary consequences will derive from an inconsistent fact. This will then require us to pose a greater control to the overall inference mechanism, thus increasing the **explainability** of the overall inference process to also ponder over conflicting data representations. This enables data in **PL**, thus including the consideration of logical connectives and negations. For brevity's sake, this chapter considers paraconsistent reasoning over programs expressed in *disjunction-free LP* where heads have no disjunctions, and tails can also express *weak negation* $\sim A$, which is false only when $A$ is explicitly in the expanded data set. These programs also contemplate the explicit detection of inconsistency by considering the universal falsehood $\bot$ also appearing in their head; no $\bot$ can appear in any of the rules' heads, thus avoiding the principle of explosion from which arbitrary consequences might be derivable. This process is guaranteed to converge as determining the consistency of such programs is in NP [33].

All the previous algorithms assumed a manually labelled set of data, from which the possibility of manually annotated data intrinsically guided specification generation, thus dealing directly with $S$ rather than an extension of $S$ with class annotation, $f : S \rightarrow \mathcal{Y}$. When considering real-world data containing conflicting and contradictory information, manually annotating vast amounts of data is often impossible. Hence, automated feature detection is often required to determine whether the data at our disposal is inherently contradictory. When this happens, and to avoid the principle of explosion, most of the machine learning algorithms will fail, for which it is then important to determine the number of contradictions in $S$ by merely relying on an environment $E$ assumed to contain trustworthy and correct information [88]. Despite several formalisations, very few implementations exist that work on real data. One of these[2] [7] is used to detect contradictory and conflicting information from the data through a preliminary knowledge expansion phase working under the *Closed-World Assumption*: given $S$ the data before the expansion phase, a subset $E \subseteq S$ is inconsistent if it is possible to derive $\bot$ via the application of the rules in $E$. Given a query-driven specification mining algorithm by matching $E \subseteq S$, the amount of inconsistency in $E$ can be expressed as the sum of the reciprocal of the size of the Minimal Inconsistent Sets (MIS) derived through the expansion phase [48]. A more conservative approach, walking in the footsteps of the aforementioned database repairs, would instead consider generating a non-contradictory specification by firstly generating the Maximally Consistent Sets (MCS) [58], for then repeating the aforementioned approach while considering approximate query answering: this measure is then required as ensuring complete consistency might discard contradictory data information. As the latter approach also blocks trivialisation, this enables subsequent probabilistic reasoning over non-contradictory data [7] with probabilistic (e.g., frequency) and fuzzy (e.g., errors) information [73].

Finally, these solutions achieve maximal **explainability** purposes by explaining contradictions through MIS, MCS, and inconsistency metrics beyond the simple violation of multivalued functional dependency constraints but by also considering data containing negations or, more generally speaking, facts that are falsifiable given a specific background knowledge. This approach is non-frequentist based, thus retaining the same guarantees from relational learning problems. Furthermore, the aforementioned methodology provides **trade-off** maximisation between query match fulfilment and consistent answering. Notwithstanding the former, no explicit relational learning algorithms with paraconsistent reasoning is known (Sect. 2.4). Furthermore, these algorithms mainly consider dealing with non-numerical data and no data predicates: Sect. 2.3.3 will discuss preliminary attempts to fill in the gaps by considering hybrid explainability, thus determining fully AI-explainable pipelines by expressing salient numerical features in **LP**.

[2] https://github.com/jackbergus/java_logicalconsistency.

Conclusions

These considerations lead to a general formulation for both mining and learning problems [28, 65]:

**Generic Learning/Mining Algorithm**

```
def genericAlgorithm(Init,stop,Quality,f,rho,Prune,n):
    Queue = Init
    Theory = []
    while not stop(Queue, Theory):
        L = Queue[:n] # extracting the first n elements
        del Queue[:n] # removing n elements
        for h in L:
            if Quality(h,f):
                Theory.append(h)
                Queue = Queue + [x for x in rho(h,f)]
        Queue, Theory, f = Prune(Queue, Theory, f)
    return Theory
```

The definition above enables **interoperability**, as many different frameworks can fit in this description. When considering algorithms with hyperparameter search, `Init` might contain all the possible preliminary specifications under all the possible parameters and just one possible initialisation otherwise.

By comparing this to the generic PPO algorithm [71], the `stop` condition considers a maximum number of iterations, `rho` will refine one specification at a time by considering at least one episode, thus including any instantiation of Equation (2); `Init` can be given as many possible initial configurations as the $n$ generated after a preliminary hyperparameter tuning search; thus, as this algorithm will constantly refine the specifications rather than producing new ones, the `Prune` function will always discard the $n$ specifications that were previously stored as specifications. No explicit `Quality` operator $Q$ is used to prune specifications in the original algorithm, as the quality search is conducted while refining the specification. When considering other simplistic reinforcement learning algorithms based upon Markov Decision Processes such as iterative policy evaluation [75], the stopping condition will correspond to the stabilisation for the state-wise value function, which variation will occur below a given value threshold $\theta$, while the specification will mainly consist of the varying value function parameters.

For mining algorithms providing a lattice search ("Mining by Quality-Driven Specification-Space Navigation"), the `Queue` is now initialised with the most general specification, while the main iteration will `stop` once the queue is empty; for these, only one specification at a time might be refined ($n = 1$), and the refinement `rho` only considers the specification's structure, thus ignoring the `f` argument; `Prune` will keep the theory unaltered, unless the algorithm is interested in retrieving the most specific specification among all the possible generated theories, thus always updating until the frontier is reached [10]. Current literature also shows how relational learning algorithms also fall into this general definition [65].

As a main **trade-off**, even though relational learning algorithms often provide the most explainable type of results, both for the machine and for the human, as they return a set of logical programs as the outcome of the learning process, these often are not well suited for considering numerical data, such as image processing or time series analysis, thus requiring to use other numerically-driven approaches for first representing this information most suitably. Notwithstanding the former, very recent approaches involving *hybrid explainability* showcase the possibility of returning such high-level approaches through the usage of preliminary specification mining phases (see "Mining by Quality-Driven Specification-Space Navigation") that serve the purpose of climbing up the representation ladder to then provide machine-readable and logical representation that these algorithms might use (Sect. 2.3.3).

#### 2.3.2.3 Ex Post

**Important**
An algorithm provides **ex post** explainability when it extracts or derives an explanation from a (previously trained) specification. An ex post algorithm is *marginally* explainable if the correlation between the input data and the target class is merely explained as a linear model from the provided input or if it just provides performance metrics for a specification. An ex post algorithm is *maximally* explainable if the correlation between the input data and the target class is also correlated with different input values.

The forthcoming subsections will introduce distinct approaches in order of increasing explainability (Fig. 2.4).

Performance Metrics

Performance metrics can be considered at the very bottom of this representation hierarchy, as they attest to the prediction abilities of a specification in terms of their ability to describe the data of interest. Given a specification $\varphi$ approximating an annotated dataset $f$, metrics such as precision and recall for binary classification tasks can be defined as follows:

$$Precision(\varphi, f) = \frac{\left|\varphi^{-1}(\mathbf{1}) \cap f^{-1}(\mathbf{1})\right|}{\left|\varphi^{-1}(\mathbf{1})\right|} \qquad Recall(\varphi, f) = \frac{\left|\varphi^{-1}(\mathbf{1}) \cap f^{-1}(\mathbf{1})\right|}{\left|f^{-1}(\mathbf{1})\right|}$$

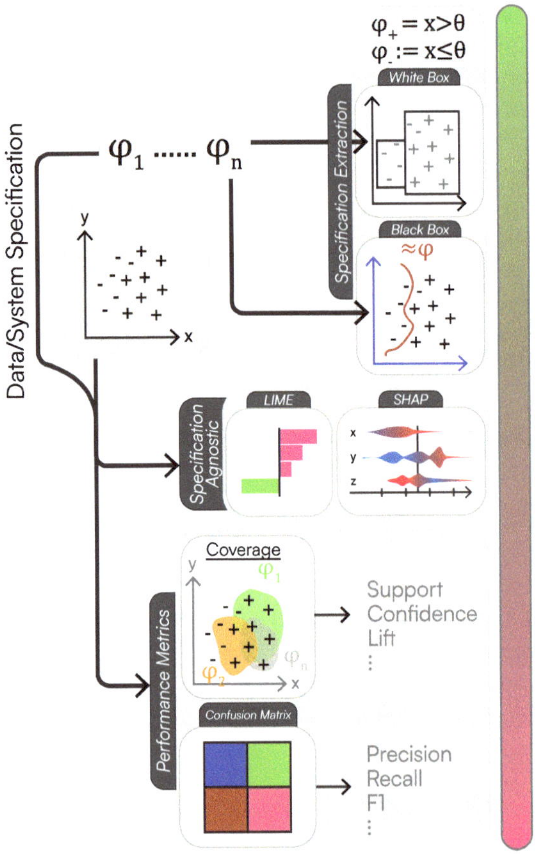


**Fig. 2.4** This chapter considers traditional explainers as ex post algorithms

Despite these being maximally **interoperable** as they can be defined for any possible specification language, they provide limited **interpretability** as they only reflect the specification's ability to capture the testing system $S$ adequately, thus providing a **trade-off** between them. As these provide no suitable explanation of how the data input is directly connected to the classification outcome, they provide a minimal numerical explanation of the specification without describing any correlation between them. The forthcoming approaches will extend this trade-off by increasingly capturing more information between input and output data and correlations between multiple dimensions.

Model-Agnostic

**Important**
Traditional **model-agnostic** explainers [32] provide a qualitative explanation of a previously-trained specification between the input variables and the expected class [69], without necessarily addressing whether a complex correlation across input values, potentially varying in time, leads to different classification outcomes.

LIME [69] offers localised interpretability concerning specific reference points; a linear specification tangent to the specification represented numerically, where such specification might also be a non-linear function. This approach has been recently extended to OptiLIME [79] to increase the former's stability in finding the best kernel width and providing the best linear fit to the trained specification while rewriting the variations in explanation outcomes due to randomisation. Differently from the former approaches, Shapely values as considered in SHAP [57] assign a value to each feature similarly to a logistic regression specification, thus indicating its contribution to a specification's output. Neither of these approaches provide correlations across dimensions, as they mainly describe the contribution of one single dimension to the final classification outcome.

As these specifications are specification-language and mining/learning algorithm agnostic, they are well suited for maximising **interoperability**, as they might be applied at the end of any specific specification-learning problem. This provides a lax notion of **interpretability**, as it allows an interpretation of any raw data information, from images [69] to textual [70] information. Despite the versatility showcased by the two previous components and providing more than just one single value as per the former approaches, these pose a major **trade-off** in explainability by preferring a simple and graphically generable numerical approximation of the specification while not capturing complex data correlations among dimensions. On the other hand, while LIME's results heavily rely on random sampling, thus potentially leading to different explanations for similar instances [2], SHAP results tend to be more stable by ensuring the principles of cooperative game theory.

It can be shown that explanations regarding correlations across variables can be extracted only by providing stronger assumptions on the specification language of interest, thus directly leveraging the specification's structure. For this, the forthcoming paragraphs will introduce the other two ways to extract explanations from the trained specifications.

Black-Box Specification Extraction

**Important**
A specification language is a **black-box** when its explanation requires further post-processing to be "fully understood" by a human.

Despite the possibility of using GNNs to decide whether a graph is considered a solution to the Traveller Salesman Problem for another graph [66], it is not directly possible to extract the decision algorithm to explain it procedurally, as having a numerical approximation of such an algorithm might not be readily interpretable by a human. Given this, the only possible explanation derivable from the trained specification (i.e., the resulting GNN) is a further black-box specification providing suitable approximations of numerical functions. While this Section will mainly consider traditional deep layer neural networks with dense layers (e.g., no dropouts), these intuitions might apply to some extent to other network architectures, such as Convolutional Neural Networks, which pipeline convolutional kernels with traditional deep learning architectures [42].

When considering the interpretation of dense layers $L_1, \dots, L_n$ under the assumption that each neuron within the same layer $L_i$ will be associated with the same activation function $\sigma_i$ [24], it is possible to derive some Python code associating to each input neuron a variable $x_i$ to each $i$-th input dimension and returns as many equations as output neurons to express any trained dense neural network as follows:

$$\sigma_n \left(W_n \left(\dots \sigma_1 \left(W_1 \mathbf{x} + \mathbf{b}_1\right)\right) + \mathbf{b}_n\right)$$

where $W_i$ (and $\mathbf{b_i}$) stands for the weights (and the bias vector) associated with the $i$-th layer. In particular, such equation can be expressed into as many equations $\varphi_j$ as output neurons within the neural network, notwithstanding the possibility of parsing each of the obtained equations using symbolic libraries such as SymPy [62] for then approximating each of these equations through Taylor expansion, the resulting equations describing the specification in its mathematical formulation might make the entire process opaque, especially when multiple input variables and multiple hidden layers are considered. The **interpretability** of such specifications heavily depends on the complexity of the specification and on the input and output size of the network, in which a specification-agnostic explanation is often preferable. The major **trade-off** with the former is that this specification provides the possibility of explicitly expressing how each input value contributes to the output of each neuron, thus clearly extending the expressive power of the former to the cost of the interpretability. Last, the **interoperability** of this specification extraction is directly dependent on the interoperability of the specification language of choice: e.g., concerning DNNs, these can be considered fully interoperable for data representable numerically.

White-Box Specification Extraction

**Important**
A specification language is a **white-box** when are "easily" interpretable by a human with minor or no post-processing.

In this scenario, the expressivity of the specification resulting from the explanation will be as expressive as the predicates of interest being used to capture the properties of the data, as well as the connectors of interest to connect such predicates, thus being directly related to the original choice for $\mathcal{L}_h$. Please observe that algorithms inducing theories (see "Relational Learning Algorithms") or reasoning paraconsistently (see "Paraconsistent Reasoning") do not necessarily need to extract or explain a specification, as they already generate specifications in **LP**, and, therefore, in a readable format. Thus, this Section will only consider specifications from which the original specification representation (e.g., a decision tree) does not necessarily match with a human representation of a specification (e.g., an **LP** formula). For this purpose, decision trees are a specific case of a white-box ad hoc algorithm from which extract a specification. When considering a finite set of classes $\mathcal{Y}$, each decision node $p_i$ is associated with a logical predicate $\lambda(p_i)$ considering the value of one single dimension at a time, while the direction of visit $dir_i$, either left or right, is assumed to refine the search by considering data examples that are satisfying or not such a predicate. Thus, given the direction decision function in the following equation:

$$\Delta(dir_i, p_i) := \begin{cases} p_i & dir_i \equiv left \\ \neg p_i & dir_i \equiv right \end{cases}$$

each path $\pi = p_1 \overset{dir_1}{\rightarrow} p_2 \overset{dir_2}{\rightarrow} \dots p_n \overset{dir_n}{\rightarrow} l_n^y$ within a decision tree from the root $p_1$ towards a leaf $l_n^y$ associated with a class $y \in \mathcal{Y}$ describes a Horn clause $H(\pi) := \bigwedge_{p_i \overset{dir_i}{\rightarrow} \dots \in \pi} \Delta(dir_i, p_i) \Rightarrow y$ . Thus, a specification $\varphi_y$ for our data as $\bigvee_{\pi \equiv p_1 \rightarrow \dots \rightarrow y} H(\pi)$ can be extracted for all the paths in such a tree leading to a leaf node associated with the class $y$ [10]. These considerations can then be further generalised with regression trees when considering $\mathcal{Y}$ being the set of the real numbers $\mathbb{R}$, where leaf nodes are associated with real values [56]. Furthermore, these predicates can be arbitrarily complex in capturing correlations between multiple dimensions, such as linear equations [84] and mathematical functions over the choice variables [4]. This extends the traditional notions of decision trees, thus allowing them to capture complex correlations within the data of interest.

The **interpretability** of such specifications depends on the specification language of choice, thus including the set of predicates of interest describing the data. Thus, novel definitions from [4, 84] capture complex correlations between numerical variables as provided by state-of-the-art black-box algorithms such as DNNs. Consequently, these specifications maximise the **trade-off** between interpretability and specification complexity by considering all the previous approaches. As decision tree predicates can also naturally support predicates over categorical data [87] without necessarily resorting to one-hot-encoding or transformer approaches, these also maximise the **interoperability** within a generic data processing pipeline.

## *2.3.3 Hybrid Explainability*

This section discusses three use case examples of hybrid explainable pipelines, which combine all the former aspects of explainability, thus discussing the interoperability of the different types of explainable algorithms to fulfil different representation aims, including the generation of specifications and specifications from the data. After introducing record deduplication pipelines keeping data in **AV** (Sect. 2.3.3.1), the section discusses two other recent examples showcasing the possibility of climbing up the representation ladder from raw data, either multivariate time series (Sect. 2.3.3.2) or full text (Sect. 2.3.3.3) where, at any point, the information is represented in **PL**. Figure 2.1 showcases the heterogeneity of these three pipelines, thus introducing the challenges for fully automating a data science pipeline for solving any generic learning task of interest, thus motivating the impossibility of directly achieving this in terms of readily available technology such as LLMs [31]. This is why GEVAI is now structured as a collection of composable services rather than an automated approach to extracting and processing information.

### 2.3.3.1 Financial Institution DeDuplication Pipeline (FIDDP)

The data deduplication pipeline for financial institutions [3] provides a domain-specific instantiation of the baseline data deduplication pipeline (BLDDP). In both scenarios, authors assume to have already integrated and cleaned data by exploiting other preliminary a priori explainable approaches. Furthermore, data is always assumed to be in the **AV** format, as this pipeline aims to transform the data further rather than climb up the representation ladder. Thus, the entire approach can clearly be described as an a priori explainable algorithm, further composed of other a priori and ad hoc explainable approaches to finalise their representation aim.

The pipeline is composed of intermediate tasks, determining the best hyperparameters and algorithms for the subsequent task, and the execution of the selected algorithm with its hyperparameters (Fine Tuning from Fig. 2.1). The main processing tasks are two a priori explanation tasks: after determining the region of records that might be potential duplicates, the authors block the most similar ones into one region, in which the pairwise record similarity is computed to determine the most similar records. As these phases mainly add information related to each single record and given that no explicit data mining algorithm is used to compute these similarities but rather similarity metrics, then these phases are rightfully considered a priori explanations of the data. Furthermore, the first phase implicitly subsumes the pruning of the dissimilar records, which do not exhibit any duplication. Last, the authors exploit the computed pairwise similarities to cluster together the similar records, thus coalescing all the records that are likely to refer to the same physical person; this naturally leads to an ad hoc explanation, as the algorithm provides a reason why the records within same region should be considered similar in terms of belonging to a group of similar records.

#### 2.3.3.2 *E*xplainable *M*ultivariat*E* Co*R*relat*I*onal *T*emporal *A*rtificial Int*TE*lligence (EMeriTAte)

EMeriTAte [14] proposes a novel Multivariate Time Series (MTS) classification task for heterogeneous temporal data assuming that, differently from current literature, each MTS is associated with a class varying in time and that, by considering all the maximal time windows where the same class holds, all of these windows will vary in length. Despite this being a fair assumption for realistic data, EMeriTAte challenged this for the first time. This work was contextualised for characterizing dyskinetic events in Parkinsonian patients, where each dimension might refer to a different motor sensor and estimated blood concentration level of a specific active principle given a pointwise drug assumption pattern. While doing this, the clinician must gain medical insight from the MTS classification, thus requiring an explainable and declarative representation of the motivations leading to the dyskinetic event.

First, authors optionally convert single pointwise events into additional time series dimensions if a specific domain-based transformation function is given. Then, they return the final MTS representation, ready to undergo the three subsequent explainability phases.

In the first a priori explanation, the authors extract for each dimension different data trends (DT) [49], describing arbitrary increase, decrease, or volatility patterns while considering them with different lengths. These discretised data trends are then transformed into durative events for a *log*, where the original raw data from the time series is also pertained. In the second and last a priori phase, such log is transformed into as many logs as the classes present across all MTSs, where each trace will represent a maximal contiguous set of events always associated with the same classification label. Last, the authors provide two pipelined ad hoc phases, where the final $\mathcal{M}$ is extracted through an *ex post* phase: first, the authors exploit efficient Business Process Mining algorithms such as Bolt2 [10] to extract relevant correlations across different DT patterns, thus providing an ad hoc explanation of the temporal correlations occurring per classification label, thus being mined in one of the aforementioned distinct logs. Next, the authors exploit an ad hoc explanation algorithm for mining the specification per classification label by exploiting a decision tree, from which it extracts an explainable specification for each classification label using an *ex post* white-box specification extractor. Preliminary results on this patient data suggest that competing state-of-the-art approaches, thus including ones considering DNNs with attention mechanisms, fail to correctly classify the resulting events due to their impossibility of explicitly deriving explicative correlations across different dimensions. This proves that achieving a better human-driven explanation can also help the machine to draw better machine specifications.

#### 2.3.3.3 *Logical*, *Structural*, and *Semantic* Text *Interpretation* (LaSSI)

LaSSI [40], to the best of our knowledge, provides for the first time a fully-explainable pipeline for full text interpretation. This algorithm starts by providing a priori explainability to first extract from the sentence contextual multi-entity units that cannot be achievable by merely considering sentence structure parsing. This approach also provides type information to the associated variables: given the sentence "*It is busy in Newcastle city centre on Saturdays*", the authors identify GPEs from GeoNames [15], in this case "*Newcastle*". Then, Stanza [67] identifies other generic geographical LOCations, such as "*city centre*", as a sub-location of Newcastle, and finally, SUTime [19] defines the specific temporal annotations, like "*Saturdays*". In parallel, authors run the specification mining pipeline, addressing the rewriting of the original full text sentence in logical form: after determining the grammatical dependencies across words, thus obtaining a graph representation of our data [21], they apply structural rewriting to provide a first intermediate representation [11] for normalising sentences in passive and active form, while associating the prepositions to the indirect objects of interest. After this step, they re-use the contextual information coming from the a priori explainability for determining possible relationships across the previous Named Entity Recognition phase: they denote a hierarchy of importance to produce a proper understanding of the structure, after which they recognise that *city centre* should be considered as a refinement of the GPE of Newcastle, and not vice-versa. So, GPE is most important, and anything below Entity is least important. After this, the authors finally derive a logical representation of this sentence. This demonstrates that a priori explanations might not necessarily happen before the specification mining phase but also during it to obtain the final specification.

Authors provide an ad hoc explainable approach explaining in which circumstances the sentences might be true by determining in which possible worlds these sentences will be true. The correlation between the two sentences and their atoms is represented tabularly: this table will contain as many columns as the atoms in each $\varphi$; authors then add two columns providing the tabular semantics for $\varphi$. After determining whether each pair of occurring atoms imply each other, are unrelated, or are mutually exclusive (Fig. 2.5) and the truth tables per sentence, the pipeline equi-join them to obtain the set of compatible worlds. This achieves simplistic paraconsistent reasoning by removing all the worlds where different atoms cannot jointly hold. This then generates a specification describing the similarity between the two

**Fig. 2.5** Tabular representation for compatible worlds implying ($a_i \Rightarrow b_j$), indifferent ($a_i ?b_j$), or mutually exclusive ($a_i \wedge b_j \Rightarrow \bot$) atom $a_i$ and $b_j$ [40]

$R_{a_i \Rightarrow b_j}$

| $a_i$ | $b_j$ |
|---|---|
| 0 | 0 |
| 0 | 1 |
| 1 | 1 |

$R_{a_i ? b_j}$

| $a_i$ | $b_j$ |
|---|---|
| 0 | 0 |
| 0 | 1 |
| 1 | 0 |
| 1 | 1 |

$R_{(a_i \wedge b_j) \Rightarrow \bot}$

| $a_i$ | $b_j$ |
|---|---|
| 0 | 1 |
| 1 | 0 |

specifications. To increase the interpretability of this tabular representation, authors provide ex post explainability by providing the ratio between the number of possible worlds where both sentences are true over the number of worlds where just the sentence selected as a premise is chosen to be true (*confidence*). Experiments showed that this explainable approach outperformed similarity scores achieved through sentence vector embedding through transformers and graph-based sentence similarity, which are unable to convey complex logical reasoning through mere structural and vector-based semantics for entire sentences.

#### 2.3.3.4 Provenance as Support for Cross-Algorithm Explainability

A data representation *natively supports* provenance information if each data record or object either provides a progressive and unique ID if coming from the original data source and provides an expression describing how it was obtained otherwise [43], thus allowing to directly derive the final uncertainty score associated to the resulting data rather than computing this continuously within the pipeline [41]. Provenance enables tracking how the provided data was produced from the original data sources by just taking into account the IDs referring to the original data while keeping track of the operations being applied to provide such results [20]. Thus, provenance support acts more as a means to achieve operational explainability, thus allowing hybrid explainability across algorithms to be supported. Current literature suggests that this approach should be exploited to achieve explainability, thus allowing us to track the rewritten data by a pipeline down to the original document and sentence from which such sentence was extracted [37].

## 2.4 Conclusions and Open Challenges

This chapter offers a compendium of the latest research on explainable and verifiable AI by showcasing how these pieces fit together in a framework to be automated in the near future. Notwithstanding the former, the codebase only provides these pieces singularly, with no current automation for orchestrating these pieces. This is also a common problem with other data science approaches [78], as they require the user to connect the components manually [22]. Similar considerations can be made for AI pipelines, which still require a huge amount of manual intervention: solving this problem requires solving a data integration task where, differently from what is currently assumed by data-aware service orchestration pipeline [9], the specification of each microservice might change dynamically according to the data elements provided as an input. Future works will attempt to automate GEVAI into a cohesive framework by considering advanced data integration algorithms to automatically orchestrate the components in the pipeline to satisfy specific user requirements. Doing so will potentially mitigate the current problems for data scientists using relational learning algorithms, which currently massively restructures the algorithm depending on the data representation of choice [28]. Such work will also verify to which extent specification mining algorithms might come into help by enriching data with contextual information. Furthermore, the description of the previous pipelines also proves that simple linear trends cannot necessarily capture data-processing pipelines, as different steps might require the ingestion of intermediate data processed at any phase of the computation. While doing so, the exploitation of explicit provenance identifiers will make it easier to reconcile different outputs from distinct phases of the algorithm.

No Relational Learning algorithm using paraconsistent logic is known at the time of the writing. As deriving specifications from inconsistent data might easily lead to the generation of inconsistent specification sets, it is therefore essential to exploit **LP** inconsistency metrics [48] to safeguard the correctness of the derived hypotheses, which are not monotonic. This issue might be solved by reason over each possible maximally consistent data set [58], for then running traditional relational learning algorithms over paraconsistently-repaired data examples. Future works will then assess the feasibility of incorporating such metrics alongside information gain metrics for specification generation. These metrics shall also be exploited to guarantee the consistency of the overall environment before expanding this, thus ensuring the correctness of the reasoning process.

## References


1. Agrawal R, Imieliński T, Swami A (1993) Mining association rules between sets of items in large databases. In: Proceedings of the 1993 ACM SIGMOD international conference on management of data, SIGMOD '93. Association for Computing Machinery, New York, NY, USA, pp 207–216
2. Alvarez-Melis D, Jaakkola TS (2018) On the robustness of interpretability methods. arxiv: abs/1806.08049
3. Andrzejewski W, Bębel B, Boiński P, Wrembel R (2024) On tuning parameters guiding similarity computations in a data deduplication pipeline for customers records: experience from a R&D project. Inf Syst 121:102323
4. Ashok P, Jackermeier M, Kretínský J, Weinhuber C, Weininger M, Yadav M (2021) dtControl 2.0: explainable strategy representation via decision tree learning steered by experts. In: Groote JF, Larsen KG (eds) Tools and algorithms for the construction and analysis of systems—27th international conference, TACAS 2021, held as part of the European joint conferences on theory and practice of software, ETAPS 2021, Luxembourg City, Luxembourg, 27 Mar 27–1 Apr 2021, Proceedings, Part II. Lecture Notes in Computer Science, vol 12652. Springer, pp 326–345
5. Baader F, Calvanese D, McGuinness DL, Nardi D, Patel-Schneider PF (eds) (2003) The description logic handbook: theory, implementation, and applications. Cambridge University Press
6. Bender EM, Gebru T, McMillan-Major A, Shmitchell S (2021) On the dangers of stochastic parrots: Can language models be too big? In: Proceedings of the 2021 ACM conference on fairness, accountability, and transparency, FAccT '21. Association for computing machinery, New York, NY, USA, pp 610–623
7. Bergami G (2019) A framework supporting imprecise queries and data. arxiv:abs/1912.12531
8. Bergami G (2024) DECLAREd: a polytime $\mathrm{LTL}_f$ fragment. Logics 2(2):79–111
9. Bergami G (2024) Towards automating microservices orchestration through data-driven evolutionary architectures. Serv Oriented Comput Appl. 18(1):1–12

10. Bergami G, Appleby S, Morgan G (2023) Specification mining over temporal data. Computers 12(9)
11. Bergami G, Fox OR, Morgan G (2024) Matching and rewriting rules in object-oriented databases. Mathematics 12(17)
12. Bergami G, Maggi FM, Marrella A, Montali M (2021) Aligning data-aware declarative process models and event logs. In: Polyvyanyy A, Wynn MT, Van Looy A, Reichert M (eds) Business process management. Springer International Publishing, Cham, pp 235–251
13. Bergami G, Maggi FM, Montali M, Peñaloza R (2021) A tool for computing probabilistic trace alignments. In: Nurcan S, Korthaus A (eds) Intelligent information systems—CAiSE forum 2021, Melbourne, VIC, Australia, 28 June–2 July 2021, Proceedings, Lecture notes in business information processing, vol 424. Springer, pp 118–126
14. Bergami G, Packer E, Scott K, Del Din S (2025) Predicting dyskinetic events through verified multivariate time series classification. In: Chbeir R, Ilarri S, Manolopoulos Y, Revesz PZ, Bernardino J, Leung CK (eds) Database engineered applications. Springer Nature Switzerland, Cham, pp 49–62
15. Bond F, Bond A (2019) GeoNames Wordnet (gnwn): extracting wordnets from geonames. In: Vossen P, Fellbaum C (eds) Proceedings of the 10th global wordnet conference, GWC 2019, Wroclaw, Poland, 23–27 July 2019. Global Wordnet Association, pp 387–393
16. Boyd SP, Vandenberghe L (2014) Convex optimization. Cambridge University Press
17. Brown B (2018) Inconsistency measures and paraconsistent consequence. In: Grant J, Martinez MV (eds) Measuring inconsistency in information, chap 8. College Press, pp 219–234
18. Camacho A, Toro Icarte R, Klassen TQ, Valenzano R, McIlraith SA (2019) LTL and beyond: formal languages for reward function specification in reinforcement learning. In: Proceedings of the twenty-eighth international joint conference on artificial intelligence, IJCAI-19. International joint conferences on artificial intelligence organization, pp 6065–6073
19. Chang AX, Manning CD (2012) SUTime: a library for recognizing and normalizing time expressions. In: Calzolari N, Choukri K, Declerck T, Dogan MU, Maegaard B, Mariani J, Odijk J, Piperidis S (eds) Proceedings of the eighth international conference on language resources and evaluation, LREC 2012, Istanbul, Turkey, 23–25 May 2012. European Language Resources Association (ELRA), pp 3735–3740
20. Chapman A, Lauro L, Missier P, Torlone R (2024) Supporting better insights of data science pipelines with fine-grained provenance. ACM Trans Database Syst 49(2):6:1–6:42
21. Chen D, Manning CD (2014) A fast and accurate dependency parser using neural networks. In: Moschitti A, Pang B, Daelemans W (eds) Proceedings of the 2014 conference on empirical methods in natural language processing, EMNLP 2014, 25–29 Oct 2014, Doha, Qatar, A meeting of SIGDAT, a Special Interest Group of the ACL. ACL, pp 740–750
22. Chen S, Tang N, Fan J, Yan X, Chai C, Li G, Du X (2023) HAIpipe: combining human-generated and machine-generated pipelines for data preparation. Proc ACM Manag Data 1(1)
23. Chen Y, Wang DZ (2014) Knowledge expansion over probabilistic knowledge bases. In: C.E. Dyreson, F. Li, M.T. Özsu (eds.) International conference on management of data, SIGMOD 2014, Snowbird, UT, USA, 22–27 June 2014. ACM, pp 649–660
24. Chollet F et al (2015) Keras. https://keras.io
25. Clocksin WF, Mellish CS (1994) Programming in Prolog, 4 ed. Springer
26. D'Agostino G, Reed CA, Puccinelli D (2024) Segmentation of complex question turns for argument mining: a corpus-based study in the financial domain. In: Calzolari N, Kan M, Hoste V, Lenci A, Sakti S, Xue N (eds) Proceedings of the 2024 joint international conference on computational linguistics, language resources and evaluation, LREC/COLING 2024, 20–25 May 2024, Torino, Italy. ELRA and ICCL, pp 14524–14530
27. Dallachiesa M, Ebaid A, Eldawy A, Elmagarmid A, Ilyas IF, Ouzzani M, Tang N (2013) NADEEF: a commodity data cleaning system. In: Proceedings of the 2013 ACM SIGMOD international conference on management of data, SIGMOD '13, pp 541–552. Association for Computing Machinery, New York, NY, USA
28. de Raedt L (2008) Logical and relational learning: from ILP to MRDM (Cognitive Technologies). Springer, Berlin, Heidelberg

29. de Raedt L, Bruynooghe M (1992) A unifying framework for concept-learning algorithms. Knowl Eng Rev 7(3):251–269
30. Detroja K, Bhensdadia C, Bhatt BS (2023) A survey on relation extraction. Intell Syst Appl 19:200244
31. Dhar V (2024) The paradigm shifts in artificial intelligence. Commun ACM 67(11):50–59
32. Dwivedi R, Dave D, Naik H, Singhal S, Omer R, Patel P, Qian B, Wen Z, Shah T, Morgan G, Ranjan R (2023) Explainable AI (XAI): core ideas, techniques, and solutions. ACM Comput Surv 55(9)
33. Eiter T, Gottlob G (1995) On the computational cost of disjunctive logic programming: propositional case. Ann Math Artif Intell 15(3–4):289–323
34. Eliason SR (1993) Maximum likelihood estimation: logic and practice. SAGE
35. Elmasri RA, Navathe SB (2016) Fundamentals of database systems, 7th ed. Pearson
36. Euzenat J, Shvaiko P (2013) Ontology matching, 2nd ed. Springer
37. Ferrucci DA (2012) Introduction to “This is Watson”. IBM J Res Dev 56(3.4):1:1–1:15
38. Feurer M, Hutter F (2019) Hyperparameter optimization. In: Hutter F, Kotthoff L, Vanschoren J (eds) AutoML: methods, systems, challenges, chap 6. Springer, pp 113–134
39. Finkel JR, Manning CD (2009) Nested named entity recognition. In: Proceedings of the 2009 conference on empirical methods in natural language processing, EMNLP 2009, 6–7 Aug 2009, Singapore, A meeting of SIGDAT, a Special Interest Group of the ACL. ACL, pp 141–150
40. Fox OR, Bergami G, Morgan G (2025) LaSSI: Logical, Structural, and Semantic Text Interpretation. In: Chbeir R, Ilarri S, Manolopoulos Y, Revesz PZ, Bernardino J, Leung CK (eds) Database engineered applications. Springer Nature Switzerland, Cham, pp 106–121
41. Fuhr N, Rölleke T (1997) A probabilistic relational algebra for the integration of information retrieval and database systems. ACM Trans Inf Syst 15(1):32–66
42. Goodfellow I, Bengio Y, Courville A (2016) Deep learning. MIT Press
43. Green TJ, Karvounarakis G, Tannen V (2007) Provenance semirings. In: Proceedings of the twenty-sixth ACM SIGMOD-SIGACT-SIGART symposium on principles of database systems, PODS ’07. Association for Computing Machinery, New York, NY, USA, pp 31–40
44. Groß A, Hartung M, Kirsten T, Rahm E (2011) Mapping composition for matching large life science ontologies. In: Bodenreider O, Martone ME, Ruttenberg A (eds) Proceedings of the 2nd international conference on biomedical ontology, Buffalo, NY, USA, 26–30 July 2011, CEUR Workshop Proceedings, vol 833. CEUR-WS.org
45. Harrison J (2009) Handbook of practical logic and automated reasoning. Cambridge University Press (2009)
46. Hastie T, Friedman JH, Tibshirani R (2001) The elements of statistical learning: data mining, inference, and prediction. Springer Series in Statistics. Springer
47. Hicks MT, Humphries J, Slater J (2024) ChatGPT is bullshit. Ethics Inf Technol 26(2):38
48. Hunter A, Konieczny S (2008) Measuring inconsistency through minimal inconsistent sets. In: Brewka G, Lang J (eds) Principles of knowledge representation and reasoning: proceedings of the eleventh international conference, KR 2008, Sydney, Australia, 16–19 Sept 2008. AAAI Press, pp 358–366
49. Huo X, Hao K, Chen L, song Tang X, Wang T, Cai X (2022) A dynamic soft sensor of industrial fuzzy time series with propositional linear temporal logic. Exp Syst Appl 201:117176
50. Jäger S, Allhorn A, Bießmann F (2021) A benchmark for data imputation methods. Front Big Data 4:693674
51. Kim B, Kim H, Kim K, Kim S, Kim J (2019) Learning not to learn: training deep neural networks with biased data. In: IEEE conference on computer vision and pattern recognition, CVPR 2019, Long Beach, CA, USA, 16–20 June 2019. Computer Vision Foundation/IEEE, pp 9012–9020
52. Koller D, Friedman N (2009) Probabilistic graphical models—principles and techniques. MIT Press
53. Kyburg HE (1961) Probability and the logic of rational belief. Wesleyan University Press, Middletown, CT, USA

54. LaForte G, Hayes PJ, Ford KM (1998) Why Gödel's theorem cannot refute computationalism. Artif Intell 104(1):265–286
55. Lerdlamnaochai T, Kijsirikul B (2004) First-order logical neural networks. In: Fourth international conference on hybrid intelligent systems (HIS'04), pp 192–197
56. Loh WY (2011) Classification and regression trees. WIREs Data Mining Knowl Disc 1(1):14–23
57. Lundberg SM, Lee SI (2017) A unified approach to interpreting model predictions. In: Proceedings of the 31st international conference on neural information processing systems, NIPS'17. Curran Associates Inc, Red Hook, NY, USA, pp 4768–4777
58. Malouf R (2007) Maximal consistent subsets. Comput Linguist 33(2):153–160
59. Maltoni D, Lomonaco V (2019) Continuous learning in single-incremental-task scenarios. Neural Netw 116:56–73
60. Megill JL (2019) Are we paraconsistent? On the Luca-Penrose argument and the computational theory of mind. Auslegung: J Philosop 27(1):23–30
61. Melnik S (2004) Generic model management: concepts and algorithms. Lecture notes in computer science, vol 2967. Springer
62. Meurer A, Smith CP, Paprocki M, Čertík O, Kirpichev SB, Rocklin M, Kumar A, Ivanov S, Moore JK, Singh S, Rathnayake T, Vig S, Granger BE, Muller RP, Bonazzi F, Gupta H, Vats S, Johansson F, Pedregosa F, Curry MJ, Terrel AR, Roučka V, Saboo A, Fernando I, Kulal S, Cimrman R, Scopatz A (2017) SymPy: symbolic computing in Python. Peer J Comput Sci 3:e103
63. Petermann A, Micale G, Bergami G, Pulvirenti A, Rahm E (2017) Mining and ranking of generalized multi-dimensional frequent subgraphs. In: Twelfth international conference on digital information management, ICDIM 2017, Fukuoka, Japan, 12–14 Sept 2017. IEEE, pp 236–245
64. Picado J, Davis J, Termehchy A, Lee GY (2020) Learning over dirty data without cleaning. In: Maier D, Pottinger R, Doan A, Tan W, Alawini A, Ngo HQ (eds) Proceedings of the 2020 international conference on management of data, SIGMOD conference 2020, online conference [Portland, OR, USA], 14–19 June 2020. ACM, pp 1301–1316
65. Picado J, Termehchy A, Fern A, Ataei P (2019) Logical scalability and efficiency of relational learning algorithms. VLDB J 28(2):147–171
66. Prates M, Avelar PHC, Lemos H, Lamb LC, Vardi MY (2019) Learning to solve NP-complete problems: a graph neural network for decision TSP. In: Proceedings of the Thirty-Third AAAI conference on artificial intelligence and thirty-first innovative applications of artificial intelligence conference and ninth AAAI symposium on educational advances in artificial intelligence, AAAI'19/IAAI'19/EAAI'19. AAAI Press
67. Qi P, Zhang Y, Zhang Y, Bolton J, Manning CD (2020) Stanza: a Python natural language processing toolkit for many human languages. In: Proceedings of the 58th annual meeting of the association for computational linguistics: system demonstrations
68. Rekatsinas T, Chu X, Ilyas IF, Ré C (2017) HoloClean: holistic data repairs with probabilistic inference. Proc VLDB Endow 10(11):1190–1201
69. Ribeiro MT, Singh S, Guestrin C (2016) Why should I trust you?: Explaining the predictions of any classifier. In: Krishnapuram B, Shah M, Smola AJ, Aggarwal CC, Shen D, Rastogi R (eds) Proceedings of the 22nd ACM SIGKDD international conference on knowledge discovery and data mining, San Francisco, CA, USA, 13–17 Aug 2016. ACM, pp 1135–1144
70. Ribeiro MT, Singh S, Guestrin C (2018) Anchors: High-precision model-agnostic explanations. In: Proceedings of the AAAI conference on artificial intelligence, vol 32(1)
71. Schulman J, Wolski F, Dhariwal P, Radford A, Klimov O (2017) Proximal policy optimization algorithms. arxiv: abs/1707.06347
72. Seshia SA, Sadigh D, Sastry SS (2022) Toward verified artificial intelligence. Commun ACM 65(7):46–55
73. Singpurwalla ND, JMB (2004) Membership functions and probability measures of fuzzy sets. J Am Statist Assoc 99(467):867–877

74. Strobl L, Merrill W, Weiss G, Chiang D, Angluin D (2024) What formal languages can transformers express? A survey. Trans Assoc Comput Ling 12:543–561
75. Sutton RS, Barto AG (2018) Reinforcement learning—an introduction adaptive computation and machine learning, 2nd ed. MIT Press
76. Tan P, Steinbach MS, Karpatne A, Kumar V (2019) Introduction to data mining, 2nd ed. Pearson
77. van der Aalst WMP (2016) Process mining—data science in action, 2nd ed. Springer
78. Verbeek E, Buijs JCAM. van Dongen BF, van der Aalst WMP (2010) ProM 6: the process mining toolkit. In: La Rosa M (ed) Proceedings of the business process management 2010 demonstration track, Hoboken, NJ, USA, 14–16 Sep 2010, CEUR Workshop Proceedings, vol 615. CEUR-WS.org (2010)
79. Visani G, Bagli E, Chesani F (2020) Optilime: optimized LIME explanations for diagnostic computer algorithms. In: Conrad S, Tiddi I (eds) Proceedings of the CIKM 2020 workshops co-located with 29th ACM international conference on information and knowledge management (CIKM 2020), Galway, Ireland, 19–23 Oct 2020, CEUR Workshop Proceedings, vol 2699. CEUR-WS.org
80. Vosoughi S, Roy D, Aral S (2018) The spread of true and false news online. Science 359(6380):1146–1151
81. Wang L, Lin P, Cheng J, Liu F, Ma X, Yin J (2021) Visual relationship detection with recurrent attention and negative sampling. Neurocomputing 434:55–66
82. von Winterfeldt D, Edwards W (1986) Decision analysis and behavioral research. Cambridge University Press
83. Xie X, Chang J, Kassem M, Parlikad A (2023) Resolving inconsistency in building information using uncertain knowledge graphs: a case of building space management. In: Proceedings of the 2023 European conference on computing in construction and the 40th international CIB W78 conference, Computing in Construction, vol 4. European Council on Computing in Construction, Heraklion, Greece
84. Yang BB, Shen SQ, Gao W (2019) Weighted oblique decision trees. Proc AAAI Conf Artif Intell 33(01):5621–5627
85. Yu Z, Chu X (2019) PIClean: a probabilistic and interactive data cleaning system. In: Boncz PA, Manegold S, Ailamaki A, Deshpande A, Kraska T (eds) Proceedings of the 2019 international conference on management of data, SIGMOD conference 2019, Amsterdam, The Netherlands, 30 June–5 July 2019. ACM, pp 2021–2024
86. Zaki MJ Jr, WM (1984) Rule-Based expert systems: the MYCIN experiments of the stanford heuristic programming project. Addison-Wesley
87. Zaki MJ, Meira W Jr (2020) Data mining and machine learning: fundamental concepts and algorithms, 2 edn. Cambridge University Press
88. Zhang T, Subburathinam A, Shi G, Huang L, Lu D, Pan X, Li M, Zhang B, Wang Q, Whitehead S, Ji H, Zareian A, Akbari H, Chen B, Zhong R, Shao S, Allaway E, Chang S, McKeown KR, Li D, Huang X, Sun K, Peng X, Gabbard R, Freedman M, Kejriwal M, Nevatia R, Szekely PA, Kumar TKS, Sadeghian A, Bergami G, Dutta S, Rodríguez ME, Wang DZ (2018) GAIA—a multi-media multi-lingual knowledge extraction and hypothesis generation system. In: Proceedings of the 2018 text analysis conference, TAC 2018, Gaithersburg, Maryland, USA, 13–14 Nov 2018. NIST (2018)
89. Zhao ZQ, Zheng P, Xu ST, Wu X (2019) Object detection with deep learning: a review. IEEE Trans Neural Netw Learn Syst 30(11):3212–3232

# Chapter 3
# Financial Explainable AI for Credit Card Scoring: Fairness Challenges and Case Study

**Mike Horia Mihail Teodorescu and Yongxu Sun**

**Abstract** Financial applications of AI, such as credit card scoring, have strict regulatory constraints and arise many ethical concerns. The chapter demonstrates how XAI techniques can be applied in credit predictions to improve both individual and group-level machine learning fairness and comply with regulations. The chapter also reviews recent literature in the field of machine learning fairness. We show that demographic factors affect the outcomes of these models, and we highlight this effect using XAI techniques in this context. A case study is used as an example, featuring visualization methods such as SHAP and LIME, as well as a detailed analysis on a publicly available lending dataset.

## Acronyms

**LIME** Local Interpretable Model Agnostic Explanations
**SHAP** Shapley Additive exPlanations
**XAI** Explainable Artificial Intelligence

## 3.1 Introduction

In recent years, decision-makers have adopted Artificial Intelligence (AI) across various critical domains of our lives, including employment, healthcare, and finance, despite the criticism regarding its explainability [1] and fairness [2–4]. Concerns of

M. H. M. Teodorescu (✉)
Responsible AI Systems and Experiences (RAISE), Information School, University of Washington, Seattle, WA, USA
e-mail: miketeod@uw.edu

Y. Sun
Information School, University of Washington, Seattle, WA, USA
e-mail: yongxs@uw.edu

 
N. Bizon and B. Appasani (eds.), *Explainable Artificial Intelligence for Trustworthy Decisions in Smart Applications*, https://doi.org/10.1007/978-3-031-97007-8_3

fairness in applying machine learning algorithms to major socioeconomic outcomes such as lending [5–7], employment [8, 9], housing [10], and health [11–13]—which can also impact societal standing and economic wellbeing, have been voiced for several years in the machine learning and information systems literature. The approach taken in the machine learning literature has been primarily to focus on statistical criteria, such as demographic parity [14], accuracy parity [15], equality in opportunity [16], and equalized odds [16, 17], to name a few (for a detailed survey of statistical fairness criteria, see [18]. This chapter will review the concepts of protected attributes, fairness criteria, briefly overview the literature on machine learning fairness, then discuss explainability within the context of machine learning applied to credit lending with an example application. For the example we use a well-known dataset in the literature, which contains documented anonymized credit card data. First, we will overview fairness criteria and the legal background for fairness criteria and protected attributes, as fundamental concepts for the forthgoing analysis.

Fairness criteria generally depend on the definition of *protected attributes*, individual characteristics that are codified in law to protect the person against discrimination, such as gender, race, age, religion, veteran status, marital status, national origin. The history of the protected attributes is long established in Federal law in the United States, for instance codified in the Civil Rights Act of 1964, as well as in the Fair Credit Reporting Act (FCRA) of 1970, though these laws pre-date the age of machine learning and algorithmic decision-making regarding employment, housing, and credit. More recent efforts in the United States which require checks against algorithmic discrimination include at the federal level the Consumer Financial Protection Bureau's oversight of lenders [19] to follow the Equal Credit Opportunity Act (ECOA), which requires creditors to provide applicants with explanations for their adverse decisions and the local law New York City Local Law 144 which compels companies to audit automated hiring algorithms and share with applicants when they are evaluated by an algorithm for employment purposes. Recent efforts in the European Union resulted in the passing of the EU AI Act "the world's first comprehensive AI act",[1] which categorizes AI risk and imposes different levels of responsibility on developers based on the potential risk to users. In previous work, the first author proposed the use of standards for AI fairness, similar to how standards regulate other technology areas such as quality management systems in manufacturing ISO or IEEE standards [20].

Credit prediction has been a long-standing focal point in the realm of human-ML augmentations. Commonly utilized techniques such as Naive Bayes, SVM, neural networks, and ensemble algorithms have found frequent applications in the assessment of credit risk and lending decisions [21, 22]. Nonetheless, ML algorithms are criticized the unsatisfactory outcomes when evaluated using various fairness criteria on real lending data [7, 23]. Even though current machine learning models can provide coefficients for each feature as an indication of the feature's impact on the output, the variability in the scale of the features can hinder the coefficients from reflecting the

[1] https://www.europarl.europa.eu/topics/en/article/20230601STO93804/eu-ai-act-first-regulation-on-artificial-intelligence, Accessed October 11, 2024.

importance of each feature [24]. Furthermore, the principles of organizational justice in the domain of business ethics emphasize the importance of algorithms being both explainable and accountable [25]. This aligns with the growing body of literature on Explainable AI (XAI) and recent legislative developments in the United States that prioritize the explainability and fairness of AI adoptions (e.g., New York Local Law 144,and the proposals in the White House Executive Order 14110 of 2023 on Safe, Secure, and Trustworthy Development and Use of Artificial Intelligence).

This chapter aims to illustrate how Explainable AI (XAI) techniques can be applied to enhance machine learning fairness by detecting biases related to legally protected attributes in the decision-making process of credit prediction. We have identified that demographic factors play a significant role in the outcomes of these models, and we highlight this effect using XAI techniques in this context. The XAI methods used in this paper are Local Interpretable Model Agnostic Explanations (LIME) and SHapley Additive exPlanations (SHAP). They will be applied to machine learning based credit prediction models using the dataset offered by UCI Machine Learning Repository,[2] previously analyzed from a fairness perspective in [7].

## 3.2 Financial Explainable AI (XAI): A Brief Overview

Among all industries, the financial sector is expected to lead the investment in AI [26], but financial AI applications also involve social, governance and economic risks [26], mitigating these risks require human intervention and specific regulations. For allowing human supervision, AI systems need to reveal the way they derived the results [26, 27]. Financial Explainable AI (XAI) refers to the use of transparent machine learning models in the financial sector to enhance understanding, trust, and accountability in AI-driven decisions. In this paper, we showcase how XAI techniques can be applied in credit predictions to improve both individual and group-level machine learning fairness.

Interpretability and explainability have different meanings in ML, with the first relating cause and effect in a data driven way that has meaning for humans, while the second is a shallower concept, namely that the system can partly fill some meaning gap between the results and the human for the human can make further reasonings and decisions [28]. Most deep AI systems are not interpretable,therefore, for being usable to decision makers, they need to provide at least some form of explanation on how the results were produced. A good notion of explainability is still lacking [28]. Explainability refers to providing the user with a description of the contributions of the various input variables to the output(s), under some specified model of the input-to-output connections, i.e., of the process. Herm et al. [28] expose various requirements for explainability and emphasize that a tradeoff has to be observed

[2] UC Irvine Machine Learning Repository, https://archive.ics.uci.edu/dataset/350/default+of+credit+card+clients.

between the complexity of the AI system performance and the degree of achievable explainability.

Recently, a flurry of frameworks for explainable AI (XAI) have been proposed for credit scoring and related fields. De Bock et al. [29] suggested a general approach to explainable artificial intelligence (XAI) for operational research, including proposals for the future. They emphasized the diverse perspectives of explainable AI (XAI), a diversity that may induce confusion. De Bock et al. distinguish between the organizational, regulatory and ethical perspectives on XAI, as opposed to the technical perspective, which seeks only the power of prediction. Most of the approach by De Bock et al. [29] may be valid for credit scoring, in particular. The lens of variance has been extended in the framework of SHAP and to several other methods, such as Sobol-based Sensitivity Analysis [30] and Shapley-Lorenz decompositions [31].

There were several attempts to apply LIME and SHAP in explaining the machine credit predictions reported in the literature, see [32–34] and references therein. Recently, [23], after reviewing the literature on AI algorithms for credit card scoring, advocated the use of Shapley values for interpretable credit scorecards and proved its benefits,they follow the scoring methodology in [35]. We exemplify in the next Sections the use of the combination of LIME and SHAP for improving the explainability.

### 3.2.1 Regulatory and Ethical Challenges of AI in Financial Applications

Financial applications of AI have stringent regulatory and ethical constraints, including accountability, auditability, and fairness, which involve transparency and explainability of model predictions [29, 32, 36]. Several bodies, such as European Commission and IEEE issued recommendations and regulations regarding this topic [37–40]. Most current regulations impose some sort of minimal transparency of the predictions made by AI for ensuring both financial inclusion and market safety, [26, 41]. The trend in the regulatory activity is to increase the coverage of various aspects and application fields of AI in finance and to extend the enforcement instruments in this respect.

The auditability of AI prediction instruments requires that human auditors are able to obtain a complete understanding of what results on can expect from the instrument, under various scenarios. In addition, the question 'how the results were obtained' has to be satisfactory answered, such that the human experts can assess the fairness of the procedure and its compliance to current regulations. On the other hand, the governance bodies of the financial institutions require a good enough understanding on the mechanisms of obtaining the predictions by the AI tools, such that those predictions can be included in strategic directions for the companies. The quality of the explanations of a negative landing result is also essential for customers' satisfaction and loyalty. One of the challenges is that these three categories of users, regulators, company governance, and customers, require different types of explanations and making explainability a three-dimensional problem. Regulators are interested in the alignment of the derivation process with respect with the laws and norms, including ethical ones. Firm governance needs an understanding of the detailed quantitative cause-effect relationships at the general, statistical level, which is closer to the interpretability meaning, while customers require explanations on the salient issues (i.e., variables) that lead to the outcome for them personally. Therefore, reducing the explainability to simple questions, such as 'what', 'why', and 'how', out of the context, is unsatisfactory: results and their derivation should be put into the form of the user's thinking model.

### 3.2.2 Lime

Local Interpretable Model-agnostic Explanations (LIME) is a model-agonistic technique that approximates the original model, regardless of its form and complexity, with an interpretable local model to explain a prediction [42]. Enhanced transparency through LIME primarily improves individual-level fairness. Credit institutions can use LIME to generate explanations for why a particular applicant was granted or denied credit. These explanations identify which features, such as income, employment history, or credit history, were most influential in the decision. If LIME reveals that certain demographic features, such as race, gender, or age, are heavily influencing the credit decision, the institution can use this information to adjust the model.

### 3.2.3 Shap

SHapley Additive exPlanations (SHAP) is a model-agnostic technique that provides local explanations by revealing the contribution of each feature to the prediction [24, 43]. SHAP improves group-level fairness by analyzing the overall impact of features across all applicants. By aggregating the SHAP values, it uncovers whether protected attributes consistently influence the predicted outcomes, providing a more holistic view that extends beyond individual-level analysis to a group-level perspective.

## 3.3 Case Study

### 3.3.1 Dataset

The "Default of Credit Card Clients" dataset [44], available from the UCI Machine Learning Repository, contains data related to default payments of credit card clients in Taiwan. The dataset comprises information on 30,000 clients, with 23 attributes for each, including demographic information (e.g., age, gender, education), credit card balance details, payment history, and billing statements over several months. The goal is to predict if an individual will default on their payment in the next month (which would be risk to the lender). This default variable is a binary target ("default payment next month" as a 1 for default and as a 0 for no default). The dataset serves as a common benchmark for classification problems, providing a rich context for exploring credit risk modeling and financial behavior patterns.

The categorical variables, including gender, education, and marriage, are transformed into categorical codes, as shown the in the following code:

```
categorical_features = ["SEX", "EDUCATION", "MARRIAGE"]

for col_name in categorical_features:
    df[col_name] = df[col_name].astype("category")
```

Then, we have used the *train_test_split* function splits the dataset (X, y) into training (80%) and testing (20%) sets. The parameter *random_state* = *919* ensures the split is reproducible for consistent results. We have also standardized the dataset before training the model by using the *StandardScaler*() function.

```
# Split the data into training and testing sets
X_train, X_test, y_train, y_test = train_test_split(
    X, y, test_size=0.2, random_state=919
)

# Standardize the training and testing data
scaler = StandardScaler()
X_train = scaler.fit_transform(X_train)
X_test = scaler.transform(X_test)
```

### 3.3.2 Explaining Neural Network Credit Predictions Using LIME and SHAP

We have built a neural network binary classifier to predict credit default outcomes. The code below defines a sequential neural network. The model consists of an input layer followed by two hidden layers and an output layer. The first hidden layer contains 25 neurons and the second hidden layer has 15 neurons. Finally, an output layer with 1 neuron is added. All layers used the ReLU activation function, which helps the model capture non-linear relationships effectively.

```
model = Sequential()

# Add the hidden layer
model.add(Dense(25, input_dim=X_train.shape[1], activation='relu'))
model.add(Dense(15, activation='relu'))

# Add the output layer
model.add(Dense(1, activation='relu'))
```

#### 3.3.2.1 LIME on Neural Network

We first initialize the *LimeTabularExplainer* using the training dataset, a list of feature names, and the possible class labels. Then, an individual instance from the test set is selected to be explained (in this case, the 11th instance). The *explain_instance* function is called, which uses the given model's prediction probabilities (model.predict_proba) to create an explanation focusing on the top 15 features that influenced the decision. Finally, the explanation is displayed within a notebook, providing insights into the model's reasoning for the specific prediction. This is particularly useful for understanding which features were most impactful for the given instance, making it easier to debug or validate model predictions in a human-readable form. LIME highlights the features that had the most significant impact on the prediction, both positively (increasing the likelihood of default) and negatively (decreasing the likelihood of default), with the most important features listed first. In this case, we will display the 15 features that impact the decision-making process most heavily according to LIME.

```
# LIME
import lime
import lime.lime_tabular

feature_names = df.columns.tolist()
feature_names.remove('default payment next month')

explainer = lime.lime_tabular.LimeTabularExplainer(
    training_data=X_train,
    feature_names=feature_names,
    class_names=['Non-default', 'Default'],
    mode='classification'
)

# Choose an instance to explain
i = 11  # explain the 11th instance from the test set

# Explanation for the chosen instance
exp = explainer.explain_instance(
    data_row=X_test[i],
    predict_fn=model.predict_proba,
    num_features=15,  # Number of features to include in the explanation
    top_labels=1
)

exp.show_in_notebook(show_table=True, show_all=False)
```

The model predicts a high probability (0.78) that the customer will default on their credit, making "Default" the predicted class. The result suggests that the payment status in the previous months (PAY_0, PAY_3, PAY_5, and PAY_2) increases the probability of default, indicating that higher values, which represent missed or delayed payments, have a strong impact. Other factors contributing to the default prediction include the credit limit (LIMIT_BAL) and specific payment amounts. A lower credit limit has a negative impact, making default more likely. From such perspective, the explanation is coherent with financial logic (Fig. 3.1).

The legally protected attributes, including gender and education, are among the 15 most influential features in the model. Specifically, in this instance, gender has a minor negative impact on the probability of default, which suggests that being female slightly decreases the likelihood of default according to the model's predictions. Conversely, having a higher education slightly increases the probability of default ($-1.08 <$ Education). However, their impact is less than payment history and credit limit. This observation suggests that, although demographic factors are accounted for in the model, they do not play a substantial role in driving the prediction relative to the more critical financial behaviors.

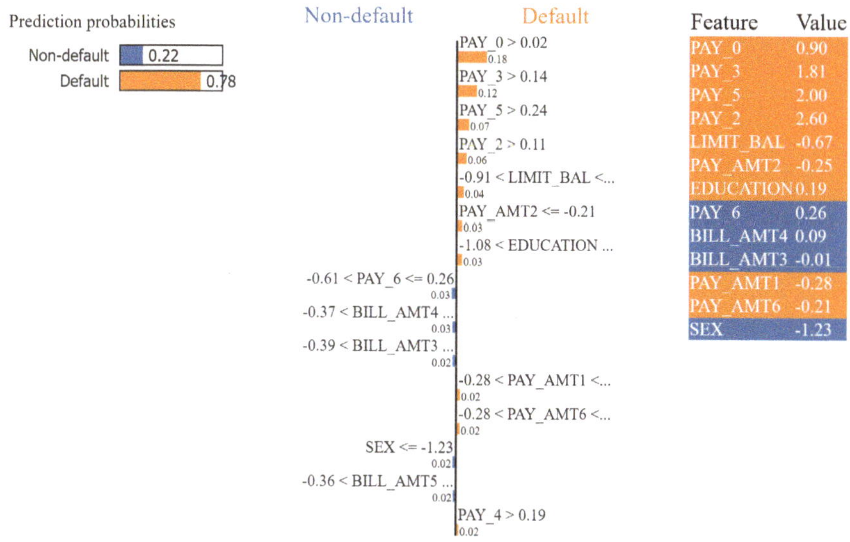


**Fig. 3.1** LIME analysis of a neural network-predicted default case

#### 3.3.2.2 SHAP on Neural Network

The SHAP analysis provides an overview of the impact of each feature on the neural network's predictions for credit default outcomes. The *shap.sample()* function is used to select a subset of the training data (*X_train_scaled*) consisting of 5 samples, which serves as the "background" dataset to simplify the computation of Shapley values. We choose 5 samples because SHAP analysis can be very computationally intensive given. The *KernelExplainer* is then used to create a SHAP explainer based on the MLP model's predictions and the reduced background dataset. The explainer computes SHAP values for all samples in the scaled test set (*X_test_scaled*). Finally, the *shap.summary_plot()* function is used to generate a summary plot, which visually represents the impact of each feature on the model's output.

```
import shap

# Reduce the number of background samples to 5 for faster computation
background = shap.sample(X_train_scaled, 5)

explainer = shap.KernelExplainer(mlp.predict, background)
shap_values = explainer.shap_values(X_test_scaled)
shap.summary_plot(shap_values, X_test_scaled, feature_names=X.columns)
```

The SHAP summary plot (in Fig. 3.2) illustrates the influence of various features on the model's output, with higher SHAP values indicating a greater impact on increasing the likelihood of default. Each point in the plot represents a SHAP value for a feature, illustrating the average magnitude and direction of the feature's contribution, thus offering an overview of feature importance and the nature of their relationships with the target variable.

The SHAP analysis reveals that the most influential factors impacting credit default predictions are related to payment history, with PAY_0 (most recent month's payment status), PAY_3, PAY_2, PAY_4, and PAY_5 significantly increasing the probability of default when values are higher, indicating missed or delayed payments.

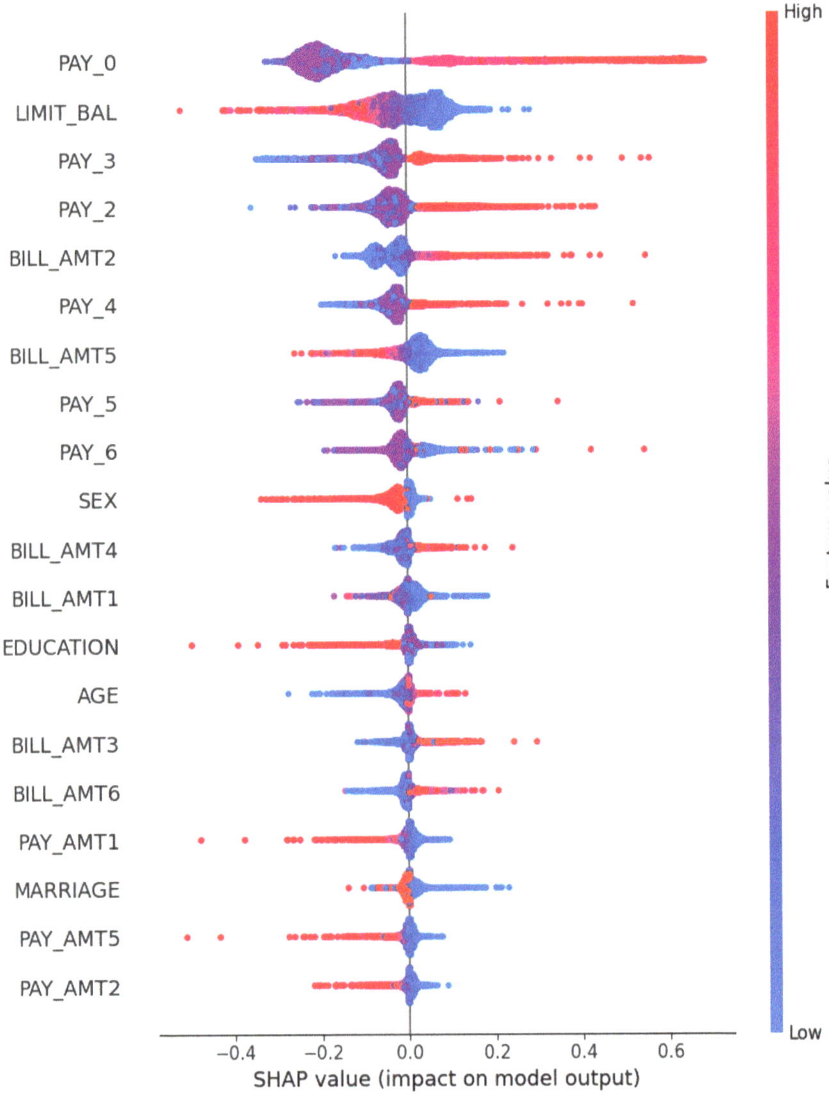


**Fig. 3.2** SHAP summary plot for neural network credit default predictions

The credit limit (LIMIT_BAL) also plays a critical role, where lower limits are associated with a higher likelihood of default. Additionally, the amount billed two months ago (BILL_AMT2) shows a notable impact, with higher billed amounts increasing the risk of default.

Demographic factors such as sex, marital status, and education ranked lower on the summary plot. Regarding the sex attribute, female applicants, represented by the red dots, clustered mainly on the negative side of the SHAP value spectrum. This indicates that females are less likely to be classified as default risks by the model.

### *3.3.3 Interpretation*

The results suggest that XAI techniques can be used to infer the fairness of a model, with LIME providing insights into individual-level fairness and SHAP offering perspectives on group-level fairness. Though LIME and SHAP offered different kinds of explanations, the explanations generated from the two kinds of XAI techniques are generally coherent: credit history is the most significant factor impacting the credit prediction outcomes, while demographic features such as gender and education also play a role, though to a lesser extent.

Interestingly, our findings suggest that being female reduces the likelihood of being classified as a default, which does not align with the traditional understanding that banks have historically lent more to men [45]. One possible explanation for this incoherence is that the local explanations offered by LIME and SHAP are typically made by approximating the decision-making process through simpler linear models. When such linear correlations do not exist or are insignificant, the XAI techniques may hallucinate and produce gravely misleading results.

## 3.4 Conclusions

This chapter explored the potential of Explainable AI (XAI) techniques, specifically LIME and SHAP, in enhancing the fairness and transparency of machine learning models used for credit default prediction in the financial sector. By applying these techniques to a neural network model, we demonstrated how XAI can provide insights into both individual-level and group-level fairness. The analysis revealed that while financial behaviors, such as payment history and credit limit, significantly influence credit decisions, demographic factors like gender and education also play a role, albeit to a lesser extent. Though we have found that the model tends to favor females by assigning them a lower likelihood of being classified as default, this contrasts with the conventional understanding that banks have historically favored males. From this perspective, the results of this paper align with existing research that the gender gaps in credit prediction are becoming smaller and could have changed over time favoring women [46]. The findings are dependent on the dataset and models used for the analysis and therefore the results may not be generalizable to a broader population.

Explainable AI (XAI) emerged to address concerns related to the limitations of AI models and their decision-making processes. However, there are challenges to implementing XAI techniques into financial credit prediction contexts. For one thing, different XAI methods offer varied explanations. LIME provides local explanations for specific instances, whereas SHAP delivers global explanations to approximate feature importance across the entire dataset. Therefore, we recommended validating the result across different XAI methodologies to testify if the explanations are accurate. For another, it is very computationally intensive to generate SHAP explanation. We reduced the background sample to 15 to speed up the process, which can pose significant limitations to the reliability of our findings. Furthermore, while credit models frequently use a binary outcome variable, decisions are rarely binary in credit lending when done by human decision-makers [47]. Factors such as down payment, repayment schedule, and interest rate add complexity beyond a simple "positive" or "negative" outcome [48]. There are also concerns that offering detailed explanations through XAI techniques risk revealing too much information about the system, which raises concerns about intellectual property rights [48]. Many of the existing XAI tools currently available in the field lack the necessary features for effectively detecting and mitigating bias [49]. This shortcoming can potentially lead to a phenomenon known as 'fairwashing,' wherein researchers may falsely believe that a model is fair whereas, in reality, the biases within the model remain undetected by the XAI tool [50].

We exposed how knowledge obtained through explainability could potentially benefit stakeholders in credit prediction. Regulators and credit applicants could leverage Explainable AI (XAI) techniques to identify patterns or practices of discrimination in credit prediction. Creditors could use this knowledge to ensure compliance with laws such as the Fair Credit Reporting Act (FCRA)—a law that precedes regular use of AI in lending as it was enacted in 1970 in the United States; the Equal Credit Opportunity Act (ECOA) passed into US federal law in 1975; and the Truth in Lending Act (TILA) which is part of US federal law since 1968, requiring lenders to provide clear explanations for their processes in credit decisions. While there are regulations by federal agencies overseeing compliance with these laws by lenders, there may be an opportunity to add context around XAI in regulations. We believe that the integration of Explainable AI (XAI) techniques offers a promising pathway toward greater fairness, transparency, and legal compliance in the financial sector. By carefully addressing these challenges and leveraging the strengths of XAI, stakeholders can enhance the integrity and inclusivity of credit lending practices, ultimately fostering a more equitable financial system.

## References


1. Deck L, Schoeffer J, De-Arteaga M, Kühl N (2024) A critical survey on fairness benefits of explainable AI. In: The 2024 ACM conference on fairness, accountability, and transparency, pp 1579–1595
2. Awwad Y, Fletcher R, Frey D, Gandhi A, Najafian M, Teodorescu M (2020) Exploring fairness in machine learning for international development. CITE MIT D-Lab. https://dspace.mit.edu/handle/1721.1/126854
3. Satell G, Abdel-Magied Y (2020) AI fairness isn't just an ethical issue. Harv Bus Rev 20. https://hbr.org/2020/10/ai-fairness-isnt-just-an-ethical-issue
4. Tarafdar M, Teodorescu M, Tanriverdi H, Robert L, Morse L (2020) Seeking ethical use of AI algorithms: challenges and mitigations. In: ICIS 2020 proceedings, vol 1. https://aisel.aisnet.org/icis2020/panels/panels/1/
5. Brotcke L (2022) Time to assess bias in machine learning models for credit decisions. J Risk Financ Manag 15(4):165
6. Spiess J (2022) Machine learning explainability & fairness: Insights from consumer lending. FinRegLab whitepaper. https://www.gsb.stanford.edu/faculty-research/publications/machine-learning-explainability-fairness-insights-consumer-lending
7. Teodorescu MH, Yao X (2021) Machine learning fairness is computationally difficult and algorithmically unsatisfactorily solved. In: 2021 IEEE high performance extreme computing conference (HPEC), pp 1–8. IEEE
8. Teodorescu MH, Ordabayeva N, Kokkodis M, Unnam A, Aggarwal V (2022) Determining systematic differences in human graders for machine learning-based automated hiring. In: Brookings institution center on regulation and markets working paper. https://www.brookings.edu/wp-content/uploads/2022/06/Determining-systematic-differences-in-human-graders-for-machine-learning-based-automated-hiring.pdf
9. Teodorescu MHM, Geiger MK, Morse L (2024) Dangers of speech technology for workplace diversity. Nat Mach Intell 6: 77–380. https://www.nature.com/articles/s42256-024-00827-5
10. Schneider V (2020) Locked out by big data: how big data algorithms and machine learning may undermine housing justice. Colum Hum Rts L Rev 52:251
11. Mhasawade V, Zhao Y, Chunara R (2021) Machine learning and algorithmic fairness in public and population health. Nat Mach Intell 3(8):659–666
12. Teodorescu MH, Taraboanta I, Andrian S, Ghiorghe AC, Teodorescu HN, Grigorie M (2023) Fairness constraints in speech-to-text applications–clinical effects of salivation. In: 2023 international conference on speech technology and human-computer dialogue (SpeD), pp 72–77. IEEE. https://ieeexplore.ieee.org/document/10314880/
13. Wang Y, Wang L, Zhou Z, Laurentiev J, Lakin JR, Zhou L, Hong P (2024) Assessing fairness in machine learning models: a study of racial bias using matched counterparts in mortality prediction for patients with chronic diseases. J Biomed Inform 156:104677
14. Pessach D, Shmueli E (2022) A review on fairness in machine learning. ACM Comput Surv (CSUR) 55(3):1–44
15. Besse P, del Barrio E, Gordaliza P, Loubes JM, Risser L (2022) A survey of bias in machine learning through the prism of statistical parity. Am Stat 76(2):188–198
16. Hardt M, Price E, Srebro N (2016) Equality of opportunity in supervised learning. In: 30th conference on neural information processing systems (NIPS 2016). Barcelona, Spain
17. Gölz P, Kahng A, Procaccia AD (2019) Paradoxes in fair machine learning. Adv Neural Inf Process Syst 32
18. Mehrabi N, Morstatter F, Saxena N, Lerman K, Galstyan A (2021) A survey on bias and fairness in machine learning. ACM Comput Surv (CSUR) 54(6):1–35
19. CFPB (2022) CFPB acts to protect the public from black-box credit models using complex algorithms. https://www.consumerfinance.gov/about-us/newsroom/cfpb-acts-to-protect-the-public-from-black-box-credit-models-using-complex-algorithms/

20. Teodorescu MH, Makridis C (2024) Fairness in machine learning: regulation or standards?, Brookings institution, center on regulation and markets. https://www.brookings.edu/articles/fairness-in-machine-learning-regulation-or-standards/
21. Bensic M, Sarlija N, Zekic-Susac M (2005) Modelling small-business credit scoring by using logistic regression, neural networks and decision trees. Intell Syst Account Financ Manag Int J 13(3):133–150
22. Nazari M, Alidadi M (2013) Measuring credit risk of bank customers using artificial neural network. J Manag Res 5(2):17
23. Hlongwane R, Ramabao K, Mongwe W (2024) A novel framework for enhancing transparency in credit scoring: leveraging shapley values for interpretable credit scorecards. PLoS ONE 19(8):e0308718. https://doi.org/10.1371/journal.pone.0308718
24. Lundberg S (2018) An introduction to explainable AI with shapley values-SHAP latest documentation. https://shap.readthedocs.io/en/latest/example_notebooks/overviews/An%20introduction%20to%20explainable%20AI%20with%20Shapley%20values.html
25. Morse L, Teodorescu MHM, Awwad Y, Kane GC (2022) Do the ends justify the means? Variation in the distributive and procedural fairness of machine learning algorithms. J Bus Ethics 181:1083–1095
26. Buckley RP, Zetzsche DA, Arner DW, Tang BW (2021) Regulating artificial intelligence in finance: putting the human in the loop. Syd Law Rev The 43(1):43–81
27. Teodorescu MH, Morse L, Awwad Y, Kane GC (2021) Failures of fairness in automation require a deeper understanding of human-ML augmentation. MIS Q 45(3):1483–1499. https://doi.org/10.25300/MISQ/2021/16535
28. Herm LV, Heinrich K, Wanner J, Janiesch C (2023) Stop ordering machine learning algorithms by their explainability! A user-centered investigation of performance and explainability. Int J Inf Manage 69(2023):102538
29. De Bock KW, Coussement K, De Caigny A, Słowiński R, Baesens B, Boute RN, Weber R et al (2024) Explainable AI for operational research: a defining framework, methods, applications, and a research agenda. Eur J Oper Res 317(2):249–272
30. Fel T, Cadène R, Chalvidal M, Cord M, Vigouroux D, Serre T (2021) Look at the variance! efficient black-box explanations with Sobol-based sensitivity analysis. Adv Neural Inf Process Syst 34:26005–26014
31. Giudici P, Raffinetti E (2021) Shapley-Lorenz eXplainable artificial intelligence. Expert Syst Appl 167:114104
32. Bussmann N, Giudici P, Marinelli D et al (2021) Explainable machine learning in credit risk management. Comput Econ 57(1):203–216. https://doi.org/10.1007/s10614-020-10042-0
33. El Qadi A, Trocan M, Díaz-Rodríguez N et al (2023) Feature contribution alignment with expert knowledge for artificial intelligence credit scoring. SIViP 17(2023):427–434. https://doi.org/10.1007/s11760-022-02239-7
34. Gramegna A, Giudici P (2021) SHAP and LIME: an evaluation of discriminative power in credit risk. Front Artif Intell 4:752558
35. Siddiqi N (2017) Scorecard development. In: Siddiqi N (ed) Intelligent credit scoring. https://doi.org/10.1002/9781119282396.ch2
36. Nguyen SP, Truong NQ (2024) An application of explainable artificial intelligence in credit scoring. In: Kreinovich V, Sriboonchitta S, Yamaka W (eds) Machine learning for econometrics and related topics. In: Studies in systems, decision and control, vol 508. Springer, Cham. https://doi.org/10.1007/978-3-031-43601-7_21
37. European Commission (2024) EU Artificial Intelligence Act. https://artificialintelligenceact.eu/
38. IEEE (2024) IEEE 2894–2024 guide for an architectural framework for explainable artificial intelligence. https://standards.ieee.org/ieee/2894/11296/
39. Nannini L (2024) Habemus a right to an explanation: so what?—a framework on transparency-explainability functionality and tensions in the EU AI Act. Proc AAAI/ACM Conf AI Ethics Soc 7(1):1023–1035. https://doi.org/10.1609/aies.v7i1.31700

40. Winfield AF, Booth S, Dennis LA, Egawa T, Hastie H, Jacobs N, Watson E et al (2021) IEEE P7001: a proposed standard on transparency. Front Robot AI 8:665729. https://www.frontiersin.org/journals/robotics-and-ai/articles/10.3389/frobt.2021.665729. https://doi.org/10.3389/frobt.2021.665729
41. Lee J (2020) Access to finance for artificial intelligence regulation in the financial services industry. Eur Bus Organ Law Rev 21(4):731–757
42. Ribeiro MT, Singh S, Guestrin C (2016) Why should I trust you? Explaining the predictions of any classifier. In: Proceedings of the 22nd ACM SIGKDD international conference on knowledge discovery and data mining, pp 1135–1144
43. Albini E, Long J, Dervovic D, Magazzeni D (2022) Counterfactual shapley additive explanations. In: Proceedings of the 2022 ACM conference on fairness, accountability, and transparency, pp 1054–1070
44. Yeh I (2009) Default of credit card clients [dataset]. UCI Machine Learning Repository. https://doi.org/10.24432/C55S3H
45. De Andrés P, Gimeno R, de Cabo RM (2021) The gender gap in bank credit access. J Corp Finan 71:101782
46. Blascak N, Tranfaglia A (2023) Decomposing gender differences in bankcard credit limits. FRB of Philadelphia working paper No. 23–30 (Federal Reserve Bank Working Paper Series). https://papers.ssrn.com/sol3/papers.cfm?abstract_id=4655583, https://doi.org/10.21799/frbp.wp.2023.30
47. Moscato V, Picariello A, Sperlí G (2021) A benchmark of machine learning approaches for credit score prediction. Expert Syst Appl 165:113986
48. Barocas S, Selbst AD, Raghavan M (2020, January) The hidden assumptions behind counterfactual explanations and principal reasons. In: Proceedings of the 2020 conference on fairness, accountability, and transparency, pp 80–89
49. Alikhademi K, Richardson B, Drobina E, Gilbert JE (2021) Can explainable AI explain unfairness? A framework for evaluating explainable AI. arXiv preprint. arxiv:abs/2106.07483
50. Shahin Shamsabadi A, Yaghini M, Dullerud N, Wyllie S, Aïvodji U, Alaagib A, Papernot N et al (2022) Washing the unwashable: on the (im) possibility of fairwashing detection. Adv Neural Inf Process Syst 35:14170–14182

# Chapter 4
# Explainable AI for Future Smart Cities: Architectures, Applications, and Challenges

**Sara Khader and Qasem Abu Al-Haija**

**Abstract** Explainable Artificial Intelligence (XAI) plays a crucial role in making AI-driven technologies in smart cities more transparent and understandable. As smart cities integrate AI to improve services like traffic management, energy distribution, and healthcare, the need for trust and accountability becomes more prominent. XAI explains AI decisions in a way that allows all stakeholders, from policymakers to citizens, to understand the reasoning behind critical AI-driven actions. This chapter discusses several architectural models that can be adopted for embedding XAI within smart city operation frameworks, namely centralized, decentralized, and layered systems. Real-world examples, like Barcelona's AI-powered traffic management system, demonstrate the practical application of XAI in enhancing urban efficiency while maintaining transparency.

## 4.1 Introduction

The idea of smart cities has developed as a response to fast-paced urban growth and 21st-century technological development [1, 2]. A "Smart city" is an urban area that uses information and communication technologies (ICT) to improve the quality of its residents' lives, utilize resources, improve governance, and enhance public services [3, 4]. Modern technologies such as big data analytics, the Internet of Things (IoT), and mainly artificial intelligence (AI) have substantially altered how cities function, creating new possibilities for sustainable development and efficient management, hence smart cities [5]. Smart cities collect data from different sources, e.g., sensors and cameras [6, 7].

S. Khader
Department of Computer Science, Princess Sumaya University for Technology (PSUT), Amman, Jordan

Q. Abu Al-Haija (✉)
Department of Cybersecurity, Jordan University of Science and Technology (JUST), Irbid, Jordan
e-mail: qsabuhaija@just.edu.jo



N. Bizon and B. Appasani (eds.), *Explainable Artificial Intelligence for Trustworthy Decisions in Smart Applications*, https://doi.org/10.1007/978-3-031-97007-8_4

Moreover, integrating AI into smart cities' sectors such as transportation, energy management, and healthcare can have the potential to transform these sectors [8–10]. In transportation, AI-driven systems may observe collected traffic data, which can effectively improve traffic signals, reduce congestion, and enhance public transit routes [11]. For instance, the AI-driven traffic flow in Barcelona has effectively enhanced traffic efficiency [4]. AI plays an important role in energy management, developing smart grids that adjust energy distribution in real-time in response to demand and supply, thereby facilitating efficient energy use, cost reduction, and a reduced environmental footprint [4, 12]. In healthcare, data from diverse sources can be collected to monitor public health trends, predict disease outbreaks, and develop personalized treatment plans [4, 8].

Integrating AI, Big Data, and IoT is essential for future smart cities [5, 9]. IoT devices such as cameras, sensors, and smart meters can generate huge amounts of data, which can be used further to improve urban infrastructure [7]. The AI algorithms analyze this information for useful insights, anomaly detection, and real-time decision-making [13]. In smart city platforms, for instance, AI can utilize the streaming data from IoT sensors to detect leaks in the water supply, predict equipment failures, or even optimize routes for waste collection [12].

### *4.1.1 Definition and Significance of Explainable AI (XAI) in Context of Smart Cities*

XAI includes a set of processes and methods that make the outcomes and operations of AI models and algorithms more understandable to humans [14]. Unlike conventional AI, which functions as a "black box" without clear decision-making functionalities, XAI explains how an AI system came to such a decision [15]. This kind of clearance is important for sectors directly impacting human life, such as health, finance, and governance of smart cities [1].

The importance of XAI within the context of smart cities can be attributed to its promoting trust, providing transparency, and increasing accountability in AI-driven systems [12]. As AI becomes more present in smart cities to improve services, so does the urgent need for more transparency in decision-making [16]. Clear explanations would make stakeholders, including policymakers, citizens, and service providers, find it easier to trust AI recommendations, especially in critical situations like emergency responses [17]. To do so, XAI plays an important role in some key aspects:

1. Providing clarity: For instance, suppose there is an AI in traffic management; then XAI helps to understand why certain routes are recommended. This keeps city planners informed and helps interpret and assess the generated decisions [17].
2. Improved Decision-Making for Domain Experts: XAI enables domain experts like doctors, financial analysts, and lawyers to make informed decisions or

change their predictions by understanding the recommendations from AI-driven decisions [13].

3. Supporting Accountability: AI systems auditing by administrators ensures that regulatory and ethical standards are complied with [18]. This has significant implications for public safety, where decisions may directly impact lives. For example, explainability in XAI would enable any potential biases from AI algorithms to be uncovered in predictive policing, making potential adjustments to assure fairness in law enforcement [4].
4. Enhancing Collaboration: By offering clear insights, XAI facilitates more informed decision-making, optimizing city management by making the decision process more interactive; this is essential for smart cities' multidisciplinary teams, including urban planners, data scientists, policymakers, and community members [2]. Regarding compliance and Ethical AI, regulators request improved transparency and accountability to address privacy, fairness, and discrimination concerns as decision-making becomes increasingly AI-driven. XAI approaches help to comply with relevant legal frameworks such as the European Union's General Data Protection Regulation (GDPR) that provide individuals the "right to explanation" regarding AI-generated decisions [19].
5. Enhancing User Experience and Personalization for Consumers: When consumers receive clear explanations for product recommendations, search results, or service suggestions, they tend to engage more positively with AI systems. These personalized explanations boost consumer satisfaction and enhance the effectiveness of AI systems by aligning them with user preferences [20].

Moreover, XAI reduces the risk of biased or unjust AI decisions by making the reasoning behind AI-generated outcomes more comprehensible. This ensures that AI applications adhere to ethical standards and legal requirements [21]. Therefore, XAI enhances AI's reliability, trustworthiness, and alignment with societal values [22].

In conclusion, the significance of XAI in smart cities lies in its ability to enhance trust, transparency, and accountability. Reflecting this growing recognition, a recent survey indicates that around 60% of smart city initiatives plan to integrate XAI by 2025 [1].

## 4.2 Foundations of XAI

### *4.2.1 Understanding Explainable AI: Definitions and Concepts*

As established in the past section, AI revolutionized the way smart cities operate by facilitating data-driven decision-making and automation across different urban sectors [8]; however, despite the advantages AI brings to smart cities, it also introduces challenges, particularly concerning transparency of the decision-making process, so XAI provided a potential solution by offering a way to enhance explainability, interpretability, and transparency of AI models, enabling stakeholders to grasp how and why decisions are made [16].

### *4.2.2 Differences Between Explainability, Interpretability, and Transparency*

Explainability, interpretability, and transparency are concepts related to machine learning and AI that are often used interchangeably but have distinct meanings [16].

**Explainability** refers to the capacity of an algorithm or model to present an understandable rationale behind its reasoning or decision-making process; it addresses the question of "why" a decision is made [8]. It aims to help users understand what factors determine the outcomes of AI models, which is essential for users to trust and accept a system. For instance, in a smart city traffic management system, if AI decides to change the timing of traffic lights during rush hour, explainability would provide the reasoning behind that decision, such as traffic congestion forecasts based on historical and real-time data [23].

**Interpretability**, however, refers to the degree to which humans can easily understand directly from the internal workings of the model why certain decisions were made. A model is interpretable if its internal processes are intuitively self-explanatory [15]. For example, a linear regression model used to predict energy consumption concerning weather and usage can be termed as interpretable because users can intuitively follow the relationships between the input and output. Decision trees are also inherently interpretable since they represent a sequential decision-making process, similar to how humans make decisions. They are, therefore, easily understandable [8]. For a smart city, an interpretable AI model may predict pollution levels based on vehicle emissions data in a manner that the town planners can easily follow and act upon [7].

**On the other hand, transparency** refers to the openness of the architecture, data, and mechanisms. A transparent system allows investigations of both external inputs and internal processing because it discloses the underlying data and methods on which decisions are based [14]. For instance, an AI system used for public safety that predicts crime hotspots based on historical crime data would be considered transparent if it explicitly shows the data sources and algorithms used [4].

Table 4.1 summarizing these concepts.

These concepts are crucial in promoting trust and accountability in AI systems within smart cities.

**Table 4.1** Explainability, interpretability, and transparency

| Aspect | Definition | Focus | Examples |
|---|---|---|---|
| Explainability | Ability to clarify why a model made specific predictions or decisions | Post-hoc clarification | LIME, SHAP |
| Interpretability | The degree to which a human can directly understand the model's decision process | Model simplicity and clarity | Decision trees, linear regression |
| Transparency | Openness about the model's structure, data, and internal mechanisms | Accessibility of internal details | Open-source models |

### 4.2.3 *Techniques and Methods of XAI*

XAI methods are techniques to understand, interpret, and trust machine learning methods. These methods allow us to uncover potential faults in models and explain how decisions are made. For example, if a model is trained to recognize certain animals, like dogs, based on their nose patterns, explainability techniques can highlight the image regions responsible for such predictions [24].

#### 4.2.3.1 Interpretable by Design Models (Ante-Hoc Models)

These models are interpretable by design and are constructed with inherent transparency. Examples include logistic regression and K-nearest neighbors. These models provide direct insights into their decision-making process by allowing the examination of parameters like weights or distances [25]. Because they are understanding how a decision is made is crucial. Ante-hoc models offer straightforward, built-in interpretability and do not require additional post-processing for explanations [22].

#### 4.2.3.2 Post-Hoc Explanations

These methods are applied after a model has been trained and used, and they can be used with any model. They explain this by generating insights into predictions of complex, often black-box models like neural networks or ensemble methods. Post-hoc explanations are critical for machine learning models in natural language processing (NLP) and computer vision, where models are inherently non interpretable [16]. These methods help bridge the gap between model complexity and the need for transparency; below is a breakdown of these methods:

1. Model Agnostic Methods: These methods can be applied to any machine learning model and focus on explaining predictions after making them.
   (a) LIME: Explains an individual prediction by fitting simple models around the local neighborhood of the prediction.

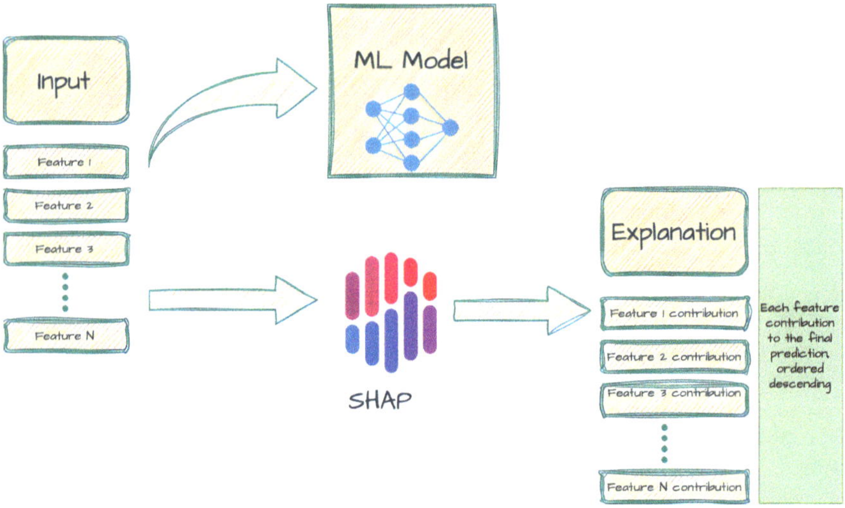


**Fig. 4.1** SHAP

(b) SHAP provides global and local importance based on shapely values, and Fig. 4.1 explains that concept.

2. Model-Specific Methods: Unlike model-agnostic methods, these techniques are tailored to specific models. For instance, gradient-based methods like integrated gradient are used in neural networks to determine the impact of individual features on predictions. These explanations are also a form of post-hoc interpretations, focusing on revealing why a specific model made a particular decision [16].
3. Adversarial Examples and Counterfactual Explanations are another category of post-hoc methods—adversarial examples test models by finding inputs that lead to misclassifications, exposing potential vulnerabilities. Counterfactual explanations, on the other hand, provide hypothetical scenarios that demonstrate what would need to change in the input to alter the outcome. Both techniques offer a post-hoc understanding of model weaknesses and decision boundaries.
4. Feature Visualizations and Activations Maximization: these methods offer a glimpse into how a model internally processes inputs, such as images or text. By visualizing the activation patterns of neurons in models like convolutional neural networks (CNNs), we can understand how different features or patterns in data influence predictions. These techniques are post hoc because they are applied after the model training to interpret the inner workings of the neural networks [16].

Figure 4.2 provides a decision-making framework for determining how to explain an AI model's decision. It begins by assessing whether the model's decision requires an explanation. If yes, the next step is determining if the model is inherently interpretable by design. This leads to simple, interpretable models such as logistic regression, K-nearest neighbors (KNN), or linear models. If not, the next question is

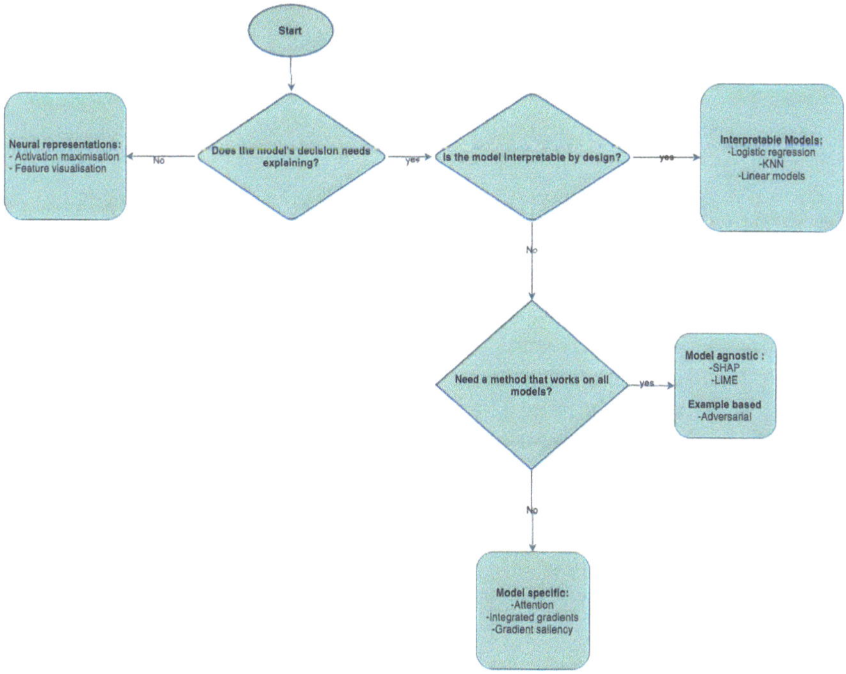


**Fig. 4.2** Choose the XAI technique

whether a method that works on all models is needed. If so, it suggests using model-agnostic methods such as SHAP or LIME or example-based techniques like adversarial approaches. If a model-specific explanation is required, techniques like attention, integrated gradients, or gradient saliency can be used. If the decision does not need explaining, neural representations, such as activation maximization and feature visualization, might be sufficient.

## 4.3 Architectural Models of XAI for Smart Cities

Smart cities rely on complex ecosystems in which AI, IoT, and edge computing collaborate to enhance urban services and infrastructure. As already noted, XAI is becoming an important tool in making AI systems transparent and explainable; this is significantly relevant when the decisions made by AI affect very important areas like transportation, healthcare, and governance. Several architectural models of XAI have thus been developed, offering different solutions depending on the centralized, decentralized, or layered AI system.

The following section describes these architecture models for implementing XAI in smart cities, considering at which layer it is possible to embed explainability within AI systems [17].

### *4.3.1 Centralized AI Architectures and Their Role in Smart Cities*

Centralized architectures are common in smart city settings, where massive volumes of data from various sources, particularly devices, are collected and then processed at a central hub, allowing for effective resource management, real-time monitoring, and large-scale decision-making. Centralized systems also support effective control and integration of smart city components, such as transport systems, energy grids, and public services. In particular, the centralization of AI decision-making is useful for such applications that have to do with large-scale data fusion and processing. While big data analytics and machine learning models can be used to perfect and optimize operations at every level in an urban environment, they must be explainable to maintain transparency and public trust in such systems. The complexity of these centralized AI systems has brought the challenge of explainability. It relates to the fact that most centralized models act like a black box that humans cannot understand for its underlying decision-making process [5].

#### 4.3.1.1 XAI Integration in Centralized Architectures

- Post-hoc XAI Methods (e.g., LIME, SHAP) are applied at the central point to explain decisions made by complex models, like neural networks, used for predicting traffic flow or optimizing energy usage.
- Feature Importance: Displays key factors (e.g., weather, traffic sensors) influencing specific decisions. For instance, why was a particular route suggested to reduce congestion?

### *4.3.2 Decentralized and Distributed Architectures*

Decentralized and distributed architectures are increasingly essential for smart cities as they allow AI processing to occur closer to where data is generated, improving responsiveness and scalability.

#### 4.3.2.1 Edge Computing and Its Impact on XAI

Decentralized architectures utilize edge computing to bring AI processing closer to the source of data generation. This method functions effectively when immediate decision-making is crucial, as it performs minor latency and improves scalability and responsiveness. Edge computing enables local devices to process data and make decisions independently, reducing dependence on a central hub. This is particularly important in smart cities with IoT networks, where data from sensors and cameras is constantly produced.

By decentralizing data processing, edge computing can improve explainability, as local models are more straightforward to understand. For example, local traffic management systems or components of smart grids can deliver real-time decisions that are more transparent to city operators. Edge computing systems typically employ smaller, domain-specific models, which can provide more precise explanations regarding decision-making processes. For instance, edge devices can clarify the reasoning behind real-time routing decisions in smart transportation systems, enhancing user trust and system transparency [23].

#### 4.3.2.2 XAI Integration in Decentralized Architectures

- Local model Explainability: Each local node (e.g., a waste sensor, an air quality monitor) makes use of interpretable models (e.g., decision trees), Or it might employ lightweight XAI methods that allow for an understanding of the decision-making process.
- Counterfactual explanations: Local systems provide "what-if" scenarios, explaining how a slight change to the input (e.g., weather conditions, local traffic patterns) would affect the outcome.

### *4.3.3 Layered AI Architectures for Enhanced Explainability*

Layered AI architectures present a hybrid model that merges the advantages of both centralized and decentralized. Multilayered AI models have several layers that perform multiple functions. These include the top layers making complex decisions in a centralized data center or facility, while the lower layers perform inference in real-time, decentralized operations from edge devices. This multi-layered approach makes for better performance and explainability by offering visibility for decision-making at various levels. An architecture using this approach would integrate explainability at every point of processing AI. In these frameworks, lower layers may handle straightforward, explainable decisions (such as traffic light management), whereas higher layers address more complex tasks such as urban planning or predictive policing. Taking smart energy management as an example, the first layer may focus on monitoring the real-time consumption of energy; the second layer may predict the use of energy shortly. In contrast, the third layer would present top-tier decisions about resource allocation. The segmentation of these roles gives the system more transparency, and stakeholders can understand every part of the decision-making process [1].

#### 4.3.3.1 XAI Integration in Layered Architectures

- Layered explainability: high-level decisions are explained using post-hoc techniques (e.g., SHAP); on the other hand, localized, real-time decision-making is explained through simpler models that are easier to understand.

### *4.3.4 Integration with Internet of Things (IoT) and Big Data*

Combining XAI with IoT and big data systems is crucial for effectively and efficiently managing smart cities. IoT devices generate vast amounts of real-time data from sensors, devices, and urban infrastructure. At the same time, big data technologies are responsible for processing and analyzing this data on a large scale. XAI is important for ensuring insights and decisions are explainable and interpretable to stakeholders. Integrating XAI into such IoT will optimize data-driven operations in smart cities through integrated IoT sensors and big-data frameworks. While being transparent to build greater confidence in AI-enabled urban environments [5] (Fig. 4.3).

### *4.3.5 Case Studies: Architectural Framework of an Explainable AI System in a Smart City [5, 8]*

**Overview**

Barcelona, Spain, leads the way in innovation among smart cities, as advanced technologies such as AI, IoT, and Big Data serve to further improve municipal services. Among the most important areas that face this city is traffic management, in which a sophisticated system was put in place to address congestion, emission, and transportation efficiencies integrates XAI within its larger smart city framework. In other words, the city has installed AI-powered traffic lights, IoT sensors, and edge computing to collect, analyze, and act on real-time data from the streets while keeping the decision-making process transparent and understandable to interested stakeholders such as city planners and the general public.

**Architectural framework**

1. IoT Sensors and Data Collection: Barcelona has installed a wide network of IoT sensors in major intersections, public transport hubs, and traffic lights that

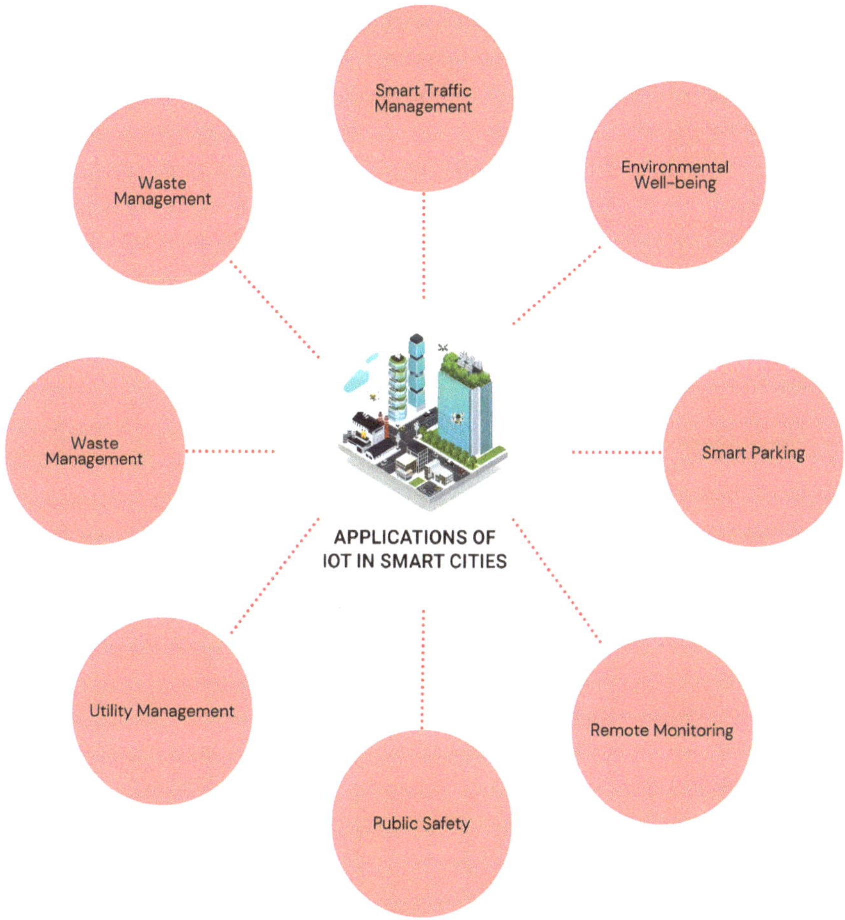


**Fig. 4.3** IoT applications in smart cities

provide real-time information about the flows in traffic, weather patterns, and levels of pollution. Besides this, data hubs are being established within the city to channel this real-time information into its AI platform.

2. AI traffic management with centralized control: Collected real-time data from the IoT sensors is piped into a centralized AI system that helps the system make informed adjustments in the traffic signals and response in case of an accident. To ensure transparency, the system incorporates XAI techniques like LIME to clarify the reasons behind specific adjustments.
3. Edge Computing for Localized Decisions: Barcelona uses edge computing nodes in various city areas, such as the primary intersections, to accelerate the response time and reduce the latency. The edge computing nodes make split-second decisions at the traffic junctions, which might involve changing the light sequences depending on the proximity and directions of the traffic. Those edge devices

promote agile responses to traffic incidents or changes without reliance on centralized AI. Furthermore, the XAI capabilities of these nodes are also available so that the city authorities can understand and explain the logic behind specific decisions.
4. Big data and predictive analysis: Barcelona uses historical traffic data, public transportation usage patterns, and demographic trends to predict traffic flow and improve the transportation network. The prediction models guide in identifying traffic congestion, adjusting public transport timing, and communicating the agenda for long-term urban development strategies.

**Explainability in Action**

1. Traffic Light Optimization: Barcelona's AI system, which is based on the data transmitted by the various transmission of IoT devices, effectively and in real-time optimizes the traffic lights. The XAI layer uses LIME to demonstrate to city officials how the timing of the traffic lights is affected by factors such as traffic density, distance to the pedestrian crosswalk, and weather conditions. In this way, the method removes jams and improves pedestrian safety.
2. Air Quality and Emissions Reduction: Barcelona prioritizes air quality improvement by first managing traffic flow to limit vehicle idling. AI systems redirect network traffic towards routes with lower pollution levels based on the dominant traffic patterns. With SHAP explanations illustrating how specific environmental factors, traffic density, and vehicle types influence the AI's decision-making process.
3. Public Dashboard for Citizens: By creating a public dashboard that allows people to be aware of traffic updates, understand the logic behind traffic management decisions, and obtain information on congestion levels in real time, Barcelona has given citizens open access to it, thereby gaining their trust. XAI is employed to explain the traffic flow's changes, ensuring that the public understands the rationale behind decisions, such as rerouting traffic due to an accident or optimizing traffic lights during peak hours.

**Outcomes and Impact**

Table 4.2 summarizing the critical outcomes of integrating AI-powered and XAI-enhanced systems in Barcelona's smart traffic management.

## 4.4 Key Applications of Explainable AI in Smart Cities

### *4.4.1 Traffic and Transportation Management [8, 17]*

While the population of cities and vehicles is increasing, some priorities are to enhance traffic flow by minimizing congestion and improving public transportation systems. It becomes very much possible with AI-driven solutions to predict

**Table 4.2** Outcome and impact

| Category | Description | Outcome |
|---|---|---|
| Traffic efficiency and reduced congestion | Barcelona's AI-powered and XAI-enhanced system has reduced traffic congestion by nearly 20% in busy areas, improving vehicle flow and shortening peak-hour travel times | Smoother traffic flow and reduced travel times during high-traffic periods |
| Environmental benefits | Improved traffic flow has also decreased vehicle emissions, contributing to Barcelona's sustainability goals. XAI tools enable environmental authorities to analyze and adjust AI models using air quality and emission data | Reduced vehicle emissions and support for the city's environmental sustainability objectives |
| Enhanced public trust and governance | Integrating explainable AI in traffic management has increased public trust by offering transparency. XAI allows citizens and city officials to understand AI decisions, improving the acceptance of AI solutions | Strengthened public trust, greater transparency, and improved governance of AI-driven urban planning decisions |

the pattern of traffic, thereby performing real-time traffic signal management and offering diversion routes to vehicles.

#### 4.4.1.1 AI Models for Traffic Prediction and Routing

AI modules that predict traffic patterns and propose optimal routes are essential for smart traffic systems. These normally come with the development of gathered historical data on traffic congestion and with updates from sensors installed in vehicles. Also, event updates and weather forecasts are included for the purposes of rerouting whenever necessary. However, one of the big challenges in the process is explaining the model's predictions and recommendations, particularly those dealing with safety and resource allocation. AI-driven systems can make a prediction of traffic congestion up to hours or days in advance. But mostly, they act like black boxes, since it is difficult to interpret their decision-making processes.

#### 4.4.1.2 Explainability in Decision-Making for Traffic Control Systems

Explainability is key to trusting the government, most especially when AI operates critical infrastructures such as traffic systems. For instance, an AI system suggested rerouting of traffic between city parts. It is important to explain how it came to that conclusion, whether due to real-time congestion information, weather conditions, or accident reports. With more transparency in decision-making, city authorities will then be able to respond better to emergencies and come up with improved traffic management strategies. The transparency will allow for good governance upkeep of regulatory standards, and it would enable the general public to have more confidence in the AI systems running city infrastructure.

## *4.4.2 Public Safety and Surveillance Systems [4, 26]*

Enhanced monitoring, real-time data analysis, and predictive analytics for prevention of crime, prompt responses in cases of emergencies, and general urban security. City infrastructure is basically incomplete without surveillance and public safety systems. Video feed analysis, real-time identification of unusual behaviors, and identifying the source of potential risks are examples of parameters used in determining when AI surveillance systems are needed for safety preservation. However, transparency is vital in any system that affects citizen privacy and safety. This is where XAI becomes essential, enabling stakeholders to grasp how decisions are made by these advanced surveillance systems.

### 4.4.2.1 AI in Crime Prediction and Emergency Response

Camera AI-driven crime prediction systems have gained much more significance in recent years; this means they offer more efficient ways for police forces to allocate resources. They can analyze historical crime information, social behavior patterns, and environmental factors before projecting areas with high chances of future criminal activities. This enables preventive policing instead of just reacting after something has already occurred. Similarly, In emergency management, they can assist in identifying regularities in situations such as accidents caused by natural forces, among others, so there can be quick reactions to minimize possible injuries.

By studying past crime trends, AI can predict where the criminal hotspots will most likely be. Therefore, this lets the law enforcement focus their attention on particular regions.

### 4.4.2.2 Challenges and Benefits of XAI in Surveillance Technologies

While AI systems benefit and improve smart city surveillance technologies, serious challenges are also involved, especially regarding privacy accountability and in public trust. The AI-enhanced surveillance systems can analyze data in real-time to determine suspicious behavior or people of interest. Still, their unclear decision-making processes raise very serious ethical and legal issues. Indeed, citizens should be guaranteed that AI surveillance systems will practice accuracy and not violate personal freedoms without justification.

XAI plays an important role in dealing with those challenges by clearly explaining how AI came to a particular conclusion. For example, if a security camera had highlighted someone, then XAI might describe some actions or conditions that led to such a conclusion, thus helping avoid wrong detections. This will be important transparency, whereby the public will more likely accept these means of surveillance when they comprehend how decisions have been made.

However, different challenges are posed in integrating XAI into biometric and surveillance systems. Probably the most critical concern is the algorithmic complexity of handling face recognition or behavioral analysis. It is not easy to clearly define how these models reach their conclusions, especially in real-time surveillance, with advanced XAI techniques that would make the system's response even slower or complicate its implementation. Besides, there is always the risk that an adversary may use the transparency provided by the XAI to manipulate or evade such surveillance systems.

## *4.4.3 Energy and Utility Management [5, 7, 23]*

Energy and utility management in smart cities focuses on optimizing resources like electricity, water, and waste by incorporating smart grids and other utility systems with AI technologies, energy demand can be predicted in advance, resource allocation optimization can be done, and wastage can be prevented. Regarding this, AI serves as the key feature in these systems. But while AI can help bring efficiency into these systems, explainability is crucial when those decisions tend to affect millions of inhabitants. For instance, AI models may decide to reroute energy from one area to avoid overloading or rerouting during peak periods to maintain efficiency, and XAI can explain such decisions in an intelligible manner even to city officials and end consumers alike.

### 4.4.3.1 Smart Grid Optimization Using AI and XAI

Smart grids are an integral part of energy management within modern smart cities. They employ AI techniques to continuously maintain the real-time balance in the supply and demand of electricity. These AI systems can, for example, analyze energy consumption patterns, detect wastage, and perform load forecasting to optimize the cost-effective and environmentally friendly deployment of resources. For instance, AI models can redistribute energy resources between residential sectors and industries according to demand forecasts, utilize alternative energy sources, and power even more—network sectors without interruption and within seconds.

Smart grids depend significantly on data from IoT sensors and AI models to oversee and regulate energy flow throughout the city. However, without XAI, the decisions made by these AI models may seem unclear or biased, particularly when they lead to rolling blackouts or other service interruptions. XAI aids utility providers in clarifying how energy resources are distributed, ensuring that AI-driven decisions are grounded in fair and justifiable criteria.

For example, in edge computing, AI tremendously helps make decisions where data is analyzed closer to the consumer. Therefore, energy needs are met quickly and much faster. In this case, XAI would provide explanations on the real-time adjustments made by the AI models, for instance, why certain regions are granted more power than others or how the power distribution is changed to utilize greener sources of energy maximally, another example of the significance of explainability in energy services related to healthcare within smart cities. For instance, hospitals and critical care units may receive prioritized power allocation during emergencies, which needs to be communicated and justified through XAI mechanisms. Ensuring transparency in energy decisions, especially in sensitive situations like healthcare, fosters trust and illustrates the fairness of AI-driven systems.

Furthermore, AI-driven energy management systems can offer utility companies and regulators valuable insights into the reasoning behind energy usage predictions and the factors that affect them. For example, suppose an AI system forecasts a significant spike in energy demand in a particular area. In that case, XAI can clarify which historical trends or external influences (weather conditions or industrial activities) led to that prediction. This enhances decision-making and enables stakeholders to review AI systems, ensuring they function within ethical and regulatory standards.

By integrating XAI into smart grid and utility management systems, smart cities can optimize energy distribution while fostering transparency and trust. This approach benefits all stakeholders—from utility companies and regulators to consumers—by helping them understand and accept the AI-generated decisions that contribute to the city's efficient operation.

### *4.4.4 Healthcare and Public Health Monitoring*

Smart cities are increasingly employing AI to improve healthcare systems and health monitoring. AI-driven solutions help advance applications such as predictive diagnosis, tracking the onset of diseases, and managing health. Due to the potential consequences of any decisions made in the healthcare sector on human life, healthcare systems must be embedded with explainable artificial intelligence systems for trust, transparency, and endorsement of AI-based outcomes.

#### 4.4.4.1 AI in Predictive Diagnostics and Public Health [12, 14, 27]

AI has played a crucial role in health cases, especially in diagnostics; one example involves developing machine learning models that can scan through large volumes of medical data to predict diseases, predict outbreaks, and give better treatment options. In public health, artificial intelligence applications are used to countercheck the epidemiology of diseases and assess and control populations' health status and healthcare delivery systems. These predictive models depend on extensive data provided by various means, such as IoT devices, electronic health records, and wearable technologies, which all aid in understanding the health patterns of individuals and the entire population. For instance, in smart cities, AI systems constantly can scan health parameters associated with blood pressure and glucose levels and provide real-time insights into health.

It could also involve public health management, such as hospital bed demand projections in pandemics or optimization of vaccine distribution strategies.

#### 4.4.4.2 The Role of XAI in Ensuring Trust in AI-Driven Healthcare Solutions

Trust is an essential element in healthcare; therefore, XAI is essential to guarantee health-based artificial systems' clarity and reliability. Patients and caregivers should have clear ideas about how AI is arriving at its decision, especially when such a decision is in life-based treatment and diagnosis. Most of the time, due to complications, AI models can be thought of as a "black box." As a result, they cannot be trusted by people without good expertise in what comes out of them. If AI is ever going to be integrated genuinely into healthcare, then XAI will be the required bridge between technical difficulty and human understanding. This includes the ability for XAI to give insights to healthcare professionals on how such a specific AI model reaches its conclusion, allowing the verification of accuracy within the medical staff recommendations from the AI before further action can take place. It can, for instance, outline how an AI system recommended one treatment to a patient over another by describing the concepts considered in making up such a recommendation; it would include patient history and physical results. It thus empowers healthcare professionals to make informed decisions.

XAI also plays a critical role in maintaining transparency in AI-driven systems that can predict infectious disease outbreaks or, on the contrary, handle healthcare resources. An example could be how, during the COVID-19 pandemic, AI models were used to forecast infection rates and recommend public health interventions. The XAI would also be useful for public health officials as it would explain the rationale behind AI-driven predictions, thus fostering trust and collaboration among government agencies, healthcare providers, and the public.

Moreover, XAI addresses the biases found in AI-driven healthcare solutions. If AI models are trained on biased datasets, they can involuntarily create biased health recommendations that might hit some populations quite hard. XAI enables finding and mitigating these biases, thus making the decision making process transparent for fairness and equity in health care.

### 4.4.5 *Environmental Monitoring and Waste Management [8, 12, 17]*

Smart cities integrate AI mechanisms in managing urban infrastructure, environmental management, and waste management to obtain sustainable solutions. AI optimizes the control of pollution, collection, and disposal of waste, and resource consumption. XAI benefited from these processes. More importantly, it has ensured that the output of such processes results in AI models whose output is transparent and interpretable for planners, elected officials, and even the public's use.

#### 4.4.5.1 AI Models for Pollution Control and Waste Management

AI models have shown great promise in addressing environmental issues in smart cities. It analyzes vast data from sensors, satellites, and IoT devices to deliver real-time insights into pollution levels, waste production, and resource usage. For instance, machine learning algorithms predict air quality by considering various factors, including tracking carbon emissions and finding hotspots of pollution in urban areas.

**Pollution control**: AI models can process data from weather stations, traffic systems, and industrial sources to predict pollution levels. These AI-powered systems are making improvements to environmental protection by proposing changes in traffic routes, industrial waste emission, or even pre-emptive closure of certain areas when levels of pollution higher than those legally permissible are predicted. This keeps cities within set regulations created to prevent pollution and protects citizens against toxic industrial emissions.

**Waste management**: AI models play an essential role in enhancing the efficiency of processes related to the collection and disposal of waste materials. For instance, smart bins could notify on-site collection services when they are full via AI sensors. This has the potential to improve route collection efforts and reduce fuel use and operational expenditure. Analytics-enabled AI can come up with historical patterns of waste production to predict innovative enhancements in waste management practices, particularly those elements that most likely can be collected cleanly at the source.

#### 4.4.5.2 Use of XAI to Improve Environmental Decision-Making

XAI will help bring this transparency to pollution control by clarifying why a particular area is highlighted as a high-risk zone due to pollution or why a measure like 'no vehicle access' is suggested. This clarity is important for trust building and assures that the actions undertaken are not only scientifically valid but also socially acceptable.

XAI, therefore, contributes to better decision-making in waste management through the explanation capability of the AI model in reasoning out route optimization, waste disposal techniques, and distribution of resources. In smart city environments, such levels of explainability enhance the overall efficiency of waste management through increased accountability of the system and, consequently, making adjustments based on feedback from human operators.

## 4.5 Challenges in Implementing Explainable AI in Smart Cities

### *4.5.1 Technical Challenges*

This section underlines some technical obstacles to implementing XAI: model complexity, scalability, and integration of different data sources.

#### 4.5.1.1 Model Complexity and Scalability Issues

Among the key technical challenges for the implementation of XAI in smart cities is the complexity and scalability of the AI models. As smart cities evolve with increasingly advanced AI models, managing urban infrastructure, transportation, and public services. These models are often built on deep learning and complex algorithms, which by nature become difficult to interpret and understand. Model performance-explainability tradeoff is really a critical challenge; in fact, more accurate models, such as neural networks, are often less transparent [11].

The lack of transparency of sophisticated models deployed in smart city applications, such as traffic forecasting and energy management, may obstruct the transparency and accountability of the decision-making processes. All these models are complicated by the need to scale across diverse transportation city systems, energy, and public safety. Making consistent explanations across multiple platforms hard to maintain [12, 28].

It, therefore, raises the issue of scalability; there can be a vast amount of data generated by systems in a smart city from sensors, cameras, and so on. Being able to process this in real-time and then convert it using scalable AIs is crucial, ones that can work with numerous inputs while giving clear and understandable explanations. Several of these models work well at small scales, but scaling them up to city-scale systems generally tends to make performance bottlenecks, which generally degrade the effectiveness of XAI solutions [23].

#### 4.5.1.2 Data Heterogeneity and Integration Problems

Another significant challenge in implementing XAI in smart cities is the issue of data heterogeneity and integration. Smart cities will rely on diverse data sources, from IoT sensors and public databases to social media and multiple mobile apps, whose data is being generated in different formats and at different speeds. This has been quite a challenge in integrating and analyzing such diverse data to allow for accurate AI-driven insights. Due to the varied nature of the data, the standardization and streamlining of AI models become complex. This, in turn, leads to problems in integration and consequently can affect the accuracy and explainability of an AI system significantly [5].

Data heterogeneity in smart cities can reduce AI systems' capability to produce reliable predictions, especially if the data is incomplete or collected at irregular frequencies. In the realm of XAI, these data-related challenges make it even more difficult to clarify AI decisions. for instance, an application powered by multiple sources of data combined to manage traffic flow or optimize energy consumption. In that case, it is impossible to trace which (part) of the data informed a particular decision, and the system needs higher transparency [8]. Furthermore, merging data from various source administrations often requires advanced middleware and frameworks and may not be compatible with the XAI models [5].

The majority of the existing XAI solutions have a limited focus on data source diversity in smart cities and thus often result in inconsistent explanations. This problem is further amplified in real-time systems where continuous streams of data must be processed, adding further challenges to the explainability and transparency of the same [13].

### *4.5.2 Ethical and Social Challenges*

XAI provides a way to enhance transparency, fairness, and accountability. The following section elaborates on the ethical and social challenges of XAI. These challenges are particularly relevant when AI systems deployed to support decision-making in smart cities enact an action in law enforcement, healthcare, and resource management among other areas [12, 16]. The following discusses some of the ethical and social challenges of XAI.

#### 4.5.2.1 Bias, Fairness, and Accountability in AI Decisions

XAI was developed to make the systems more transparent and fairer, although such goals are hard to achieve. One major problem is that the explanations it gives might still reflect biased interpretations.

XAI systems can inherit biases from the AI models on which they are based. XAI can increase the explainability of the AI decisions, but no guarantee exists that those explanations are going to be fair, considering the fact that the model may have been trained on biased data. Thus, XAI can also imply clear but faulty reasons for the decisions taken, which may lead to perpetuating unfairness with the semblance of transparency [8, 12].

XAI aims to focus on explaining how AI models came up with their decisions, but some of the models are extremely hard to explain, especially for people who do not have deep technical backgrounds. This may lead to a situation whereby the operators of AI systems fail to comprehend the complicated details of the explanation given by AI. The operators may risk uncertainty as to where to lay responsibility when issues arise [13, 29].

#### 4.5.2.2 Acceptance and Trust in AI Systems

In smart city contexts, people can be suspicious of AI systems that oversee and control public services-mostly when the provided explanations are either too technical or too complicated. Success in gaining the trust of XAI is related to the explanations actually reaching out to the public and whether they will be further empowered by the transparency provided. If explanations are not accessible and meaningful, XAI may fail to fulfill its potential to enhance societal trust in AI systems [12, 13].

### *4.5.3 Privacy and Security Concerns*

Integrating XAI in smart cities opens up opportunities for more transparency and accountability; nevertheless, significant privacy and security concerns come along with this. These are basic concerns since a huge amount of data is collected and analyzed on a continuous basis in urban environments to manage service delivery, which includes traffic, healthcare, and public safety. XAI models face a challenge in the balance between the potential to effectively apply insight while maintaining the security and privacy of data requirements [8, 12].

#### 4.5.3.1 Ensuring Data Privacy in AI Systems' Models

Many personal and sensitive data can be collected from different stakeholders through various sources, including IoT devices, cameras, sensors, and citizen interactions. To provide explanations for their decisions, the XAI systems usually require access to this kind of data; unveiling sensitive information in the explanation process could unintentionally cause a breach of privacy [12, 30].

A possible risk is that, for example, XAI-generated explanations in healthcare or law enforcement might inadvertently leak personally identifiable information (PII) or sensitive information about single entities or groups. The key challenge is to find a balance between providing understandable explanations and protecting sensitive information. In general, it is technologically challenging to develop XAI models capable of anonymizing or masking sensitive data while producing meaningful, understandable explanations [29]. In smart cities, such data is usually interrelated from various sources and interconnected, hence opening more channels of privacy vulnerabilities. The more related this data is, the more difficult it is to ensure that the XAI models don't unintentionally reveal private information when explaining [13].

#### 4.5.3.2 Security Implications of Explainable AI Systems

While XAI works on increasing the level of transparency in AI-driven decisions, this may also become a source of weakness in its security. Comprehensive and revealing explanations about the workings behind an AI model can inadvertently leak systems-level information to malicious actors. Thus, giving away the internal details of an AI model may enable attackers to discover prospective vulnerabilities or ways through which systems can be adversarially attacked or manipulated [12, 17].

For example, an XAI model explaining its decisions in a smart city traffic management system could inadvertently reveal information about specific algorithms or data patterns that might be leveraged to manipulate the traffic signals into changing state. Similarly, an explanation for a surveillance system might show the parameters for defining suspicious behavior, providing attackers with the knowledge needed to avoid detection [1]. Moreover, there is a possibility of "adversarial explanations" by which the attackers can manipulate the explanations themselves, which would lead the XAI system to result in misleading or deceptive interpretations of AI decisions. Such types of attacks may collapse the confidence in AI systems and threaten the security of smart city infrastructure [23].

In summary, while XAI fosters transparency and accountability in AI systems for smart cities, it also brings significant privacy and security challenges. Safeguarding sensitive data and securing AI models from exploitation while providing meaningful explanations is a prerequisite challenge for the effective implementation of XAI in urban settings [29].

### *4.5.4 Regulatory and Policy Challenges*

Various regulatory and policy challenges emerge as XAI becomes more integrated into smart city systems, which range from modifying existing legal frameworks to cater to the needs of XAI to creating new regulations for the well-considered use of XAI. All these issues will become of great importance in view of the ethical and efficient implementation of XAI technologies [8, 12].

#### 4.5.4.1 Current Legal Frameworks and Their Limitations

Current legal frameworks struggle to keep pace with the rapid speed at which AI and XAI evolve. Data protection and privacy laws, notably the General Data Protection Regulation (GDPR) within the European Union, were designed in the absence of explicit consideration of such unique XAI requirements. While these frameworks consider data privacy and security, they may fail to consider the complexities of the issues involved with XAI systems, especially about how they manage or explain personal data [1].

For example, GDPR grants a right to an explanation of automated decisions made by individuals. Applying this right in the context of complex AI systems can be tricky. Current legal frameworks often do not detail how XAI shall work or how much detail in an explanation is needed without infringing on privacy issues [12, 13]. Furthermore, Current regulations also lack clear definitions of the technical aspects of how AI systems should be designed and explained, creating uncertainty for organizations deploying XAI systems [30].

#### 4.5.4.2 Need for New Regulations and Standards

Given the limited capabilities of existing legal frameworks to deal with XAI in smart cities, new regulations and standards focused on XAI are urgently called for. Such regulations shall address both the technical and ethical dimensions of XAI deployment [29]. New rules could include specific provisions for:

- Explainability Requirements: Clearly state what will be considered an adequate explanation and how this will be communicated to all concerned parties, including the public and any regulating agencies [13, 29].
- Data Handling Practices: Set standards on how XAI systems shall manage data in a manner that anonymizes information to protect privacy while at the same time offering meaningful explanations [12].
- Accountability Mechanisms: Design frameworks that allow accountability from organizations when the system fails in transparency or commits ethical violations [29].

The regulatory bodies need to work with developers of AI on technically feasible standards that are in harmony with ethical principles. In this respect, a cooperative approach would enable new regulations to be practical and effective in allowing responsible XAI practices [13].

In summary, while existing legal frameworks lay the groundwork for data protection and privacy, they often fail to address the specific requirements of XAI in smart cities. Developing new regulations and standards considering XAI's technical, ethical, and practical aspects is essential to ensure these systems are deployed responsibly and effectively [29, 30].

## 4.6 Frameworks and Strategies for Effective Explainable AI Implementation

### *4.6.1 Design Principles for XAI in Smart City Contexts*

#### 4.6.1.1 Human-Centric AI Design

Human-centered AI design in smart cities concentrates on the needs and experiences of the residents, making the AI system intuitive and transparent, with alignment with human values. This means creating interfaces and interactions that are intuitive for people. For example, AI systems controlling urban traffic should be able to give good, understandable explanations for the decisions they make about traffic management, thus enabling residents to understand how their commutes may or may not be affected. On the other hand, a human-centered approach considers many different users' points of view during the design phase and ensures solutions using AI meet real-world demands. They are accessible to all users with disabilities and less technology-savvy individuals [8, 12].

DoReMi—Human-centered XAI design framework, emphasizing making AI explanations useful, actionable, and understandable for the users: [13, 16]. It is based on three guiding principles:

- **Do**: Explanations should help users take action based on the AI's decisions. For example, in healthcare, AI recommendations should guide doctors toward the next steps.
- **Re**: Explanations should be reflective, allowing users to understand the reasons behind AI decisions and helping them trust the model's choices.
- **Mi**: Explanations should mitigate confusion by aligning with the user's mental model of how the AI works, ensuring clarity and fostering trust.

The DoReMi framework ensures the explanation is user-centered, clear, and actionable, with links to real-world tasks [29, 30] (Fig. 4.4).

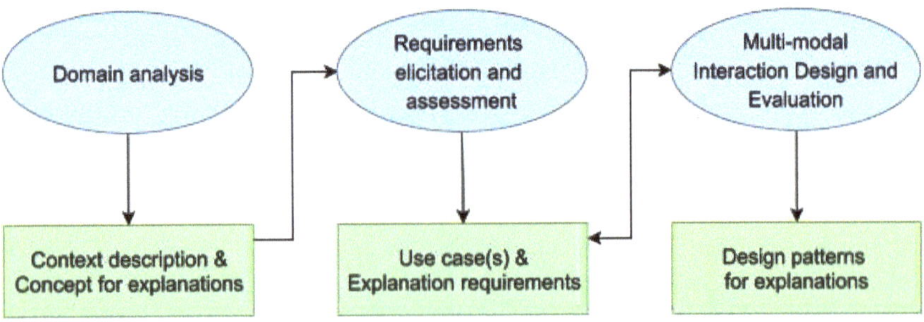


**Fig. 4.4** DoReMe framework

### 4.6.2 *Ethical AI Frameworks for Urban Environments*

Ethical AI frameworks in urban settings ensure that the technologies operate fairly and responsibly. It enables guidelines on data privacy, bias, and accountability issues [13, 29].

### 4.6.3 *Privacy-Preserving and Secure XAI Approaches*

#### 4.6.3.1 Federated Learning and Differential Privacy

Federated learning and differential privacy are the most critical components in improving privacy in AI systems. Federated learning will make it possible to have models that can be trained on several decentralized devices without data sharing, hence allowing for the preservation of individual privacy. Differential privacy injects noise into the data and thus protects the privacy of individuals while, at the same time, gaining insights of tremendous value from the information. These techniques are crucial in smart cities where personal data is gathered heavily, and associated privacy issues are considered quite critical [12, 13]. With this integration, AI systems in smart cities could balance data utility and privacy protection. The Table 4.3 shows how these concepts can be integrated with XAI to ensure privacy, trust, and transparency [8, 29].

**Table 4.3** Federated learning integrated with XAI

| Concept | Benefits to XAI | Example |
|---|---|---|
| FL | – Preserve privacy while contributing to the global model<br>– Enables explanations at both local and global levels | Smart city traffic systems are trained on local data without sharing individual travel details, but they provide city-wide insights |
| DP | – Protects individual data in model explanations<br>– Ensures that sensitive information isn't leaked through explanations | Healthcare AI explains diagnostic decisions without revealing patient-specific data |
| FL + DP for XAI | – Provides transparent, trustworthy AI<br>– Ensures user data privacy while maintaining the explainability of AI decisions | AI models in finance explain loan approvals while protecting personal financial information |

#### 4.6.3.2 Blockchain-Based Approaches for XAI

XAI Blockchain-based Methods: Blockchain enforces the development of transparent and safe artificial intelligence systems. Blockchain-based approaches could provide unalterable records of AI decisions and interactions, making the system highly accountable and traceable. Furthermore, Blockchain can help enable privacy preservation techniques and secure data sharing or management practices. These could foster trust in AI systems used within smart cities by ensuring that data integrity is upheld and users' privacy is protected [31, 32].

## 4.7 Future Directions and Emerging Trends in XAI for Smart Cities

Future XAI will be much improved with new, innovative algorithms, considerations for ethics, and incoming emerging technologies such as quantum computing; below is a glance at these advancements.

### *4.7.1 Advances in AI Algorithms and Models for Explainability*

Traditional machine learning models, especially the deep learning models, have been regarded as "black boxes" and sometimes impenetrable regarding how they make certain decisions. Interpretable machine learning does just the opposite. It builds much-needed transparency and trust in AI systems [9, 28].

The importance of using inherently interpretable models is particularly evident in critical fields like healthcare, public safety, and urban management. These models are built to be transparent from the beginning, removing the need for post-hoc explanations that can be flawed or incomplete [29]. At the same time, advancements in post-hoc explanation methods such as LIME and SHAP are helping to clarify the predictions made by complex AI models. These tools let users see which factors influenced a model's decisions in a particular instance [16].

Hybrid models also combine the strengths of inherently interpretable and post-hoc explainable models. These approaches embed interpretable decision trees or rule-based models with complex machine-learning algorithms. The derived explanation balances simplicity and predictive accuracy [29]. Future research is likely to concentrate on enhancing the scalability of these methods for large-scale smart city initiatives while ensuring that the explanations remain accessible and pertinent to stakeholders in various urban sectors [9, 28].

### 4.7.2 Impact of Quantum Computing on Explainable AI

Quantum computing is a groundbreaking technology that can change the landscape of AI, particularly in XAI systems. With its superior processing power, quantum computing can tackle complex problems that classical computers struggle with. This advancement could lead to more robust and efficient XAI systems [32]. Quantum computers, processing enormous flows of information obtained from transport, energy, and public services, can enormously improve the efficiency of smart city systems. The possible applications of quantum computing in AI indicate that it can speed up machine learning algorithms and enhance their scalability. However, the intricate nature of quantum algorithms may introduce new challenges for explainability, as these systems could add layers of complexity that necessitate innovative XAI strategies to clarify [32].

Quantum machine learning (QML) is a leading recent domain where quantum computing intersects with XAI. QML could shorten model training times and yield more profound insights into data trends. The challenge remains in crafting quantum algorithms that are understandable to human users. If the algorithms employed by quantum computers to generate predictions are not transparent, the transparency issues that affect classical AI systems could become even more pronounced [32].

Incorporating XAI into quantum computing will likely demand the development of new methods and frameworks. Future research should concentrate on creating explainability tools tailored for quantum contexts, helping clarify the decision-making processes of quantum algorithms. This could involve designing visualizations and interactive explanation techniques for quantum systems, enabling users to grasp how quantum computations impact urban decision-making [32].

### 4.7.3 Research Opportunities and Open Challenges in XAI

As XAI continues to develop, various challenges still need to be tackled, particularly concerning the domain of smart cities. Perhaps the biggest challenge remains finding that spot between explainability and model complexity; simple, hence more interpretable models are much easier to understand but can usually fall behind in predictive power compared to more complex models, such as deep neural networks. Future studies are leaning more toward hybrid models, which can achieve high accuracy with interpretability.

Another significant challenge is creating XAI systems that are understandable to a wide range of users, not just AI specialists but also city planners, policymakers, and the general public.

Applying XAI in an urban environment will introduce even more research challenges. AI systems in an intelligent city must frequently operate in dynamic environments, processing a huge amount of data in real-time. A key area of research is ensuring that those systems can provide timely and accurate explanations along with performance. Additionally, as AI systems gain more autonomy, it raises questions about how to effectively audit and monitor these systems to ensure they remain transparent and accountable over time.

Finally, there is a significant growing area of research into the societal implications of XAI. As systems of XAI become integral to smart cities, they will impact urban living in everything from public safety to transportation and healthcare. How such systems shape and influence different demographic groups and ensure fairness and equity will be critical issues for future research.

## 4.8 Conclusion

This chapter explored XAI's critical role in smart cities, balancing model complexity with interpretability, creating user-friendly explanations, and addressing technical challenges in real-time urban environments. XAI is essential for ensuring transparency, fairness, and accountability, particularly in public safety, transportation, and healthcare sectors, while hybrid models aim to balance predictive power and interpretability. As smart cities evolve, XAI is pivotal in fostering public trust and promoting ethical, sustainable, and socially responsible AI systems. Researchers and practitioners should focus on scalable XAI solutions accessible to non-experts, enhance real-time system performance, and align with ethical standards. Collaboration across disciplines will ensure that AI-driven smart cities prioritize inclusivity, equity, and transparency.

## References

1. Gandomi AH, Rahmani Z, Gandomi MH (2022) Artificial intelligence for smart cities: problems, challenges, and opportunities. CRC Press
2. Townsend A (2014) Smart cities: big data, civic hackers, and the quest for a new Utopia. W.W. Norton & Company
3. Batty M (2013) The new science of cities. MIT Press
4. Zhao J, Wang T, Wang X (2023) Urban traffic flow prediction with explainable deep learning. Transp Res Part C Emerg Technol 147:103793
5. Zhang H et al (2020) Blockchain for smart cities: the challenges and opportunities. Front Blockchain 3:25
6. Shi W, Cao J, Zhang Q, Li Y, Xu L (2016) Edge computing: vision and challenges. IEEE Internet Things J 3(5):637–646
7. Wang H et al (2020) Convergence of edge computing and deep learning: a comprehensive survey. IEEE Commun Surv Tutor 22(2):869–904
8. Camero A, Alba E (2019) Smart city and information technology: a review. Cities 93:84–94
9. Guidotti R et al (2018) A survey of methods for explaining black box models. ACM Comput Surv (CSUR) 51(5):93
10. Kitchin R (2014) The real-time city? Big data and smart urbanism. GeoJournal 79(1):1–14
11. Batty M (2018) Inventing future cities. MIT Press
12. Dubey A, Malik P (2022) Explainable AI in smart cities: a step towards trustworthy AI. J Big Data 9(1):1–28

13. Chamola V, Hassija V, Gupta V, Guizani M (2023) A review of trustworthy and explainable artificial intelligence (XAI). IEEE Access
14. Samek W, Montavon G, Vedaldi A, Hansen LK, Müller K-R (2021) Explainable AI: interpreting, explaining and visualizing deep learning. Springer
15. Molnar C (2022) Interpretable machine learning: a guide for making black box models explainable. Lulu.com
16. Arrieta AB et al (2020) Explainable artificial intelligence (XAI): concepts, taxonomies, opportunities, and challenges toward responsible AI. Inf Fusion 58:82–115
17. Wirtz BW, Weyerer JC, Geyer C (2019) Artificial intelligence and the public sector—applications and challenges. Int J Public Adm 42(7):596–615
18. Russell S (2020) Human compatible: artificial intelligence and the problem of control. Viking
19. Kitchin R (2016) The data revolution: big data, open data, data infrastructures, and their consequences. Sage Publications
20. Zeng Y et al (2021) Fostering trust in AI: the role of explainability in the perception of AI trustworthiness. Ethics Inf Technol 23(4):581–592
21. Dwork C, Roth A (2014) The algorithmic foundations of differential privacy. Found Trends® Theor Comput Sci 9(3–4):211–407
22. Rudin C (2019) Stop explaining black-box machine learning models for high-stakes decisions and use interpretable models instead. Nat Mach Intell 1(5):206–215
23. Gilpin LH et al (2018) Explaining explanations: an approach to evaluating interpretability of machine learning. In: 2018 IEEE 5th international conference on data science and advanced analytics (DSAA), pp 80–89
24. Gilpin LH et al (2018) Explaining explanations: an overview of interpretability of machine learning. In: IEEE 5th international conference on data science and advanced analytics (DSAA)
25. Doshi-Velez F, Kim B (2017) Towards a rigorous science of interpretable machine learning. arXiv:1702.08608
26. Gunning, D., & Aha, D. (2021). DARPA's Explainable Artificial Intelligence (XAI) Program: A Retrospective. MIT Press.
27. Florida L (2019) The logic of information: a theory of philosophy as conceptual design. Oxford University Press
28. Florida L (2014) The fourth revolution: how the infosphere is reshaping human reality. Oxford University Press
29. Ribeiro MT, Singh S, Guestrin C (2016) Why should I trust you?: explaining the predictions of any classifier. In: Proceedings of the 22nd ACM SIGKDD international conference on knowledge discovery and data mining, pp 1135–1144
30. Ortigossa ES, Gonçalves T, Nonato LG (2024) EXplainable artificial intelligence (XAI): from theory to methods and applications. IEEE Access
31. Gunning D, Aha D (2019) DARPA's explainable artificial intelligence (XAI) program. AI Mag 40(2):44–58
32. Valli A, Cakmak AS, Soni N (2022) Quantum computing in AI: a game-changer for future research directions. J Artif Intell Res 67:99–117

# Chapter 5
# XAI for Entertainment, Education, and Environment: Case Studies and Trends

**Valeriu Manuel Ionescu, Madalin Ciprian Enescu, and Cristina Claudia Bizon**

**Abstract** Artificial Intelligence (AI) is becoming present in all Information Technology aspects as an automated component with human like behavior. Understanding the decision process is important for increasing the trust in AI tools. This Chapter creates an overview of the current state of research for introducing explainability in AI models for Entertainment, Education, and Environment. Aspects like the use of generative AI tools and content personalization in entertainment, the enhancement of educational tools with the potential of introducing bias and the ability to detect patterns and trends in the massive Internet of Things generated data for Environment are investigated. Based on this analysis, this Chapter outlines future trends in this domain.

**Keywords** Artificial intelligence (AI) · Explainable AI (XAI) · Explainability · Artificial intelligence · Entertainment · Education · Environment

## Abbreviations

| | |
|---|---|
| AHA–XDNN | Artificial Hummingbird Algorithm–eXplainable Deep Neural Networks |
| AI | Artificial Intelligence |
| BERT | Bidirectional Encoder Representation from Transformers |
| CL-XAI | Cognitive Learning supported by XAI |

V. M. Ionescu (✉)
National University of Science and Technology POLITEHNICA Bucharest, Pitesti University Centre, Pitesti, Romania
e-mail: valeriu.ionescu@upb.ro

M. C. Enescu
Political Sciences National School of Political and Administrative Studies Bucharest, Bucharest, Romania

C. C. Bizon
Doctoral School, Bucharest University of Economic Studies, București, Romania

 
N. Bizon and B. Appasani (eds.), *Explainable Artificial Intelligence for Trustworthy Decisions in Smart Applications*, https://doi.org/10.1007/978-3-031-97007-8_5

| | |
|---|---|
| DNN | Deep Neural Networks |
| EDM | Educational Data Mining |
| EMGCD | Explainable Multi-View Game Cheating Detection |
| GPT | Generative Pre-trained Transformer |
| GXAI | Gaming Explainable Artificial Intelligence |
| IDE | Integrated Development Environment |
| IoT | Internet of Things |
| IT | Information Technology |
| LIME | Local Interpretable Model-Agnostic Explanations |
| LLM | Large Language Models |
| LMS | Learning Management Systems |
| ML | Machine Learning |
| NLP | Natural Language Processing |
| NPC | Non-Player Characters |
| PDP | Partial Dependency Plot |
| SHAP | SHapley Additive exPlanations |
| SoC | System on a Chip |
| XAI | Explainable Artificial Intelligence |
| XAID | EXplainable AI for Designers |
| XAI-ED | Explainable Artificial Intelligence in EDucation |

## 5.1 Introduction

The need for development of software that performs tasks which usually require human presence was felt when most activities had an Information Technology (IT) component and became interconnected via networking. The society is at the moment where this goal (and necessity) is starting to become a reality.

Artificial Intelligence (AI) has appeared on the basis of machine learning, deep learning and predictive analytics as a means to mimic the human intelligence and even exceed it due to its capacity to store and process existing data with great speed and efficiency. As any major change in human history (electricity, internal combustion engine, flight, telephone, nuclear power), it is regarded with mistrust and fear [1]. The problem relies not with the scientists that work within the AI domain (even if some scientists have expressed concerns with unregulated development of AI technologies that can lead to pitfalls [2]), but mainly with the general public that does not understand how this technology works.

Some AI technologies were gradually pushed to the general public and generated acceptance in a short time, for example: in industry [3]; in cybersecurity [4]; for the environment (such as sorting materials [5]), in smartphones for improving the picture quality and for recognizing text [6] or voice; in games for managing non-player characters, level generation or player modeling [7], etc. On the other hand, AI that directly touches sensitive areas such as human rights and opportunities was met with resistance because of the black box nature of many of these systems and reports of AI implementations that have interacted with the general public and exhibited "hallucinations"—generating content that is has no basis in the common sense or does not respect privacy, copyright, etc.

One way to reduce general public resistance and to dispel the "magic" behind the AI algorithms (their black box design) is to introduce explainability to the AI.

This Chapter novelty consists of:

- An analysis of current research in the domain of XAI for Entertainment, Education and Environment with the purpose of identifying which areas have more research performed and the possible reasons for this;
- An investigation in the current state of national and international regulations for AI development, their commonalities and the influence on XAI implementation;
- Identifying future trends for this applying XAI in Entertainment, Education and Environment based on the performed analysis.

The reminder of the Chapter is structured as follows: Sect. 5.2 details the context that led to the need of introducing explainability in AI models and the attempts to regulate the AI domain and applications; the following sections present how explainability methods are used in Entertainment (Sect. 5.3), Education (Sect. 5.4) and Environment (Sect. 5.5) in order to increase trust in AI models; finally 5.6 will present the conclusions of the study and identify future trends in the XAI for Entertainment, Education and Environment.

## 5.2 Explainable Artificial Intelligence (XAI)

In order to understand what XAI is, it is necessary to identify where AI originated, how it evolved and why scientists have problems understanding the decisions made by their own AI models.

The idea of machines endowed with human thinking existed long before the age of computers (for example the Mechanical Turk [8]), probably motivated by the prospect of a human-like entity with endless life that can learn and improve itself. The concepts behind machine learning can be traced in the 20th century with the development of the neural network in 1943 [9] (a machine learning algorithm for of binary classifiers), the perceptron, the Turing machine and the Turing test [10], but the reduced interest, hardware availability and power has kept the development limited until the end of that century. At the beginning of the 21st century computing machines have become widespread, having reasonable computational power to run new algorithms that have been developed (such as training artificial neural networks through backpropagation of errors [11], neocognitron [12], convolutional neural networks [13], Long Short-term Memory recurrent neural network [14], the generative adversarial network [15] and the transformer architecture in 2017 [16] which are the basis of the Generative AI (which include the well-known Large Language Model-LLM—chatbots like ChatGPT).

The increase in development speed was also due to the creation of complex frameworks to guide the development flow for the creation of models operating on large datasets. The results offered by these models, however, became hard to explain even by their creators. The paper [17] makes the case that, while a scientist working with neural networks in 1991 would have been able to follow the decision path and understand their reasoning (because of its simple structure: a feedforward multi-layer perceptron, 960 inputs from a $30 \times 32$ pixel video image, 30 outputs, steering direction and 4 hidden layers [18]), this was no longer the case in 2016 when the data that needs analysis has increased exponentially and similarly the algorithm complexity (for example deep learning) and technology in a number of applications. These make the decision process a black box. For example, Grok-0 LLM was trained with 33 billion parameters, Grok-2 with 600 billion parameters, LLaMA 2 has approximately 70 billion parameters, Generative Pre-trained Transformer (GPT) 4 has 1.76 trillion, GPT-3.5 has 175 billion, Gemini Flash 1.5 by Google with an estimate of 32 billion parameters (with model distillation—transferring knowledge from the 137 billion parameters of Gemini 1.5 Pro), and Bard 137 billion parameters [19]. Lower parameter models are faster, run better on lower specification hardware and are easier to understand but are less flexible and target specific use-cases (Fig. 5.1).

In this context has appeared the need for introducing methods that can explain how different parameters affect the end result and how the model arrives to certain conclusions. Explainable Artificial Intelligence (XAI) implements methods that provide, for humans, insights into the reasoning behind the decisions taken by the AI. The purpose is to increase trust in AI for developers, users and regulators, and to avoid (intentional or unintentional) errors in the models.

Defense Advanced Research Projects Agency (DARPA) initialized in 2015 a four-year program to enable users to understand, trust and manage AI systems [20]. Because of this, 2015 is identified as a significant turning point for XAI because of the development of new algorithms (such as deconvolutional networks that can visualize the neural network layers) and LIME (Local Interpretable Model-Agnostic

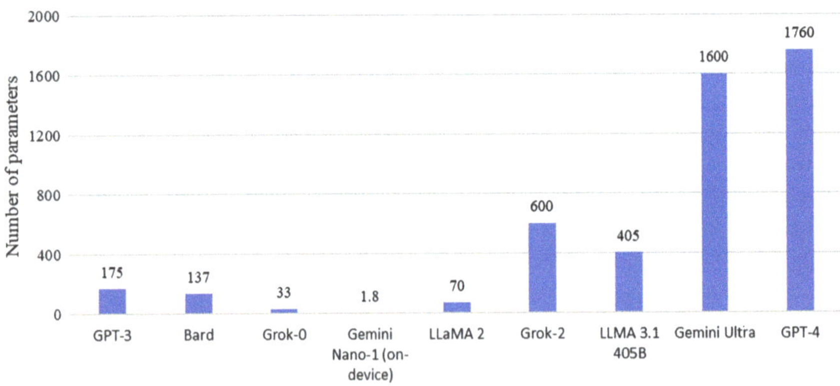


**Fig. 5.1** Number of parameters in some of the large language models used in 2024

Explanations) [21]. One of the results of the program was the Explainable AI toolkit [22] with interface software modules that can be used for the development of XAI systems.

The components found to be taken into consideration for explainability are transparency, fairness, robustness, privacy and interpretability as outlined in 2015 paper [20].

- Transparency: represents the ability to understand the decision process in order to be able to investigate and fine tune the model.
- Fairness: the AI algorithm does not exhibit a discriminatory behavior to a specific group.
- Trust: placing the human in the loop front methods that allow monitoring the decision process increasing trust.
- Robustness: this is an important component because, as we have shown, the most energy is consumed not on the initial training but on the tuning to new data.
- Privacy: this is another sensitive topic as most AI models keep the data source hidden from the public, creating privacy issues (such as crating content based on a person's private data placed online) and copyright issues [23] or missing and incomplete licensing [24].
- Interpretability: documenting a system is critical as it can speed up development or, by missing, block development for long periods of time.

The explainability methods should offer human understandable information for monitoring the AI decision process. The article [25] finds that there are two categories of model interpretability:

- model-specific methods for specific categories of models like linear regression or decision trees;
- model-agnostic methods that can be used in any machine learning models and focus on analyzing the input–output pair (like LIME or SHapley Additive exPlanations (SHAP) [26]).

As AI technology is evolving and getting more complex to the point of making researcher understanding problematic (and therefore making difficult the process of improving the existing algorithms and processes), the authors in [25] identify several trends in XAI: to simplify the model in order to improve the implementation of explainability (which can lead to performance loss) or implement the explainability only for important aspects of the process, therefore making it hard for researchers to understand a supposedly "explained" model. The conclusion is to place more accent on developing algorithms (specific or general) that can be used to implement explainability especially for complex AI models but also to integrate ethics focused algorithms directly into the development process.

In regard to explainability, three development directions were identified [27]:

- white-box or glass-box model: development AI models that are inherently explainable for machine learning experts. Only some simpler models can be used for this, for example Decision trees, and have less accuracy and therefore applicability;

- gray-box: are systems where Deep explanation approaches can be applied to produce explanations in addition to predictions, for example Fuzzy Rule-Based Systems or Bayesian Networks;
- black box: are usually the most accurate but the least trustworthy, systems that need model induction techniques to create approximate explainable models, for example in the case of Deep neural networks.

It is notable that measuring the efficacy of explainability method results usually involves psychological experiments [28].

If we look at the explainability components found, from the developer's perspective, only some present interest: transparency; robustness and interpretability. The remaining aspects are critical to the public and not to the entities performing the development. In order to make developers put effort for implementing fairness, trust and privacy, national entities need to get involved by providing regulations.

Unfortunately, until 2024 there are only few attempts at international level for regulation in this domain, the main being the European Union AI Act which was adopted into law on 13 March 2024. Other than that, declarations of compliance made by large AI developers (as a response to public pressure and lawsuits started for privacy or copyright infringement) are generic or even try dismiss these claims [29] of lack of fairness, trust and privacy. This led to increased skepticism for AI adoption in sensitive areas for the general public.

There are discussions about AI regulation to be dependent on the domain of applicability, with different levels of restrictions. There are IT domains where AI is already present for years and there is little regulation needed or only voluntary regulation on how the user data is used. For example, Android application's access to sensors, components and features is categorized depending on the risk of accessing sensitive user information (classified in Android as: normal/dangerous/signature/knownSigner/signatureOrSystem) and permissions are necessary to be obtained from the user [30]. In this context, access to camera or location data needs strict control, while other sensors (such as the light sensor or the accelerometer) are less inclined to give Personally Identifiable Information (PII) or Protected Health Information (PHI).

It is a native human desire to be in control of the access to PII or PHI information (and it can be extended to intellectual property and financial information) for applications [31], however, it is also desirable to exercise this control in a simple manner. This is why many users are using a smart help system to manage their data, but this system, in turn, needs access to the large amount of information generated by the user. This can lead to problems, as the users are the worst enemy of cybersecurity when complex software is used because (due to lack of knowledge or commodity) they often find ways to circumvent security measures [32, 33], meaning that all the information gathered by the AI system in order to improve user experience can be exposed and extracted by an attacker.

For this, in 2024, the European Parliament adopted the Artificial Intelligence Act (AI Act) [34] after a draft was presented for over a year, this being the first framework in the world on regulating the AI use [35]. This is part of the EU strategy on integrating digital technologies in all aspects of society and industry, but also for analyzing the impact of such technologies in all domains including telecommunications, health care and education. In European Union, in 2023, 42% of people lack basic digital skills [36] and this strategy aims to reduce this value and make people able to understand and leverage the IT advantages in regard to education (where problems were outlined during the COVID-19 health crisis, when people were encouraged to study and work from home).

The risk levels proposed by the EU document are:

- Unacceptable risk—covering prohibited practices such as: subliminal manipulation, social classification, non-discriminatory biometric motoring for law enforcement and vulnerability exploitation. As they are excluded, no regulation is needed.
- High risk—this is the main purpose of the propose legislation, identified as having simultaneously two conditions: AI systems intended to be used as safety component of products and stand-alone AI systems that have fundamental rights or health and safety implications.
- Low or Limited risk: these systems should only specify that they are AI systems and the user will decide if the interaction will continue, for example chatbots or 'deep fakes.

In the United States of America (USA), standards for safe development of AI where privacy, advances equity and civil rights are protected were also introduced. The USA congress presented a bill called "Blueprint for an AI Bill of Rights" that outlines these aspects with accent on: Safe and Effective Systems; Algorithmic Discrimination Protections; Data Privacy; Notice and Explanation and Human Alternatives, Consideration, and Fallback. One aspect of this bill, should be noted, applies to "automated systems that have the potential to meaningfully impact the American public's rights, opportunities, or access to critical resources or services" [37], meaning that private automated systems that do not affect these rights are not subject to these considerations.

Due to lacking legislation, some companies have set their own rules and limitations. For example, current smartphones have integrated AI tools that are used for smart assistants, Virtual Reality, scene detection, image processing with black box design that are not the target of the public's fear or concern making the developers save money by not implementing fairness, trust and privacy methods. For example, Samsung introduces the Mobile AI term for the S24 series [38].

Google uses the Tensor hardware System on a Chip (SoC) in the Pixel phones that is a ML hardware capable of processing audio, photo and video [39]. In this case, due to the closed nature of the hardware, only regulatory laws will make explainable methods available to public developers. An exception in this domain would be an open-source environment like Raspberry Pi, even if the Broadcom chip is proprietary, with TensorFlow, Keras, PyTorch, OpenCat and OpenCV being available [40] (or Arduino platform for a less powerful system, with less libraries like AIfES, OpenCat). Apple has similarly started introducing explainability to its customers through podcasts, articles and books, the incentive being not only gaining the thrust of the people, but also improving their own products [41].

In other domains, that affects public's rights, opportunities, or access to critical resources or services, a black box approach to AI without offering explainability methods can discourage the broad adoption of AI model.

The paper [27] creates an overview of the XAI assessment methods, the available tools and XAI datasets. The authors identify four directions of classification for XAI components: data explainability, model explainability, post-hoc explainability, and assessment of explanations. The conclusion of the article is that the explainability of an AI model should be customized to specific user types to reduce development costs.

AI can be used in the legal domain because the cases are based on existing experiences and facts that only an AI algorithm can summarize and explore in reasonable time. Explainability in legal domain is critical therefore the need for XAI methods to be implemented for AI operating in this domain, along with the inherent trust issues. The paper [42] links different forms of legal inference to specific forms of algorithmic decision-making, concluding that AI development in this domain should be heterogenous in design, as each branch of the legal domain s (administrative, criminal, civil, etc.) has its particularities and different models fit each category and require different levels of explainability.

The paper [43] performs a survey on XAI and identifies four complementary axes for the need to implement explainability: to justify (in order to comply with authority requests but also to explain unexpected decisions), to control (to prevent errors and to correct existing ones), to improve (for the development purpose) and to discover (to learn new methods or facts that have escaped so far to human understanding).

In conclusion, there is a large number of research papers and books that point out that introducing explainability in AI models is a necessity for increasing trust in AI for the general public, because national and international regulation has remained behind the current production implementations. Only recently legislation has been adopted to define notions and require companies to take measures for transparency implementation in order to verify model compliance. Europe Union is leading in this domain with the laws adopted in 2024.

Introducing explainability in a model can be done in multiple moments of the AI system's lifecycle, each moment having tradeoffs between efficiency and implementation difficulty. That is why many AI models operate already with little or no explainability, as this aspect is left to the latitude of companies. This, however, is changing even if for different reasons: explainability allows understanding the model decisions and can speed up future development. In depth analysis of current research and implementations in the domains of entertainment, education, and environment will be presented in the following sections. An investigation of the online activity in each domain was made using Google Trends, a service offered by Google to identify the interest level for a topic at local or global level when searching for: "Explainable AI" + the name of the topic. The results, shown in Fig. 5.2 in logarithmic scale, show that the top interest was found for social media, followed by education, environment and politics.

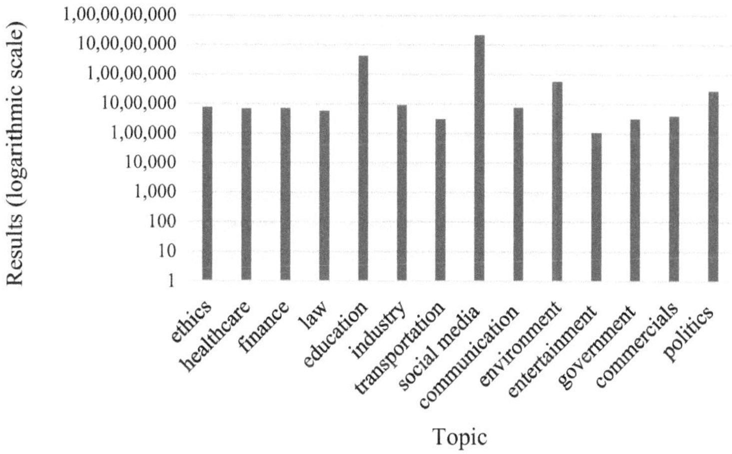


**Fig. 5.2** Interest found in google trends for explainable AI for different topics

## 5.3 XAI in Entertainment

Entertainment is worldwide business that generates important revenue (14.81 billion dollars in 2022) [44, 45]. Increasing profit by using AI models is a trend that has started at the beginning of the 21th century [46] (with an estimated increase in 2023–2030 of 26% in North America), driven mainly by the ability to generate virtual content and personalize the user experience. Big companies have implemented AI to determine the user preferences and offer content that is closer to their preferences. For example, Netflix [47] uses AI for content personalization, thumbnail generation, streaming quality to reduce congestions in specific regions, ranking [48], advertisement, etc. The information used by the algorithm includes time, location, genres, favorites but also if the user usually watches multiple episodes from a series in sequence or not. YouTube [49] is also using AI ranging from suggestions, stream recommendations to YouTube's generative AI products. YouTube also intends to introduce a system to warn users when a content uses AI [50], with special attention being paid for elections, ongoing conflicts, public health crises, or public officials. In this case the warning will be placed in a more visible fashion. Also, YouTube uses AI in automatic detection of copyright and of content that does not conform with the publishing rules [51], but this has created problems and still requires real people to supervise the results of the process.

Entertainment has multiple domains, such as music, film, advertisement, etc. In the reminder of this section, each domain will be investigated in relation to AI tools and the existence of XAI methods to implement explainability.

***Music and movies***

In music industry the purpose of AI is to generate content, recommend similar songs and enhance existing content (live, historical, etc.) by eliminating imperfections.

The question is if XAI is relevant for this domain. The paper [52] shows that the introduction of explainability methods in music generation is less explored but has a potential in case people want to use the AI tools to the best of their potential by adjusting components and monitoring the changes in real time or by offering new features. The code provided was developed in Python and uses the PyTorch package. Paper [53] again observers the lack of applications in music and presents a framework for explainability in music performance where the explainability requirements are dictated by the target audience. The more subjective the target is, the higher is the need to explain the AI model parameters effect in order to better target the audience.

Similarly, movies and television are highly subjective areas, and have potential benefits from explainability. AI is used for script generation (during the film writers guild strike in 2023 AI was used to supplement the lack of humans' writers as follows: a human author would lay the plot, characters and the key aspects of the script, while Large Language Models will generate the rest of the script), film production with actors de-aging, virtual scenery generation, removing unwanted sounds and images from the film, automating scene selection and for box office predictions (like Cinelytic AI, ScriptBook, Merlin or the Korean DRECE (Dimension REduction, Clustering, and classification for Explainable artificial intelligence [54]).

LLMs can be used for summarization of large texts, for example movie reviews, that do not give direct conclusions or scores. Explainability methods can be used to complement the process and make us understand how the AI model [55] concluded that a movie is good or bad based on the analysis of multiple reviews. In this case LIME Algorithm was used to understand the reason of misclassifications of some movies. Paper [56] presents a comprehensive investigation of XAI and AI in general in audio-visual media, touching: big data and its origins, AI use in news and media, cultural diversity, copyright, advertisement, deepfakes and ghost acting and finally concluding with proposals for a regulatory framework in audio-visual industries in Europe.

***Gaming***

AI in gaming is often used in order to control the non-player characters (NPCs) and make the world that the player is in more believable and look alive, to adapt the game difficulty automatically based on the player skill and to make game recommendations.

Paper [57] investigates the game cheating detection using XAI and proposes the EMGCD (Explainable Multi-View Game Cheating Detection) framework that combines cheating explainers to cheating classifiers from different views to generate individual, local and global explanations. In [58] the authors introduce an explainable AI workflow, which combines multiview data sources and multiview black-box models. Both papers target NetEase Games.

In [59] the authors suggest using the technique called Gradient-weighted Class Activation Mapping [60] (Grad-CAM used for creating a localization map for showing detected regions of interest of any Convolutional Neural Network model) in order to visualize the actions of AI players who had undergone Deep Reinforcement Learning training. Their application explains how a black box AI model plays a game by identifying the areas of interest that the AI uses as the basis for taking action.

### *Advertising*

Advertising is a business that is present in all entertainment components because it brings attention to a specific product to increase its revenue. It is also the domain where many market research practices are becoming borderline criminal, commercializing personal data, spying on user activities and, in general, infringing on the privacy rights. Some activities where AI is involved include advertisement personalization, predicting consumer trends, content generation and social media analysis. AI is used to combine information from media taken from multiple sources that are liked to a single account (for example: Meta Platforms, Inc. company that includes Facebook, Instagram, WhatsApp and Threads; Alphabet with Google search engine, Chrome, Google Ads, YouTube, Android operating system and cloud business; Microsoft with Windows, Azure, Microsoft Office/365 and Bing search engine; Amazon with e-commerce platform and video streaming, Twitch and Kindle e-reader; Tencent with the WeChat, TencentQQ and the gaming platform; ByteDance Ltd. with TikTok/Douyin platform; X Corp. with X (formerly Twitter) and Tesla) in order to understand the user and propose relevant advertisement across multiple platforms.

The number of papers investigating advertisement and XAI is low when compared to other domains. One possible cause is that rarely the companies developing AI products for advertisement have interested to allow external developers to understand, use or influence their products or to create a tool open for the competition to use.

In [61] the aspect of content generation by game designers with the help of AI/ML tools—called eXplainable AI for Designers (XAID)—is investigated. Key challenges of the proposed framework are discussed where, for a better usage of the tools, the properties of the AI tools need to be exposed and the user particularities need to be understood in detail.

In [62] an explainability model is proposed to analyze and classify the advertisements based on their exaggeration of truth in regards to food products.

Paper [63] focuses on implementing XAI for hate speech detection, with an implementation using Python. The datasets used were HateXplain and Google Jigsaw and the explainability was described using BERT (Bidirectional Encoder Representation from Transformers) and LIME (Local Interpretable Model—Agnostic Explanations). Platforms used to extract data were Twitter, Gab and Wikipedia.

### *Books*

For book publishing the uses of AI tools are multiple (such as simplify the paper evaluation process). In this context Scopus, Elsevier's publishing company citation database, has introduced in 16 January 2023 the Scopus AI tool [64] that can be used to synthesize abstracts of relevant documents and is able to create a summary of related authors work based on the information in the Scopus database (which is a curated peer-reviewed source). The trust in the generated results is increased by the generated concept map of each query. Data source trust and process transparency are keywords here in order to avoid any "hallucinations" of other existing summarization tools.

AI tools can also check paper for clarity, consistency (especially for large documents) and plot structure (it is documented that authors lose a lot of time trying to keep track of what is happening in large books or related book series with multiple authors).

The problem in this domain derives from the unlawful use of copyrighted content [65] as is the case of the Books3 dataset comprising pirated books that was used for training by the Meta AI tools.

Content creation is another point of dispute. A simple search on the internet shows that there are even guides on how to quickly script, write and grammar proof a material (and even proof it against plagiarism checkers). Here are the steps that can be followed for fast content generation:

- Any of the existing ChatBot for suggesting ideas for writing a book: ChatGPT, Bard, etc.
- Sudowrite or Jenni AI for writing a material draft with protection against plagiarism checking;
- If images are needed as book illustrations the text to image generative AI can be used, for example: Midjourney (in 2024 Midjourney takes about 1 min to generate an image in FAST mode), OpenAI's Dall-E, Playground AI or Stable Diffusion;
- Finally, Grammarly or QuillBot for checking grammar and content and making corrections.

These steps allow the creation of synthetic content really easy, but—at this point in time—it will be remarkably close to the materials that the model was trained on (leading to plagiarism) and humans will still be necessary to proof the result. The fear that AI will replace the human authors was one of the causes of a 148 days strike in USA in 2023. Explainability could prove in this case that the generated content does not infringe copyright laws and it not biased.

***Consumer experience***

This represents the context of offering a specific service, whether it is easy to use and looks good. Chatbots [66] are applications that use AI to simulate a human presence and discuss with people (often in text format) in order to give them information that can help them. ChatGPT is one example [67] that uses the Natural Language Processing (NLP) model, presented to the public in November 2022 by OpenAI, for open-ended conversations, based on GPT 3.5 and GPT-4 for paying customers. Other applications like text summarization, language translation, and content creation were sidelined by the chat potential. There are many chatbots implemented today: Bard, Bing, Jasper, GitHub Copilot, OpenAI Playground, etc. However, the most attention in scientific literature was generated by ChatGPT because it was the first to be mass used by the general public. Comparisons were made in different papers Plevris et al. [68] and Bhardwaz and Kumar [19] concerning understanding, result accuracy, consistency, response time, content relevance and overall effectiveness. Results showed that there are significant differences between them and there are tradeoffs that have to be considered, with no overall clear winner. For example, ChatGPT-4 has the advantage of accuracy and content relevance, ChatGPT-3.5 has a short response time while Google Bard offers best responses if the responses are already on the internet.

These algorithms were found to:

- have security issues due to the commercial black box nature of their development [69]. This is one of the reasons why ChatGPT was banned for a period in Italy [70, 71].
- give erroneous answers and lie, often intentionally to provoke or extend a conversation [72, 73].
- have copyright infringement problems [74] because the companies do not disclose the data sources used to train the AI algorithms. The authors of [19] propose to apply formal methods in all the stages of development of chatbot systems that would help eliminate errors.
- can be used especially in social engineering attacks [75], one of the most common attacks in cybersecurity (46–60%) [76]. ChatGPT can even be used to generate code for executing a cyber-attack. Even if safeguards are in place, if the human will formulate the request in a specific manner, it is possible to circumvents the protections [77].

Some researchers believe that this is an inherent characteristic of Large Language Models because humans use knowledge that is not written down and common sense rooted in realty. Therefore, human logic is used as feedback in protecting the AI generated data from aberrations called "hallucinations" that are present even in the newer models.

In other cases, these are generated by the fact that the model does not have enough relevant data to be trained on. In the context of image generation faces are often present in pictures used for training, while hands and feet are not. These lead to the generation of anatomically wrong images or blurry content as seen in Fig. 5.3.

## 5.4 XAI in Education

Education is another domain where AI has an important impact and is on a continuous ascending trend. The students are those that will use the current AI tools and develop in the future new AI tools, therefore they need to understand the need for embedding explainability. Paper [78] identified three areas for explainability in educational AI tools: to enhance educational tasks and lower the costs because the tools will be available for all stakeholders; to address system-level risks and anxiety about potential future risks such as student surveillance fear and teacher fear of replacement; to prevent the use of biased tools for hiring teachers and to offer personalized content to students in order to prevent the creation of a knowledge gap between them.

Many papers have investigated the use of AI in education and the need for explainability. Some of the domains where AI tools are currently present but need

(a) (b)

**Fig. 5.3** Images with errors generated by Gemini 1.5 flash on an android phone: (**a**) has hands with multiple fingers, (**b**) has distorted cars and random placement of support pillars

explainability are: scientific paper checking, evaluation and summarization; automatic test generation and checking (even for questions with free form answers) [72]; the creation of tutorials for students and for managing student progress.

The appearance of generative AI in education is forcing teachers, already confronted by copy/paste plagiarism and ghost writing to switch to other evaluation techniques that require a hands-on/experience approach which cannot be replaced easily by AI tools. AI can generate fake content that is believable by a less knowledgeable person, as is the case described in [79] where ChatGPT generated academic content with fake references.

Figure 5.4 shows how some existing LLMs helped grading a free form answer to an Operating Systems question: *Please grade on a scale from 1 to 10 the answer: "The types of synchronization and communication are signal, pipe, fifo and socket." for the question "Which types of synchronization and communication between processes are use in operating systems?"*. Some give the grade with few explanations of the process while others are explaining in detail the reasoning and even give hints on how to improve the answer. This can help both teachers and students.

Khosravi et al. [80] presents the term XAI-ED (or XAIED) for Explainable Artificial Intelligence in EDucation and identifies concerns regarding the aspects of Fairness, Accountability, Transparency, and Ethics for educational use of Artificial Intelligence algorithms. Education has specific needs and a framework is introduced with six aspects in relation with explainability: the benefits to the stakeholders; the delivering of explanations; the models used in education; the human centered design of AI to support explanations; the pitfalls of providing explanations. The pitfalls are identified as: the needless complexity of models used for teaching; inaccurate and incomplete explanations when the model is oversimplified; misconceptions that rise

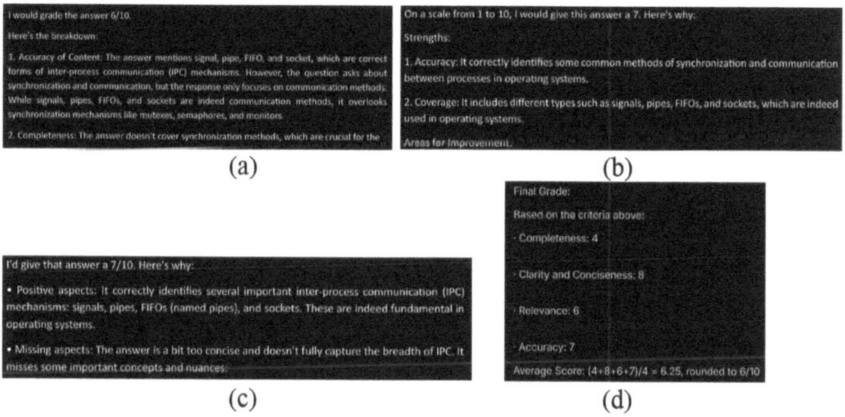


**Fig. 5.4** Grading of a free form answer by different LLMs: (**a**) ChatGPT; (**b**) Copilot; (**c**) Gemini 1.5 Flash; (**d**) Grok

from too much data that overwhelms the student, and finally promoting dysfunctional behavior where students learn that there are "shortcuts" to attain a specific target (such as multiple plagiarism checks until the paper passes the check). The case studies investigated show how explainability can be applied to AI tools used in education (for example targeting the RiPPLE Adaptive educational system).

Paper [81] expands on the proposed XAIED and makes a review of the literature and concludes that explainability is a relative term because of the perspectives of different stakeholders (for example students want to understand the model, managers want to monitor the process, teachers need automated assessment tools, etc.). The authors point out that total transparency is also disruptive as, for example, by knowing the algorithm used for evaluation, students will engage in algorithm manipulation and put less accent on actual learning.

The overview paper [82] points out the increased use of AI models and systems in LMS (Learning Management Systems) and EDM (Educational Data Mining) tools but, similarly to previous analyzed papers, identifies the distinct aspects of AI used in education that requires explainability for students as one of the stakeholders in the domain.

Paper [83] shows that standard XAI approaches are ineffective across different domains and can lead to biased results. The authors use cognitive psychology to outline a strategy on explainability functions interface design with educational value, where a balance must be found between simple and cognitively engaging explanations.

In [84] an AI framework is presented to analyze the impact of students' learning input on their learning effectiveness. The framework uses explainability methods to provide information and explain why the model identified some students as having potential to fail the class. The framework was implemented by using Microsoft Power BI tools to demonstrate the analysis outcomes. The same dropout prediction is made in [85] with XAI tools being used to show the main factors for fall in grades for high school students and university students. This paper also outlines the importance of offering a good user interface to the tools and their results, because the management staff are the ones using them and they do not have the time and knowledge to understand a complex visualization output.

Paper [86] investigates Exemplar-based XAI and proposes an explainability framework that analyses the trade-off between the example complexity and the fidelity the human understanding. The framework generates the examples used to learn until is finds the right balance between example complexity and human understanding. A paper related to this [87] shows that implementing explainability is not a certain recipe for understanding a model's decisions. Oversimplified answers or highly complex explanations usually lead to the same conclusion of not understanding the AI models that they were applied to. The case is made that the psychology of the student should be considered where it is equally important to understand the AI behavior but also to check that the correct, complete information was understood and was not oversimplified.

The paper [88] emphases the importance of feedback in the understanding of AI through explainability. It introduces an intelligent system called Cognitive Learning supported by XAI (CL-XAI) for understanding how humans comprehend the operation of AI models that have incorporated explainability methods, that also allows evaluating the effectiveness of such tools through human feedback.

Brdnik et al. [89] performed a study to understand the perceived impact of XAI and which is the most effective way to present the explainability, through feedback on Slovenian students of Information Technologies and Computer Science. The authors observed a difference between master students and bachelor students, as former preferred global feature explanation while the latter were satisfied by both local feature explanation and global feature explanation (explained at their level of understanding).

In [90] student success prediction though AI is investigated. Because the authors found that the current systems were limited, they implemented the system explainability as a combination of AI and social science methods. The propose a three-level interpretability system that has LIME in the first step, a novel method called Minimal Counterfactual-LIME that finds the smallest number of changes necessary to change a prediction in the second step and a final step of uncovering causal relations for similar students.

Tutoring is investigated by Conati et al. [91] where an AI gives hints and feedback adapted to the student level. The system was complemented with explainability that increased trust in hints and perceived usefulness. Similar to other papers, it was found that explanations implementation should consider context, task criticality, and user needs.

## 5.5 XAI for Environment

Artificial intelligence plays a crucial role for the environment because it can analyze the large amount of data generated by Internet of Things (IoT) devices and make predictions for the evolution of the environment [92]. AI helps investigate the impact of certain actions on the environment through visualization tools that have much larger impact when compared for example with text representations [93]. Unfortunately, AI has also a negative impact on the environment, because:

- it requires a lot of energy to train and process data [94] and this has increased exponentially since 2012, doubling the amount of computing power required by AI models every few months and it is estimated that, the emissions by 2040 the IT industry will reach 14% of the global emissions [95]. Some researchers [96] have found ways to reduce the needed energy by promoting energy efficient models, energy efficient model tuning, and improved software packages that can interact with deep learning frameworks such as PyTorch and TensorFlow. Authors argue that if the software is not optimized, the developers will default to other easy to use packages that are not energy efficient.
- it generates waste of IT hardware. This is a general problem in IT and a complete solution was not yet found, as not all components can be recycled;
- using AI for predictions in agriculture might place accent on economic growth and endanger biodiversity.

The advantages, on the other hand, are significant, as presented in the United Nations Environment Program concerning the use of AI for the benefit of the environment [97]:

- weather monitoring and prediction: processing big data generated by the environment monitoring tools, predicting weather evolution in short term and identifying dangerous patterns and changes in the long term. For example, the paper [98] shows that weather prediction algorithms are getting more complex and, while accuracy in increased, trust is decreasing because the specialists cannot explain why a certain prediction was made. Explainability is needed in this domain to fine tune the weather models architecture.
- disaster prevention: identify natural disasters form monitoring stations (such as seismic sensors) and generate early warning with integrated communications systems;
- tracking pollution: identify pollution and suggest solutions to mitigate the problem;
- carbon neutrality: use algorithms to increase efficiency of complex processes that have a significand carbon footprint, reduce supply chains and increase the efficiency of renewable sources;
- optimize food production and minimize waist generation.

As in other domains there is a need to increase the trust in the AI models and better tune them for environment related predictions. In this regard, the paper [99] develops an explainability framework for environmental management research that can reduce the computation cost for training by eliminating object categories that are not relevant to the domain monitored. The framework is monitoring the outputs for interpretability and robustness to avoid making false predictions with a high confidence level. For the code implementation Python is used with the PyTorch library.

The paper [100] presents a XAI approach called AHA–XDNN (Artificial Hummingbird Algorithm—eXplainable Deep Neural Networks) for water quality prediction. The results of an optimized DNN model are explained using the explainability method SHAP, which increases user's trust in the algorithm.

Authors of [101] aim to evaluate the AI algorithms used in Climate Science by implementing multiple explainability methods that assess robustness, faithfulness, randomization, complexity, and localization. For the stated purpose, the authors have found that the Integrated Gradients, Layer-wise relevance propagation, and Input Gradients XAI methods were the best in term of robustness and faithfulness. Finally, the results help users that use explainability in climate science models to choose the suitable methods for a specific Climate Science AI implementation.

Paper [102] introduces a framework for human in the loop Intelligent and Sustainable Maintenance of AI algorithms. Most assets develop problems toward the end of the life cycle. By allowing cyber assisted maintenance activities that implement explainability methods, the lifecycle of the product can be extended in order to achievement of a net zero carbon emission economy. This is especially the situation for AI technology training that was found to have a large carbon footprint [103, 104] because it requires energy to process all the initial data and an even larger amount of energy for tuning the model to a new dataset or performing the research for model development.

Paper [105] targets the flight of drones and the influence of weather on their flight capability. Machine learning algorithms were trained on ten years of sail data in order to predict the flight duration and the influence of thermals on the flight soaring potential. SHAP approach was used to understand the predictions given by the models. The conclusion was that because of the explainability implemented, human knowledge can be used to understand the logic and avoid potential prediction errors.

Authors of [106] introduce a XAI framework for investigating wildfire occurrence in Italy. While the results are good, the authors place accent on the importance of XAI implementation success on existing AI algorithms with the purpose of improving trust in the results. The authors conclude that a framework with integrated decision support systems allows forest managers to prevent disasters and develop new strategies for fire management and response. Similarly [107] uses XAI techniques to increase the trust in the results of the AI algorithms investigating climate data and to improve the algorithms beyond what the designers envisioned.

Paper [108] investigates climate changes effect on seagrass development in Hainan area China. For a model with 15 variables, the contribution is outlined with the help of explainability methods. The SHAP and Partial Dependency Plot (PDP) analysis methods were used to inspect how the model variables affect the predictions.

In [109] implements a decision-making system for landfill selection that uses XAI with an optimized bagging (bootstrap aggregating—training multiple models on random data subsets and aggregating the results) ensemble model to understand how a specific choice was made. SHAP and LIME were used to provide better understanding of the global and local determinants of landfill site suitability section.

Paper [110] uses machine learning models with thousands of variables to predict yield models with good accuracy. Implementing explainability allows a detailed interrogation of these models to determine crop behavior and complex interactions in yield prediction.

## 5.6 Conclusions and Future Trends for XAI

In order to outline the future trends for explainability tools (which are freely offered to the developers in most cases) it is necessary to investigate the history regarding other projects that have started as open-source projects and evolved into companies. In [111] an analysis is made where examples like Red Hat (offering paid support for open-source software), MySQL with its dual license model, Kafka that became commercial as Confluent, Android with its blend of open-source principles and monetization through licensing, Google Play Store purchases and advertising, etc. The steps of this evolution are:

- present the solution as open-source or as free of charge and thus obtain community engagement and project validation;
- develop commercial services (often for enterprises) on top of the open-source platform to provide value for paying customers. While open-source code is easy to migrate from, the black box, commercial code is harder to migrate from [112] authors coin the term as "stickiness".

This is exactly what is happening with XAI, even if the domain is still in an early stage of development, most of the existing explainability tools are available for free and are open source. They usually use Python as a development language, as a front-end API for building applications or by creating wrappers for underlying languages such as C++, R, Swift and Java. Outside the development language, standardization is still an issue, making it difficult for algorithms to be implemented consistently. The developers of these tools are keen on attracting developers as their number is low due to the high-level requirements of competences. However, most implementations are dependent on the framework used and it is hard to migrate from on framework to another.

It is not hard to imagine that in the near future, when the number of frameworks and companies has been reduced, some of the tools and algorithms will become black boxes themselves in order to keep the developers and companies locked in a specific software environment.

It can be concluded that the main problems are the drive and the financing behind the XAI development. In regard to these two aspects aspect, the developers can be grouped as:

- National or international organizations: if there is interest in the development of a certain area and financing is offered, a viable XAI framework can be created. The problem is continuous financing and industry adoption.
- Private businesses: they may introduce explainability in their AI models in order to comply with existing legislation requirements, but they may opt to implement further modules as black boxes based on the Google/Android model;
- Individuals: innovative ideas for explainability are often designed by individual researchers or small teams, as part of research projects. However, in order to have financial support in long term, they can:
  - Switch the development model and attempt to find financing for bringing a commercial product to the market;
  - Sell the implementation to private business to be used in a closed environment therefore defeating the openness of XAI;
  - Obtain national or international organization backing which can give the proposed platform a chance to extend its developer and user base.

Concerning the actual model implementation, the trend is to use model agnostic algorithms for explainability that can be applied to any machine learning model, regardless of its structure or type. Even if they were found to give less reliable information on the model decisions than creating the model with integrated explainable methods and are prone to bias, but are easier to develop with, can work with black box models and are being constantly improved.

In regard to the evaluation methods there is no consensus on a general evaluation method for explainability due to the lack of metrics, subjectivity and model complexity. Finally, better visualization tools and techniques are necessary for supporting the decision-making process and to allow quick understanding of the presented explanation results.

In conclusion, the Chapter presented a survey of papers that propose using explainability methods for AI models used in education, environment and entertainment. As the development of an AI model (explainable or not) is a task that involves multiple stakeholders: private companies, governments as well as individuals, the need for introducing explainability for a model can be differentiated into two categories in respect to general human rights and opportunities:

- AI models that do not influence them in an explicit way, and as a result will implement explainability methods only for improving the development process;
- AI models that affect these rights and that will be subject to regulation at national and international level, and therefore will have to implement explainability to be able to cope with the legislation.

In regard to legislation for regulating AI and the need for XAI, European Union and United States of America have already presented the basis for future legislation in the domain at international level. The proposed legislation targets applications and not the AI models in order not to limit their development.

Concerning the educational research and implementation, the general ideas that can be extracted from the analyzed papers were:

- having XAI models is important but the particularities of educational systems must be considered in their development;
- as there are many stakeholders in this domain with various degrees of understanding, the trained persons satisfaction with the tools should be monitored continuously as wrong conclusions can be understood by the user;
- AI needs to be integrated into the education and this needs explainability for understanding and using the AI tools at their true potential and reduce the chances of introducing bias;
- oversimplification of explanations is as bad as giving too complex explanations;
- the academic impact is large as it becomes difficult to identify original papers and AI ghostwritten papers, therefore models should implement with safeguards that can be monitored.

In regard to the environment research and implementation, the general ideas that can be extracted were:

- there is a positive potential of AI models used for Environment that offsets the current carbon footprint that these create. Methods and solutions to reduce this footprint have already proposed;
- explainability for AI is necessary to increase trust but also to be able to fine tune the AI models.

Finally, concerning the Entertainment research and implementation, the general ideas that can be extracted were:

- a balance is yet to be found between potential harms and potential benefits of AI use;
- XAI models are important as they increase trust in their predictions;
- special attention must be paid to the datasets used for training the AI models in order to prevent privacy and copyright issues. Here explainability can show that the generated output does not infringe these aspects.

## References

1. Nowak PL, Horodecki P (2018) Assessing artificial intelligence for humanity: will AI be the our biggest ever advance? or the biggest threat [Opinion]. IEEE Technol Soc Mag 37(4):26–34. https://doi.org/10.1109/MTS.2018.2876105
2. Russell S, Dewey D, Tegmark M (2015) Research priorities for robust and beneficial artificial intelligence. AI Mag 36:105–114. https://doi.org/10.1609/aimag.v36i4.2577

3. Peres RS, Jia X, Lee J, Sun K, Colombo AW, Barata J (2020) Industrial artificial intelligence in industry 4.0—systematic review, challenges and outlook. IEEE Access 8:220121–220139. https://doi.org/10.1109/ACCESS.2020.3042874
4. Kaur R, Gabrijelčič D, Klobučar T (2023) Artificial intelligence for cybersecurity: literature review and future research directions. Inf Fusion 97:101804. ISSN 1566-2535. https://doi.org/10.1016/j.inffus.2023.101804
5. Krishna G, Sharma A (2023) An improved system built on artificial intelligence for sorting and handling plastic materials. In: 2023 IEEE international conference on integrated circuits and communication systems (ICICACS), Raichur, India, pp 1–7. https://doi.org/10.1109/ICICACS57338.2023.10100302
6. Bernardino R, Lins RD, Barboza RDSA (2023) Quality, size and time assessment of the binarization of documents photographed by smartphones. J Imaging 9(2):41. https://doi.org/10.3390/jimaging9020041
7. Xia B, Ye X, Abuassba AOM (2020) Recent research on AI in games. In: 2020 international wireless communications and mobile computing (IWCMC), Limassol, Cyprus, pp 505–510. https://doi.org/10.1109/IWCMC48107.2020.9148327
8. Jay R (2000) The automaton chess player, the invisible girl, and the telephone. Jay's J Anom 147:147–162
9. McCulloch W, Pitts W (1943) A logical calculus of ideas immanent in nervous activity. Bull Math Biophys 5(4):115–133. https://doi.org/10.1007/BF02478259
10. Turing AM (1950) Computing machinery and intelligence. Mind LIX(236):433–460. https://doi.org/10.1093/mind/LIX.236.433
11. Werbos PJ (1994) The roots of backpropagation: from ordered derivatives to neural networks and political forecasting. New York: John Wiley & Sons. ISBN 0-471-59897-6
12. Fukushima K (1980) Neocognitron: a self-organizing neural network model for a mechanism of pattern recognition unaffected by shift in position. Biol Cyber 36:193–202. https://doi.org/10.1007/BF00344251
13. LeCun Y, Bottou L, Haffner P (1998) Gradient-based learning applied to document recognition. Proc IEEE 86(11):2278–2324. https://doi.org/10.1109/5.726791
14. Hochreiter S, Schmidhuber J (1997) Long short-term memory. Neural Comput 9:1735–80. https://doi.org/10.1162/neco.1997.9.8.1735
15. Goodfellow I et al (2014) Generative adversarial nets. In: Proceedings of the international conference on neural information processing systems (NIPS 2014), pp 2672–2680
16. Vaswani A et al (2017) Attention is all you need. Adv Neural Inf Process Syst 30
17. Davide C (2016) Can we open the black box of AI? Nature 538(7623):20–23. https://doi.org/10.1038/538020a
18. Pomerleau DA (1993) Neural networks for intelligent vehicles. In: Proceedings of the intelligent vehicles'93 symposium, Tokyo, Japan, pp 19–24. https://doi.org/10.1109/IVS.1993.697291
19. Bhardwaz S, Kumar J (2023) An extensive comparative analysis of chatbot technologies—chatgpt, google BARD and microsoft bing. In: 2023 2nd international conference on applied artificial intelligence and computing (ICAAIC), Salem, India, pp 673–679, https://doi.org/10.1109/ICAAIC56838.2023.10140214
20. Gunning D, Vorm ES, Wang J, Turek M (2021) DARPA's explainable AI (XAI) program: a retrospective. Appl AI Lett 2. https://doi.org/10.1002/ail2.61
21. Ribeiro MT, Singh S, Guestrin C (2016) Why should I trust you? Explaining the predictions of any classifier. In: Proceedings of the 22nd ACM SIGKDD international conference on knowledge discovery and data mining
22. DARPA (2021) XAITK—an open-source, explainable AI toolkit built for analytics and autonomy applications.https://xaitk.org/
23. Banciu D, Cîrnu CE (2022) AI ethics and data privacy compliance. In: 2022 14th international conference on electronics, computers and artificial intelligence (ECAI), Ploiesti, Romania, 2022, pp 1–5, https://doi.org/10.1109/ECAI54874.2022.9847510

24. Edd G (2023) Public AI training datasets are rife with licensing errors. IEEE Spect. https://spectrum.ieee.org/data-ai
25. Zoumana K (2023) Explainable AI—understanding and trusting machine learning models. https://www.datacamp.com/tutorial/explainable-ai-understanding-and-trusting-machine-learning-models
26. Lundberg SM, Lee SI (2017) A unified approach to interpreting model predictions. In: Proceedings of the 31st international conference on neural information processing systems (NIPS'17). Curran Associates Inc., Red Hook, NY, USA, pp 4768–4777
27. Ali S et al (2023) Explainable artificial intelligence (XAI): what we know and what is left to attain trustworthy artificial intelligence. Info Fusion 99:101805. ISSN 1566-2535. https://doi.org/10.1016/j.inffus.2023.101805
28. Linder R, Mohseni S, Yang F (2021) How level of explanation detail affects human performance in interpretable intelligent systems: a study on explainable fact checking. Appl AI Lett. 2(4):e49
29. Blake B (2023) Meta tells court AI software does not violate author copyrights. Reuters. https://www.reuters.com/legal/litigation/meta-tells-court-ai-software-does-not-violate-author-copyrights-2023-09-19/
30. Google LLC (2024) Android permissions. https://source.android.com/docs/core/permissions
31. Mehrnezhad M, Makarouna C, Gray D (2022) Risks of mobile ambient sensors and user awareness, concerns, and preferences. In: 2022 European symposium on usable security (EuroUSEC 2022), September 29–30, Karlsruhe, Germany. ACM, New York, NY, USA, p 21. https://doi.org/10.1145/3549015.3554171
32. Rawat DB, Doku R, Garuba M (2021) Cybersecurity in big data era: from securing big data to data-driven security. IEEE Trans Serv Comput 14(6):2055–2072. https://doi.org/10.1109/TSC.2019.2907247
33. Hoffer G (2023) Ease of use, ease of integration encourages data protection. https://www.cyberdefensemagazine.com/complexity-is-still-the-enemy-of-security/
34. European Commission (2021) Regulation of the European parliament and of the council laying down harmonised rules on artificial intelligence (Artificial Intelligence Act) and amending certain Union Legislative Acts. https://eur-lex.europa.eu/legal-content/EN/TXT/?uri=celex%3A52021PC0206
35. European Parliament (2023) EU AI act: first regulation on artificial intelligence. https://www.europarl.europa.eu/news/en/headlines/society/20230601STO93804/eu-ai-act-first-regulation-on-artificial-intelligence
36. European Parliament (2021) Shaping the digital transformation: EU strategy explained. https://www.europarl.europa.eu/news/en/headlines/society/20210414STO02010/shaping-the-digital-transformation-eu-strategy-explained
37. The White House (2023) Executive order on safe, secure, and trustworthy artificial intelligence. https://www.whitehouse.gov/briefing-room/statements-releases/2023/10/30/fact-sheet-president-biden-issues-executive-order-on-safe-secure-and-trustworthy-artificial-intelligence/
38. Samsung (2024) Enter the new era of mobile AI with samsung galaxy S24 series. https://news.samsung.com/global/enter-the-new-era-of-mobile-ai-with-samsung-galaxy-s24-series
39. Gupta M (2021) Google tensor is a milestone for machine learning. https://blog.google/products/pixel/introducing-google-tensor/
40. Lehman M (2023) Raspberry Pi AI integration: guide to building intelligent mobile assistants. https://www.gmihub.com/blog/raspberry-pi-ai-integration/
41. Holtzman A, Buys J, Du L, Forbes M, Choi Y (2019) The curious case of neural text degeneration. arXiv:1904.09751
42. Nielsen A, Skylaki S, Norkute M, Stremitzer A (2023) Effects of XAI on legal process. In: Proceedings of the nineteenth international conference on artificial intelligence and law (ICAIL'23). Association for computing machinery, New York, NY, USA, pp 442–446. https://doi.org/10.1145/3594536.3595128

43. Adadi A, Berrada M (2018) Peeking inside the black-box: a survey on explainable artificial intelligence (XAI). IEEE Access 6:52138–52160. https://doi.org/10.1109/ACCESS.2018.2870052
44. Heath DR (2023) All in on AI: how smart companies win big with artificial intelligence. J Inf Technol Case Appl Res 25(3):200–204. https://doi.org/10.1080/15228053.2023.2228675
45. Grand View Research, Inc (2023) AI in media & entertainment market size, share & trends analysis report by solution (Hardware/Equipment, Services), by application (Gaming, Personalization), by region, and segment forecasts, 2023–2030. Report ID: GVR-4-68039-944-3. https://www.grandviewresearch.com/industry-analysis/artificial-intelligence-ai-media-entertainment-market-report
46. Forbus KD, Laird J (2002) AI and the entertainment industry. IEEE Intell Sys 17(4):15–16. https://doi.org/10.1109/MIS.2002.1024746
47. Netflix (2024) Machine learning. https://research.netflix.com/research-area/machine-learning
48. Devashree HG, Sharma N, Mangla M (2022) Analysis and visualization of netflix shows. In: 2022 4th international conference on artificial intelligence and speech technology (AIST), Delhi, India, pp 1–6. https://doi.org/10.1109/AIST55798.2022.10065331
49. Suresh SK, Srivastava M (2023) YouTube and movie recommendation system using machine learning. In: 2023 Second International Conference on Electronics and Renewable Systems (ICEARS), Tuticorin, India, pp 1352–1356. https://doi.org/10.1109/ICEARS56392.2023.10084999
50. O'Connor JF, Moxley E (2023) Our approach to responsible AI innovation. https://blog.youtube/inside-youtube/our-approach-to-responsible-ai-innovation/
51. Rosenstein D (2016) Improving content ID for creators. https://blog.youtube/news-and-events/improving-content-id-for-creators/
52. Bryan-Kinns N et al (2023) Exploring XAI for the arts: explaining latent space in generative music. arXiv:2308.05496
53. Privato N, Armitage J (2023) A context-sensitive approach to XAI in music performance. In: The 1st international workshop on explainable AI for the arts (XAIxArts), ACM creativity and cognition (C&C) 2023. Online, p 3. ACM, New York, NY, USA. https://doi.org/10.48550/arXiv.2309.04491
54. Leem S, Oh J, So D, Moon J (2023) Towards data-driven decision-making in the Korean film industry: an XAI model for box office analysis using dimension reduction, clustering, and classification. Entropy (Basel) 25(4):571. https://doi.org/10.3390/e25040571. PMID: 37190359; PMCID: PMC10137647
55. Ahmed F et al (2022) Interpretable movie review analysis using machine learning and transformer models leveraging XAI. In: 2022 IEEE asia-pacific conference on computer science and data engineering (CSDE), Gold Coast, Australia, pp 1–6, https://doi.org/10.1109/CSDE56538.2022.10089294
56. European Audiovisual Observatory (2020) IRIS special 2020-2 artificial intelligence in the audiovisual sector. Strasbourg ISBN 978-92-871-8806-9
57. Tao J et al (2020) XAI-driven explainable multi-view game cheating detection. In: 2020 IEEE conference on games (CoG), Osaka, Japan, pp 144–151. https://doi.org/10.1109/CoG47356.2020.9231843
58. Tao J et al (2023) Explainable AI for cheating detection and churn prediction in online games. IEEE Trans Game 15(2):242–251. https://doi.org/10.1109/TG.2022.3173399
59. Joo HT, Kim KJ (2019) Visualization of deep reinforcement learning using grad-cam: how AI plays Atari games? In: IEEE conference on games (CoG), IEEE, pp 1–2. https://doi.org/10.1109/CIG.2019.8847950
60. Selvaraju RR et al (2017) Grad-CAM: visual explanations from deep networks via gradient-based localization. In: 2017 IEEE international conference on computer vision (ICCV), Venice, Italy, pp 618–626. https://doi.org/10.1109/ICCV.2017.74
61. Zhu J et al (2018) Explainable AI for designers: a human-centered perspective on mixed-initiative co-creation. In: 2018 IEEE conference on computational intelligence and games (CIG), Maastricht, Netherlands, pp 1–8, https://doi.org/10.1109/CIG.2018.8490433

62. Park Y et al (2023) Exaggerated advertisement inspection system for judging the suitability of advertisements in social media environment. In: 2023 international conference on information networking (ICOIN), Bangkok, Thailand, pp 778–781, https://doi.org/10.1109/ICOIN56518.2023.10048933
63. Mehta H, Passi K (2022) Social media hate speech detection using explainable artificial intelligence (XAI). Algorithms 15:291. https://doi.org/10.3390/a15080291
64. Elsevier (2024) Scopus AI: trusted content. Powered by responsible AI. https://www.elsevier.com/products/scopus/scopus-ai
65. Reisner A (2023) These 183,000 books are fueling the biggest fight in publishing and tech. https://www.theatlantic.com/technology/archive/2023/09/books3-database-generative-ai-training-copyright-infringement/675363/
66. Albayrak N, Özdemir A, Zeydan E (2018) An overview of artificial intelligence based chatbots and an example chatbot application. In: 2018 26th Signal Processing and Communications Applications Conference (SIU), Izmir, Turkey, pp 1–4. https://doi.org/10.1109/SIU.2018.8404430
67. Abdullah M, Madain A, Jararweh Y (2022) ChatGPT: fundamentals, applications and social impacts. In: 2022 ninth international conference on social networks analysis, management and security (SNAMS), Milan, Italy, pp 1–8. https://doi.org/10.1109/SNAMS58071.2022.10062688
68. Plevris V, Papazafeiropoulos G, Jiménez Rios A (2023) Chatbots put to the test in math and logic problems: a comparison and assessment of ChatGPT-3.5, ChatGPT-4, and Google Bard. AI 4 (4):949–969. https://doi.org/10.3390/ai4040048
69. Gupta M, Charankumar A, Kshitiz A, Elisabeth P, Lopamudra P (2023) From ChatGPT to ThreatGPT: impact of generative AI in cybersecurity and privacy. IEEE Access 11:80218–80245
70. Saracco R (2023) How smart is it to block ChatGPT? https://cmte.ieee.org/futuredirections/2023/04/02/how-smart-is-it-to-block-chatgpt/
71. Wu X, Duan R, Ni J (2024) Unveiling security, privacy, and ethical concerns of ChatGPT. J Info Intell (2):102–115
72. Ionescu VM, Enescu MC (2023) Using ChatGPT for generating and evaluating online tests. In: 2023 15th international conference on electronics, computers and artificial intelligence (ECAI), Bucharest, Romania, 2023, pp 1–6. https://doi.org/10.1109/ECAI58194.2023.10193995
73. Smith CS (2023) Hallucinations could blunt ChatGPT's success. IEEE Spectrum. https://spectrum.ieee.org/ai-hallucination
74. Wang Y, Pan Y, Yan M, Su Z, Luan TH (2023) A Survey on ChatGPT: AI–generated contents, challenges, and solutions. IEEE Open J Comput Soc 4:280–302
75. zvelo, Inc. The role of AI in social engineering. https://zvelo.com/the-role-of-ai-in-social-engineering/
76. Bakhshi T (2017) Social engineering: revisiting end-user awareness and susceptibility to classic attack vectors. In: 2017 13th international conference on emerging technologies (ICET), Islamabad, Pakistan, pp 1–6. https://doi.org/10.1109/ICET.2017.8281653
77. Thatha P (2023) ChatGPT refused to write cyber-attack code for us, so this is what we did next. https://www.radware.com/blog/security/chatgpt-refused-to-write-cyber-attack-code-for-us-so-this-is-what-we-did-next/
78. U.S. Department of Education, Office of Educational Technology (2023) Artificial Intelligence and future of teaching and learning: insights and recommendations, Washington, DC
79. Hillier M (2023) Why does ChatGPT generate fake references? https://teche.mq.edu.au/2023/02/why-does-chatgpt-generate-fake-references/
80. Khosravi H et al (2022) Explainable artificial intelligence in education. Comput Educ Artif Intell 3:100074. ISSN 2666-920X, https://doi.org/10.1016/j.caeai.2022.100074
81. Farrow R (2023) The possibilities and limits of XAI in education: a socio-technical perspective. Learn Media Technol 48(2):266–279. https://doi.org/10.1080/17439884.2023.2185630

82. Rachha A, Seyam M (2023) Explainable AI in education: current trends, challenges, and opportunities. In: SoutheastCon 2023, Orlando, FL, USA, pp 232–239. https://doi.org/10.1109/SoutheastCon51012.2023.10115140
83. Simkute A et al (2022) XAI for learning: narrowing down the digital divide between "new" and "old" experts. In: Adjunct proceedings of the 2022 nordic human-computer interaction conference (NordiCHI '22). Association for computing machinery, New York, NY, USA, Article 39, pp 1–6. https://doi.org/10.1145/3547522.3547678
84. Chou T (2023) Apply an integrated responsible AI framework to sustain the assessment of learning effectiveness. In: Proceedings of the 15th international conference on computer supported education (CSEDU 2023) vol 2, pp 142–149. https://doi.org/10.5220/0012058400003470
85. Nagy M, Molontay R (2023) Interpretable dropout prediction: towards XAI-based personalized intervention. Int J Artif Intell Educ. https://doi.org/10.1007/s40593-023-00331-8 .
86. Håvardstun B et al (2023) XAI with machine teaching when humans are (Not) informed about the irrelevant features. In: European Conference on Machine Learning, ECML
87. Ferri C, Hernández-Orallo J, Telle JA (2019) Teaching explanations by examples. MI21-HLC machine intelligence workshop on human-like computing
88. Suffian M et al (2023) Toward enriched cognitive learning with XAI. https://arxiv.org/abs/2312.12290
89. Brdnik S, Vili P, Boštjan Š (2023) Assessing perceived trust and satisfaction with multiple explanation techniques in XAI-enhanced learning analytics. Electronics 12(12):2594. https://doi.org/10.3390/electronics12122594
90. Cohausz L (2022) Towards real interpretability of student success prediction combining methods of XAI and social science. In: Proceedings of the 15th international conference on educational data mining, pp 361–367, Durham, United Kingdom
91. Conati C, Barral O, Putnam V, Rieger L (2021) Toward personalized XAI: a case study in intelligent tutoring systems. Artif Intell 298:103503. https://doi.org/10.1016/j.artint.2021.103503
92. Bamdale R, Shelar S, Khandekar V (2021) How to tackle climate change using artificial intelligence. In: 2021 12th international conference on computing communication and networking technologies (ICCCNT), Kharagpur, India, pp 1–7. https://doi.org/10.1109/ICCCNT51525.2021.9579674
93. Luccioni A, Schmidt V, Vardanyan V, Bengio Y (2021) Using artificial intelligence to visualize the impacts of climate change. IEEE Comput Graph Appl 41(1):8–14. https://doi.org/10.1109/MCG.2020.3025425
94. Li P et al (2023) Towards environmentally equitable AI via geographical load balancing. https://doi.org/10.48550/arXiv.2307.05494
95. Kanungo A (2023) The green dilemma: can AI fulfil its potential without harming the environment? https://earth.org/the-green-dilemma-can-ai-fulfil-its-potential-without-harming-the-environment/
96. Strubell E, Ganesh A, McCallum A (2020) Energy and policy considerations for modern deep learning research. Proc AAAI Conf Artif Intell 34(09):13693–13696. https://doi.org/10.1609/aaai.v34i09.7123
97. United Nations Environment Program (2024) How artificial intelligence is helping tackle environmental challenges. https://www.unep.org/news-and-stories/story/how-artificial-intelligence-helping-tackle-environmental-challenges
98. Mamalakis A, Ebert-Uphoff I, Barnes E (2022) Explainable artificial intelligence in meteorology and climate science: model fine-tuning, calibrating trust and learning new science. In: Holzinger A, Goebel R, Fong R, Moon T, Müller KR, Samek W (eds) xxAI—beyond explainable AI. xxAI 2020. Lecture notes in computer science, vol 13200. Springer, Cham. https://doi.org/10.1007/978-3-031-04083-2_16
99. Arashpour M (2023) AI explainability framework for environmental management research. J Environ Manag 342:118149. ISSN 0301-4797. https://doi.org/10.1016/j.jenvman.2023.118149

100. Ezzat D et al (2023) An optimized explainable artificial intelligence approach for sustainable clean water. Environ Dev Sustain. https://doi.org/10.1007/s10668-023-03712-0
101. Bommer P et al (2023) Finding the right XAI method—a guide for the evaluation and ranking of explainable AI methods in climate science. https://doi.org/10.48550/arXiv.2303.00652
102. Turner C, Okechukwu O, John O (2022) XAI sustainable human in the loop maintenance. IFAC-PapersOnLine. 55:67–72. https://doi.org/10.1016/j.ifacol.2022.09.185
103. Hutson M (2022) Measuring AI's carbon footprint. https://spectrum.ieee.org/ai-carbon-footprint
104. Strubell E, Ganesh A, McCallum A (2020) Energy and policy considerations for modern deep learning research. Proc AAAI Conf Artif Intell 34(09):13693–13696. https://doi.org/10.1609/aaai.v34i09.7123
105. Schnieder M (2024) Using explainable artificial intelligence (XAI) to predict the influence of weather on the thermal soaring capabilities of sailplanes for smart city applications. Smart Cities 7(1):163–178. https://doi.org/10.3390/smartcities7010007
106. Cilli R et al (2022) Explainable artificial intelligence (XAI) detects wildfire occurrence in the Mediterranean countries of Southern Europe. Sci Rep 12:16349. https://doi.org/10.1038/s41598-022-20347-9
107. Barnes E (2022) Explainable AI (XAI) for climate science: detection, prediction and discovery. Monday. https://www.imsi.institute/videos/explainable-ai-xai-for-climate-science-detection-prediction-and-discovery/
108. He B, Zhao Y, Mao W (2022) Explainable artificial intelligence reveals environmental constraints in seagrass distribution. Ecol Indic 144:109523. ISSN 1470-160X. https://doi.org/10.1016/j.ecolind.2022.109523
109. Awadh MA, Mallick J (2024) A decision-making framework for landfill site selection in Saudi Arabia using explainable artificial intelligence and multi-criteria analysis. Environ Technol Innov 33. ISSN 2352-1864. https://doi.org/10.1016/j.eti.2023.103464
110. Newman SJ, Furbank RT (2021) Explainable machine learning models of major crop traits from satellite-monitored continent-wide field trial data. Nat Plants 7:1354–1363. https://doi.org/10.1038/s41477-021-01001-0
111. Oh JB (2024) The evolution of open source projects into successful companies. https://ohjungbin.medium.com/the-evolution-of-open-source-projects-into-successful-companies-04708f649c2c
112. Paul FC et al (2023) Deep reinforcement learning from human preferences. https://arxiv.org/pdf/1706.03741.pdf

# Chapter 6
# LLMs and XAI: Use Cases, Dependency and Challenges

**Vishnu S. Pendyala and Kargi Chauhan**

**Abstract** This chapter explores the transformative potential of Large Language Models (LLMs) across various critical sectors, including healthcare, finance, education, and autonomous systems. While LLMs offer unprecedented capabilities in data analysis, prediction, and automation, their increasing use in high-stakes domains brings forth critical concerns around accountability, fairness, and transparency. The necessity of Explainable Artificial Intelligence (XAI) is emphasized to address the "black box" nature of LLMs, ensuring that their decisions are understandable and trustworthy for professionals, regulators, and end-users. The chapter discusses the application of XAI techniques like SHAP, LIME, and counterfactual reasoning to enhance the interpretability of LLMs in these domains. It also examines the unique challenges and ethical considerations in each sector, advocating for ongoing research and collaboration to develop transparent, fair, and reliable AI systems.

## List of Abbreviations

| Acronym | Description |
|---|---|
| AI | Artificial Intelligence |
| GPT | Generative Pre-trained Transformer |
| LLM | Large Language Model |
| XAI | Explainable Artificial Intelligence |
| LIME | Local Interpretable Model-Agnostic Explanations |
| SHAP | Shapley Additive exPlanations |
| MRI | Magnetic Resonance Imaging |
| NLP | Natural Language Processing |
| EHRs | Electronic Health Records |

V. S. Pendyala (✉)
San Jose State University, One Washington Square, San Jose, CA, USA
e-mail: vishnu.pendyala@sjsu.edu

K. Chauhan
Sunnyvale, CA, USA



N. Bizon and B. Appasani (eds.), *Explainable Artificial Intelligence for Trustworthy Decisions in Smart Applications*, https://doi.org/10.1007/978-3-031-97007-8_6

| | |
|---|---|
| IoMT | Internet of Medical Things |
| GDPR | General Data Protection Regulation |
| ECOA | Economic Community Agreement |

## 6.1 Introduction

With the rapid innovations in artificial intelligence (AI), working, learning, and even healing have improved significantly. One such powerful innovation is the Large Language Model (LLM), in particular OpenAI's Generative Pre-trained Transformer (GPT-3 and GPT-4), which has been leveraged in a variety of industries across varied domains such as healthcare, finance, education, and autonomous systems. LLMs can analyze gigantic volumes of data and provide analyses some of which were previously only possible in science fiction.

They help doctors recognize complex patterns in a patient's history, assist financial analysis in detecting potential fraud, and provide bespoke learning provisions for students, among many other applications. However, increasing use of them in high-risk domains has raised questions of accountability, fairness, and transparency, emphasizing the significant and urgent demand for Explainable Artificial Intelligence (XAI). How do we ensure that the AI-driven systems make decisions fairly, and can explain how they arrived at a particular choice?

This strategy is broadly covered under a growing field of AI called Explainable AI (XAI), which aims to fill in the gaps in the interpretability and usability of AI decision mechanisms. Traditional AI models often function in a"black box" fashion, making the decision-making process behind such systems impossible to understand. Picture a doctor employing AI to recommend a treatment regimen, with no insight into the reasoning behind the suggestion. Would they also trust it and feel confident in doing so? One would not like to sit in a self-driving car if one cannot understand why the car took a U-turn or stopped in the middle of the road. Likewise, in finance, if an AI marks a transaction as suspicious, analysts need an explanation.

By providing transparency, XAI facilitates trust and also helps to minimize the risk of biased or erroneous decisions. Over time, researchers have introduced various novel approaches to improve the interpretability of LLMs' decision-making. These techniques not only respond to the critique of their opacity but also provide relevant insights regarding the behavior and functioning of such systems. In healthcare, for example, doctors can explore in more detail why an AI model suggested a certain diagnosis, enabling them to validate and trust the recommendation. In finance, analysts can investigate flagged transactions so the systems of fraud detection are accurate and fair. When this fact relates to education, interpretability allows us to make better choices, enabling us to validate and trust the recommendation. In finance, analysts can investigate flagged transactions so the systems of fraud detection are accurate and fair. When this fact relates to education, interpretability allows us to make certain that AI tools serve all requirements of various students reasonably, allowing inclusivity in different classroom environments.

Multiple explanation tools specific to LLMs have been created to aid with the accountability and transparency of an internal system. Similar to the few illustrations on healthcare and finance mentioned above, LLMs help with education to make it equitable and fruitful for different learners. Popular interpretability methods, like Shapley Additive Explanations (SHAP), Local Interpretable Model-agnostic Explanations (LIME), and counterfactual reasoning, are essential for developing trust and responsible AI applications. Counterfactual reasoning encourages fairness [1] and inclusiveness as it analyzes how small changes in the input can affect the outcomes. Incorporating such methods into LLM usage aids in aligning human values and legal frameworks, causing a ripple of change. There are methods among these that help the attribute model decisions to relevant input features, creating focused justifications that interpret AI predictions. Mechanistic interpretability is another promising path largely in financial services [2].

Mechanistic interpretability is about understanding the structure and function of a model's internal components, providing transparency into how the model works internally. This level of interpretability is especially important for improving model reliability, fairness, and trustworthiness in complex systems, and for identifying potential problems or biases within a model. Mechanistic interpretability is akin to reverse engineering the decisions made by a complex deep learning model such as an LLM. Understanding the behavior of an LLM at a mechanistic level will enable controlling the model at the neuron level. Individual neurons can then be programmed to alter the behavior of the LLM to better align with ethical principles, regulations, and established values. The initial work in mechanistic interpretability was more focused on computer vision tasks but is fast expanding into more application domains. Model interpretation techniques such as SHAP and LIME focus on attributing the decisions of the model to specific input features to provide localized explanations that can render AI predictions interpretable. For vision-based tasks, saliency maps are employed to visualize areas of an image that contribute to making a decision. Counterfactual reasoning identifies how small changes in input can lead to different outcomes and encourages equity and diversity. Coupling these models with techniques used when employing LLMs guarantees that all outcomes correspond with human values and laws, making a difference in the decision-making process and increasing robustness and ethics. But as LLMs become more deeply embedded in social and industrial functions, explainability crosses the threshold from technical desideratum to moral imperative. This is especially important as LLMs are susceptible to hallucination, or producing false information. Whether it is in life-critical healthcare, high-stakes financial transactions, personalized education, or autonomous transportation, explainability, and trust are the foundation of successful AI adoption and the reason XAI is now a necessity.

XAI explains the opaque processes involved in LLMs that map well onto the intersection of advanced AI research, which is at the core of what LLMs exemplarily showcase, with critical domains like healthcare. Techniques like LIME and SHAP are effective in explaining the decisions made by LLMs in detecting misinformation

[3] and fake images [4]. Recent work shows that explainability improves correctness too, by surfacing errors or bias, which leads to iterative improvements [5]. Using XAI, AI models can provide accurate, reliable, and ethical insights, especially for regulated fields [6] such as healthcare.

This chapter gives an overview of important application domains, such as healthcare, finance, and education, where LLM applications are developed and used, and describes the need and current implementation of explainability in these domains. XAI, particularly in the context of LLMs is an evolving area of research. The following discussion may not entirely be specific to LLMs but the ideas and scenarios presented may as well be applicable when the AI models used are in fact LLMs. For instance, so far, LIME and SHAP have been mostly used with AI models that are not LLMs. However, as indicated above, these XAI techniques have recently been used with LLMs as well [3]. The chapter also discusses the challenges encountered in achieving explainability and suggests prospective pathways to incorporate XAI into critical domains.

## 6.2 Healthcare

LLMs like GPT-3 and GPT-4 are going to play a big role in healthcare. These models have changed medical practices by helping in diagnosing diseases, predicting outcomes of patients, as well as effective communication between health professionals and patients. As healthcare professionals are trying to provide optimal support, it becomes important to have an interpretable and explainable AI system. In sensitive domains such as mental health, trust in AI recommendations and diagnoses relies on clinicians and patients grasping their rationale [7].

This is where XAI comes into play, a trend that sits at the junction of cutting-edge AI research and medical practice, a methodology developed to help make sense of the processes that underpin LLM-generated insights that is akin to reading from a black box. Explainability is not only a technical challenge. It is an ethical and moral imperative. In high-stakes situations such as healthcare, where human lives and well-being are at stake, the transparency of AI decisions influences both the trust that clinicians place in these systems as well as the trust that patients place in their care. While LLMs for clinical applications must ensure regulatory compliance, emerging research higights the need for explainability in improving performance and reliability.

It is the continued pursuit of improved patient outcomes and delivery of care that is pushing the boundaries of innovations in LLMs. Excellent in processing multi-modal data, these models are trained on large datasets of text, images, and code and can be applied in a variety of healthcare domains [8]. LLMs can ingest millions of pages of medical literature, clinical notes, and research findings, enabling physicians to find differential diagnoses and clinical outcomes. LLMs drive more precise and informed decision-making to enhance patient care by revealing hidden patterns in complex medical datasets. Outside of diagnostics, LLMs have incredible potential to improve patient communication. By creating custom patient education resources aligned with specific needs and literacy levels, they can help patients participate actively in their healthcare pathways [8]. LLMs help patients better understand their diagnoses and treatment options by translating clinical jargon into everyday English. Breaking down these barriers leads to better communication and trust between healthcare professionals and patients, and as a result, greater efficiency in the delivery of care.

Although the promise of LLMs in healthcare is compelling, there are challenges to overcome. Transparency is a key requirement, as clinicians and patients must trust the reasoning behind LLM-generated suggestions. Equally important, LLM training data can perpetuate existing bias, resulting in unequal or even false outcomes. These biases can be mitigated through careful selection of the data and rigorous model construction. Privacy and security issues are major concerns as well; LLMs may need access to sensitive patient data in massive volumes. Ensure maximum security: protect this data from unauthorized access to ensure patient privacy is paramount [9].

Despite these challenges, LLMs offer promising opportunities in healthcare. In particular, their use in medical image analysis may lead to earlier and more accurate diagnoses, and enhanced patient care [8]. With LLMs continuing to evolve, the importance of maintaining a clear ethical framework so that we can ensure transparency, accountability, and fairness should not be lost. To realize this vision, collaboration between AI researchers, healthcare professionals, and policymakers will be essential. LLMs could fundamentally transform the future of medicine, moving AI from being a "black box" to a trusted partner at the bedside. This can help build trust in AI-powered systems and support their integration into clinical practice, ultimately improving patient care. This step will allow it to connect the world of technological innovation to patient care. This step will allow it to connect the world of technological innovation with the world of human trust, where LLMs can be an integral part of the world of personalized, efficient, and equitable patient care [10].

### *6.2.1 Challenges in Applying LLMs to Healthcare*

LLMs provide a significant preview of the future of the healthcare revolution. However, their greatness will only be achieved if we overcome substantive challenges of implementation, trust, and ethics. AI systems introduce their own"pitfalls" that require careful consideration [11]. Some of the main challenges with the usage of LLMs in the healthcare industry are:

- **The Interpretability Enigma**: The potential mechanisms of LLMs could be as baffling as human intuition. Such a"black box" nature presents a challenge in healthcare, where transparency and trust are key. Clinicians need to understand the reason for LLMs advice to integrate it into their decision-making processes successfully. This highlights the demand for an XAI solution capable of rendering the reasoning behind the LLM outputs understandable, thereby developing trust within healthcare professionals engaged in responsible artificial intelligence application [9].

- **Bias in Training Data**: LLMs are trained on large text and code datasets. Yet, these datasets often contain the biases of society from which they are taken, and which can result in the discrimination of AI outputs. For instance, if an LLM is biased due to its training data being skewed toward younger patients, it may not be able to effectively identify diseases in older adults. Graph-augmented LLMs have been designed to help ameliorate biases of the model and offer more tailored health information feed by integrating heterogeneous sources of information [12].

This approach also increases interpretability by relating outcomes to major drivers [13].

- **The Data Security Tightrope**: LLMs often need access to huge volumes of patient private data. Ensuring data security and maintaining patient privacy are crucial challenges. Strong security protocols, as well as ethical data governance standards, need to ensure against the abuse and misuse of this data [14].
- **The Human Element Dilemma**: LLMs are intended as tools to supplement and not supplant existing expertise. The trick is to extract optimal efficiency out of LLMs without taking away the empathy, intuition, and contextual knowledge that healthcare workers possess [15]. LLMs are meant to complement decisionmaking, not to replace human judgment.
- **The Regulatory Minefield**: LLMs are a newer technology in healthcare and regulatory frameworks are still emerging. There is a lack of specific criteria to enable safe and effective clinical implementation of these tools. It is only with collaboration among developers, medical professionals, and policymakers that we can earmark regulations that ultimately protect the patient while promoting innovation.

These challenges are substantial, but not insurmountable. This means they have the power to transform the healthcare landscape to provide better patient outcomes along with an improved system of healthcare delivery through adherence to transparency, ethical principles, and ongoing enhancement of all AI-based systems.

### *6.2.2 Explainability Techniques in Healthcare*

Due to the high-stakes nature of medical decision-making, where a misinterpretation or a miscalibration of a model may have life-altering consequences, explainability remains a critical property of AI systems for transparency, trustworthiness, and ethical compliance. As deep learning models (like LLMs) also become more central to personalized medicine, therapy planning, and diagnostics, it will become important for us to reduce the difference between machine outputs and human.

#### 6.2.2.1 Importance of Explainability in Healthcare

The main use case of explainability in healthcare is to build trust for the end-users, such as patients or physicians. Medical practitioners need explainability Medical practitioners must come to grips with how AI models make decisions before we can expect them to blindly trust diagnostic and treatment recommendations. In contrast to other domains, healthcare decisions are required to be described and explained, given that clinicians are personally accountable for patient outcomes. Explainability allows AI to be consistent with medical knowledge, medical standards, and clinical processes, enabling ethical AI in medical care [16].

Because of their "black box" nature, LLMs and deep learning models also have their challenges. While these models are excellent at analyzing complex data sets such as patient histories and medical imaging, their internal workings are often opaque. To solve such a challenge, explainability approaches like SHAP and LIME give post hoc information about model predictions. For example, SHAP and LIME can show how individual input features, like lab results or MRI scans, affect AI-driven diagnoses, so clinicians can check and have confidence in the recommendations [17].

#### 6.2.2.2 Techniques for Model Explainability in Healthcare

Several techniques can improve the explainability of AI systems in healthcare:

- **Feature Significance Analysis**:This allows discovering which variables influence a model the most in terms of predictions. The influence of age, blood pressure, and cholesterol levels, for example, can be determined using XAI techniques [10]. This allows doctors to check if the model's reasoning conforms to known medical facts.
- **Visualization Tools**: Methods like saliency maps identify portions of diagnostic images that strongly impact model outputs. This gives radiologists a visual explanation with which to compare the model's predictions versus their expertise. Tools such as VBridge [18] build on this idea by connecting model decisions to patient history or image characteristics, thus simplifying model outputs into more understandable features.
- **Attention Mechanisms**: In NLP models, attention layers tell the model which parts of a text to pay attention to when making a prediction. Domain knowledge is needed to provide this level of explanation to predicate true scientific meaning on the explanation, subsuming a bigger scientific part—this is very needed in domains like oncology, where treatment-related decisions very much hinge on nuanced clinical histories and test results [16].

By providing interpretable and explainable insights, these techniques contribute to the establishment of trust in AI systems and their adherence to ethical and legal frameworks inherent in healthcare. This transparency and interpretability can help healthcare professionals incorporate AI-based solutions into their practice with greater confidence, ultimately enhancing patient care and outcomes.

#### 6.2.2.3 Trust and Ethical Considerations

Trust underlies AI acceptance in healthcare. In contrast to other industries, where mistakes can cost money or time, errors in healthcare AI can have dire consequences: misdiagnoses, wrong treatments, or even death. As such, explaining strategies is paramount to maintaining ethics and transparency. Such audits allow medical practitioners and healthcare providers to question or confirm recommendations made by AI, which ultimately leads to transparency, justice, accountability, and fairness [17].

User-centric explainability in healthcare is about finding an explanation that makes sense from the perspective of the medical professional. It brings everything back to a direct, clear, and relevant understanding, rather than an explanation that follows all the logical and mathematical rigor of the algorithm. Doctors are overwhelmed with data. Based on the data, AI models deliver simple information that fits in with what they already know. For instance, decision trees and case-based reasoning techniques provide interpretable models highlighting how certain inputs, like patient symptoms or medical history, drive certain outcomes. These approaches guarantee that AI modes collaborate with human decision-making frameworks, advancing clinical trust and applying them in practice.

Emergent explainability methods can be particularly useful in the expanding domain of the Internet of Medical Things (IoMT). AI-based IoMT systems handle plenty of big data generated in real-time by medical devices, wearables, and other health monitoring tools, ensuring continuous monitoring and predicting patients' needs. As these systems become more complex, explainability can no longer simply be a technical consideration. It has to include how usable or comprehensible that information is for different stakeholders like clinicians. In many cases, this means providing or prescriptively recommending actionable recommendations that improve patient care, which are potentially best suited for human understanding. Recent work has focused on the development of dynamic interfaces and contextualized decision-support systems to aid stakeholders in interpreting AI-generated outputs and incorporating them into solutions.

In the future, explainability must shift from simply making an AI system transparent to ensuring it is usable and accessible to a wider set of stakeholders. Explainability techniques that provide trust, accountability, and collaboration between AI systems and medical practitioners and patients and vice versa should play a crucial role in adopting ethical and effective AI in medical practices.

### *6.2.3 Case Study: Explainability in Patient Communication*

Visualization and explainability are both key—this goes beyond the fact a clinician may know the outcome an AI model can reach. Generalized AI technologies must be interpretable and trustworthy, not just for clinicians, but also for patients, who are often non-technical experts, to be fully integrated into healthcare. This case study illustrates how the use of explainability strategies led to improved trust and transparency in the patient-clinician dialogue.

This application of explainability is most clear in AI which is used for the diagnosis of cancer. Models, such as those that evaluate MRI, and other medical imaging data, classify tumors as benign or malignant. However, the "black box" aspect of these models presents challenges to oncologists, who must make sense of how these predictions are made and what factors are relevant, to ensure their consistency with clinical knowledge. The use of SHAP in oncology [17] shows how this can be done. For detecting tumors, for example, SHAP values capture MRI features such as the size, shape, and contrast of tumors, and indicate how the model's final prediction was impacted by those features. It allows oncologists to"ground" AI outputs with corresponding clinical expertise, which establishes trust and confidence.

Explainability is just as important for chronic conditions such as diabetes. AI models evaluate patient-specific parameters, medication regimens, lifestyle habits, and blood glucose levels to suggest individualized therapies. LIME is one such technique. It allows doctors to understand the influences that create treatment plans, such as daily exercise, insulin dosage, and so on. This increased transparency allows clinicians to modify AI recommendations and convey them reliably to patients, thereby increasing trust and adherence.

Counterfactual explanations are another innovative method that enhances AI transparency. These scenarios make it possible for clinicians to sense what might change with relatively small shifts in patient data, which ultimately leads to different treatment recommendations. So, for example, in cases where AI indicates that medication is appropriate for heart disease instead of surgery, a counterfactual explanation could tell us if cholesterol levels a little higher than measured would lead to the recommendation of surgery. This enables clinicians to make better decisions and to be able to communicate the reasoning to the patient.

Other recent developments are dynamic visualization tools that allow for real-time interactivity with AI-generated insights. It enables patients and clinicians to explore the decision-making process together, which, in turn, improves trust and communication [18]. These techniques of explainability offer insight into complex decision-making processes in a way that is human-readable, ensuring that AI systems can play an effective role in personalized and transparent patient care.

### *6.2.4 Challenges and Future Directions*

LLMs have great potential to revolutionize healthcare, especially in the areas of diagnosis, treatment recommendations, and patient communication. LLMs are capable of digesting a lot of medical academic literature, clinical notes, and patient data quickly and efficiently to suggest the most personalized treatments. For such recommendations to be trusted, however, decisions driven by AI need to be explainable. Using explainability methods such as SHAP and LIME, health-centered LLMs can offer insights into how individual data points contribute to diagnoses or treatment recommendations. These techniques, for instance, enable the explanation of why some lab results are given precedence over others or how a diagnosis has been made, helping close the loop between the findings derived from AI and how the clinical decisions would be made.

While the application of LLMs in healthcare shows great promise, several challenges remain:

- **Interpretability and Explainability**: The"black-box" nature of LLMs is a significant obstacle, as it prevents healthcare professionals from understanding how decisions are made. Lack of transparency leads to distrust and poses ethical concerns, especially in high-stakes environments [19].
- **Bias in Model Predictions**: The presence of social, financial, medical, or any sort of bias in the training dataset may result in biased or discriminatory prediction outcomes. For instance, LLMs trained on populations of young patients can be inaccurate in diagnosing older adults. Proper data selection must go through a rigorous process to lock in bias, and models must be regularly validated.
- **Data Privacy and Security**: Data privacy and security challenges are crucial in training to protect patient data. Access to large volumes of sensitive data increases the risk of privacy breaches. Adherence to strict security protocols and privacy regulations is crucial to protect patient data.
- **Regulatory Challenges**:The existing regulatory frameworks are insufficient in navigating the complexities brought about by using LLMs in clinical applications. These AI systems require updated frameworks that communicate similar standards of safety and update innovative processes in healthcare.
- **Ethical Considerations**: Striking a balance between AI-based decision-making and human compassion is an ongoing dilemma. LLMs are valuable tools for supporting decision-making; however, they should augment rather than supplant clinical expertise in healthcare contexts [20].

Future research and development efforts should address these challenges to ensure the safe and effective integration of LLMs into healthcare:

- **Improved Explainability**: New Explainable AI (XAI) techniques like SHAP and LIME need to be leveraged to make LLM outputs interpretable. Dynamic visualizations and contextual explanations are more fuel to improve trust and usability.
- **Lower Bias**: Continuous monitoring and inclusion of a diverse set of datasets can drive down bias in LLM outputs. Graph-augmented LLMs that combine multimodal information types generate more personalized and equitable health insights while boosting model interpretability.
- **Enhanced Security**: Studies dedicated to safety frameworks for LLMs training and deploying. Methods like encryption data storing, dissimilarity acknowledgment, and following terms will guarantee data sustainability while enhancing innovation [14].

**Table 6.1** LLM use cases in healthcare, explainability methods, and challenges

| Use case | Description | Explainability methods | Challenges addressed |
|---|---|---|---|
| Diagnosis support | Analyzes medical assist in records to diagnoses | SHAP, LIME | Bias, lack of inter-pretability |
| Patient communication | Translates medical jargon for patients | NLP-based explanations, attention | Clarity without oversimplification |
| Medical imaging | Interprets X-rays or MRIs | Saliency maps, SHAP | Complex image features |
| Trial design | Matches patients to trials | SHAP, attribution | Bias, regulatory concerns |
| Treatment reco | Gives personalized suggestions | LIME, counter-factuals | AI versus clinician expertise |
| Health records | Analyzes EHR patterns | Attention, importance | Privacy, sensitive data |

- **Human-AI Collaboration**: Future AI systems should enhance human decision-making instead of replacing it. LLMs trained to be integrated alongside clinicians can drive expertise, outcomes, and patient trust.

Addressing these challenges and focusing on transparency, security, and collaboration, LLMs have great potential to not only transform healthcare delivery but also to do so in an ethical manner. This will ensure improved outcomes for patients along with a more equitable healthcare system. Table 6.1 summarizes the XAI techniques used in some of the healthcare-related use cases.

## 6.3 Explainability in Finance

The financial services industry has quickly embraced AI and ML models to improve processes such as fraud detection and credit scoring. Though these models increase accuracy and efficiency, their complexity often prevents users and regulators from understanding how they make decisions. This lack of transparency has amplified the need for explainability in the financial domain to guarantee that AI models are transparent, fair, and in line with legal needs.

In domains like fraud detection and credit scoring, where decisions can have farreaching consequences, explainability becomes a critical feature. Fraud detection systems need to offer clear reasons for flagging suspicious transactions, and lenders have to appreciate why a credit score was assigned to them or a loan application was approved or denied. It enhances trust, limits biases, and ensures regulatory compliance, thereby enriching a transparent financial system.

### 6.3.1 Explainability in Credit Scoring

One very common use of AI, especially in the finance industry is credit scoring which determines whether a loan is disbursed, whether further interest is charged, etc. Classic credit scoring models use indicators such as payment history and credit utilization, while modern machine learning scoring models have expanded to accommodate a wider variety of data, which can pose challenges to both predictive accuracy and interpretability.

Transparency of credit scoring is crucial to guarantee equity in financing choices. Regulatory requirements generally require lenders to give reasons for adverse credit decisions. XAI enables regulatory compliance by explaining how a consumer's credit score is determined and what decision-driving features they are comprised of. Model behavior explanations based on SHAP, for example, provide feature attribution that demonstrates how factors like credit card balances and payment history contribute to an individual's credit score. This serves a dual purpose of not only enabling lenders to justify their decisions but also assisting consumers in understanding how to enhance their creditworthiness [21].

Feature attribution techniques elucidate the importance of the different criteria in credit scoring. For example, SHAP values can help understand why a good-scoring customer with balanced finances has a higher score than one with a high debt-to-income ratio. Counterfactual explanations also serve an important purpose by providing"what-if" scenarios. These scenarios can showcase how small shifts in behavior, such as paying credit card bills on time, could raise a credit score. A counterfactual explanation, for instance, might show that a bank customer's credit score would have been boosted by 20 points had they not missed a payment three months ago.

Newly developed tools like FinGPT [22] and other open-source financial large language models (LLMs) show how AI can rapidly improve the transparency and personalization of credit scores. FinGPT can assess consumer habits and preferences so that credit suggestions are personalized, promoting trust and understanding between lenders and consumers. Even with these advancements, obstacles have yet to be overcome. Interpreting complex machine learning models becomes more challenging with larger and more complex datasets. Techniques such as LIME create model-agnostic frameworks that enable financial institutions to comply with regulations while achieving higher levels of accuracy.

### 6.3.2 Explainability in Fraud Detection

While AI models are integral to fraud detection in the financial sector, the balance between accuracy and transparency in developing the system remains a challenge. Anomaly detection, a common fraud detection technique, identifies unusual patterns in transactions that may indicate fraud. But in real-time systems, it is important to identify anomalous transactions and also to deliver clear explanations of the reasons why certain transactions are flagged.

Explainability in fraud detection helps financial institutions improve consumer trust, reduce false positives, and enhance the accuracy of their models. False positives can disrupt customers and hinder legitimate transactions, leading to dissatisfaction and financial hardship. By explaining why certain actions are flagged as abnormal, institutions can refine their detection strategies and improve customer satisfaction [21].

In fraud detection, techniques such as LIME and SHAP are commonly used to promote explainability. For example, SHAP values can show which features of the transaction (size, location, frequency) led to a fraud alert. LLMs trained with episode embeddings have already been successfully applied to financial time-series data to spot suspicious activity in client accounts and to explain differences in transaction patterns, thereby aiding analysts and consumers who want to understand a suspended transaction. Visualization methods (e.g., graphs that show transaction groups or trends per period) can also help fraud analysts and customers interpret reported activities.

FinGPT and other advanced tools further improve fraud detection by sifting through enormous datasets of historical transactions and identifying patterns of behavior that may signal when fraud has occurred. For example, an LLM may notice an aberrant spike in high-value transactions in a normally inactive account and offer a granular explanation for its suspicion. Furthermore, LLMs are useful for real-time graphics of sentiments over financial markets and can detect potential fraud or market manipulation by scanning the social media content from Reddit and so on.

Despite this progress, ethical concerns remain. A balance between transparency and security is key. Overly transparent fraud detection models expose what you might call the matrix of vulnerabilities to bad actors, while overly opaque systems can become trust-absorbing. But it raises ethical issues too—for instance when legitimate transactions are wrongly flagged, leading to needless financial stress for customers.

### *6.3.3 Regulatory Requirements and Ethical Implications*

Explainability is important for both fraud detection and credit rating to be compliant with regulations. Financial institutions must comply with the General Data Protection Regulation (GDPR), which requires that automated decisions be transparent to consumers in the European Union (EU) and the United Kingdom (UK), including decisions produced by AI models. Explanations also ensure that AI models do not discriminate against certain demographics unfairly, which helps the institutions comply with anti-discrimination laws.

### 6.3.4 *Future Trends in XAI for Finance*

As AI matures, explainability will have growing significance in financial services. Tools like FinGPT and other LLM models are already being used to make financial advice more transparent and accessible. Future LLMs could be used to build fully interactive XAI systems where users can ask questions and get direct, clear, and comprehensible explanations of why certain financial decisions should be made. However, new visualization paradigms are being introduced that allow users to interact with XAI models through graphical or dashboard interfaces. These advancements will not only boost user confidence in AI-enabled financial systems but also allow them to make informed financial decisions.

In finance, explainability is paramount to ensure that AI models are fair, transparent, and compliant. This has enabled LLMs, such as FinGPT, and XAI methods, such as SHAP and LIME, to help financial institutions win their clients' trust in credit scoring and fraud detection. Future applications of AI will uncover different methods and models to augment the sustainability and transparency of financial institutions to create a just and equitable financial system.

### 6.3.5 *Market Sentiment Analysis and Financial Decision Making*

Additionally, LLMs are increasingly being used to analyze market sentiment [23], a critical component of financial decision-making. By scouring sites like Reddit, LLMs can get a sense of public sentiment regarding specific stocks or market trends. Trained on a time snapshot of data, one will be able to predict the market and tell when to invest and what not to invest. LLMs, for instance, can find weak financial sentiment labels from Reddit posts to examine the underlying sentiment surrounding specific stocks or overall economic conditions. These models often outperform traditional supervised learning methods when used with semisupervised learning methods, making them a strong option for financial analysts [24].

However, using LLMs for sentiment analysis gives rise to ethical concerns, notably in data privacy and the prospect of automated judgment influencing financial markets. Without explainability, there is no cure for these problems. Financial institutions need to back up their recommendations with evidence, e.g., if they are requesting an automated trading system to make a decision based on market sentiment. Yes, the processes of explainability can help in disentangling the impact of individual aspects, be it social media sentiments or company operating performance metrics, enabling accountability and foresight in those decisions.

### 6.3.6 *Ethical Considerations and Regulatory Compliance*

The growing adoption of LLMs within the finance sector is generating some major ethical and legal questions. This poses privacy and bias risks further compounded by LLMs using vast amounts of consumer data, from financial histories to public opinion. For financial institutions, there is also a need to ensure these models comply with regulatory provisions like the EU's General Data Protection Regulation (GDPR) and the US Economic Community Agreement (ECOA), which require transparency and fairness in automated decision-making.

Explainability is key to alleviating these concerns. Providing clear, comprehensive, and factual explanations for AI-driven financial decisions within an institution can ensure that the institution's financial decisions made using AI comply with ethical or legal standards. For instance, explainability approaches can be leveraged to surface and correct biases in a credit scoring model that disproportionately denies loans to individuals in certain demographic groups [25].

In addition, financial firms that are explainable can gain their customer's trust. Customers demand greater transparency as AI-driven decisions have proliferated in the financial sector. By providing insight into the decisions behind their models, regardless of the application, firms can build client trust and avoid the guilt by association that can arise from murky decision-making processes, making the implications for financial institutions potentially far-reaching.

For example, we have been seeing major advancements in areas such as market sentiment research, where LLMs have been applied to analyze trends and forecast prices or trends, which ultimately helps in making investment decisions. However, explication of the ethical aspects of data use is important for transparency, equality, and counterfactual. Table 6.2 illustrates the use of XAI in the finance domain.

We are already witnessing some impressive applications. LLMs have begun to find their way into the financial world as businesses leverage them not only for fraud detection but also for risk assessment, credit scoring, and algorithmic trading. Institutions will be able to make more precise predictions. Decision-making will be supported by increasingly autonomous systems and LLMs trained on complex datasets. Nevertheless, the financial industry is highly regulated and therefore requires that AI systems produce transparent and explainable decisions to comply with legal standards and uphold public trust. For instance, for fraud detection, LLMs learn transaction pattern identifications and deviations; even then, they need to come up with the proper kind of explanations on why a transaction was flagged off. Likewise, if an LLM is employed in any credit score, then the reason for which the LLM determines, weighs, and combines different aspects like income, credit history, and spending habits cannot be arbitrary, just so that the derived score is not biased or discriminatory.

Various explainability tools such as feature attribution and LIME can assist financial institutions in exhibiting AI-powered decisions in a more clear-cut way that improves both regulatory adherence and customer faith in automated procedures. If in the past, the historical financial data reflects the echoes of societal inequalities and biases, then the biggest challenge in finance is when real-time decisions need to

**Table 6.2** LLM use cases in finance, explainability methods, and challenges

| Use case | Description | Explainability methods | Challenges addressed |
|---|---|---|---|
| Fraud detection | Detects transaction anomalies for fraud detection | SHAP, feature importance | Handling imbalanced data, minimizing false positives |
| Risk assessment | Assesses financial risk by analyzing market and company data | LIME, counterfactual explanations | Balancing interpretability with accuracy |
| Credit scoring | Analyzes financial history to calculate credit scores | Feature attribution, SHAP | Bias in data, fairness in scoring |
| Algorithmic trading | Makes trading decisions by analyzing market trends and news | Attention mechanisms, feature importance | Real-time decision-making versus interpretability |
| Customer service automation | Automates customer queries related to accounts, transactions, etc | NLP-based explanations, attention mechanisms | Understanding queries with clarity and accuracy |
| Regulatory compliance | Helps companies adhere to financial regulations | SHAP, legal XAI techniques | Compliance with regulations, avoiding overfitting |

be made in fast-paced financial markets with interpretability preserved. As financial systems rely more on AI, progressing in these areas means ongoing work toward minimizing bias and speeding up and clarifying the decisions made by LLMs. The future of LLMs in finance lies in making systems that will be efficient and yet moral, and understandable, and affirm monetary security and fairness for all their users.

## 6.4 Explainability in Education

Educational technology is another area where LLMs are making a substantial impact, with their potential to provide personalized learning experiences and automate grading systems, among other applications [26]. In this environment, XAI is a necessary measure to ensure accountability and transparency when LLMs are used for decision-making. This analysis focuses on the applications, benefits, and challenges in an LLM-supported classroom setting.

The increasing complexity of education data [5] necessitates advanced analytical techniques such as LLMs. LLMs can process vast quantities of data and may also identify trends and provide insights to promote better learning outcomes. XAI gives explainability of LLM-driven decisions, ensuring fairness and accountability that enables instructors to decipher the outcomes. The automation of grading, made possible by LLMs, can help to create fairness in education by increasing transparency, explainability, and the availability of knowledge [27]. LLMs can evaluate the accuracy, content omission, and risk of harm in responses to medical education topics [25]. An LLM-driven system, for example, can give personalized recommendations to a student based on their learning methods. Such a strategy has been used in education where LLMs can support personalized learning pathways [26].

Value judgments in computerized assessments refer to the implicit or explicit decisions made about what is important, relevant, or appropriate in the evaluation process. These judgments are often encoded within AI models that drive the assessment system. For instance, an AI-based educational tool might prioritize certain types of student behaviors or responses over others, reflecting underlying assumptions about what constitutes successful learning or desirable skills. Agency refers to the capacity of individuals or entities to act independently and make choices. In the context of computerized assessments, agency relates to how much control is given to the system versus the users such as students, job applicants, or patients. Automated systems can either augment human agency by providing insightful recommendations or constrain it by making decisions without sufficient human oversight or understanding. In both Value Judgments and Agency Computerized Assessment, current academic literature corroborates the importance of explainability in each case.

### *6.4.1 Personalized Learning*

LLMs have been discovered to be extremely good at delivering personalized learning experiences by tailoring learning content for every student. Personalized learning refers to customizing lessons, tasks, and evaluations to the strengths and weaknesses of each student to ensure fairness and effectiveness in the educational process utilized by these models.

Personalized learning models and customized learning path models, need to be amended according to the specific needs, learning preferences, and talents of every student. Educators should also tailor how LLMs generate educational pathways for students to ensure equity. Studies [5] highlight the importance of explainability to ensure learning recommendations are based on educational value, not inadvertent biases in model training datasets.

A systematic study analyzing learning liaison professionals using LLM in the context of medical education found LLMs to be able to change course materials during a lesson based on success rates of students that day [5]. Such uses also point out that without insight into the underlying algorithms, students and educators may not understand how certain recommendations are made, potentially leading to a lack of trust in the system. In other words, the application of XAI has the potential to bridge this gap by providing educators with a general understanding of what is driving various aspects of the decision-making process about learning adaptions.

The impact of LLMs on education as well as an explanation of how the adoption of LLMs is gradually gaining popularity and how personalized education provides information to parents. It can analyze the student's learning history and suggest recommendations based on the strength of the read of a student. LLMs can model the assessment of student understanding and provide recommendations. To ensure that the recommended paths to learn and evaluate young learners are consistent with developmental and cognitive standards, the AI model behavior needs to be [26] endorsed using XAI tools.

In this context, it is important to document the pains of personalized learning. There are still many companies and universities that use LLMs in personalized study experiences. One of the biggest hurdles is ensuring that models can adapt without biasing against already-held prejudices or disadvantageous groups of students. One of the researchers states [28] LLMs may inadvertently help the already better programming students when computing science students are aided by these tools. This can make the feedback and recommendations of these models opaque, particularly for less skilled students who may not understand why their learning trajectory is different from that of their peers.

Saliency maps and attention visualization are types of XAI methods that can aid teachers in understanding which parts of the student input the model is leveraging to identify how to personalize the curriculum. This awareness is crucial to minimize prejudice and ensure that all students receive equal opportunities to benefit from tailored teaching. Make sense of a grading mechanism—by automating the grading process, an LLM-based grading system can relieve educators of the burden of grading. The process of grading must be made clear to both the teacher and the student. Systems such as these with explainability-enabled automated grading [29] provide students with detailed feedback about the marking process, allowing them to understand what they made mistakes on and how to improve on it. Without this kind of openness, it may be difficult for students to trust the reliability of assessments that are machine-generated.

### 6.4.2 Automated Grading

Another important application of LLMs in education concerns those that deal with grading students' work and leaving comments. Such computerized grading systems should, however, be transparent to instructors and students so that marks can reflect true and fair assessments of students' performance.

By automating grading, LLM-based intelligent grading systems lighten some workload for educators. Teachers and students have a right to know the process behind the determination of grades. Explainability of Automated Grading Systems is key to increased adaptability of the systems. Explainability gives feedback to students in the marking process, and they should thus understand and learn from their mistakes. Without such transparency, it would be hard for students to believe the accuracy of the assessments that automatically pop out of machines. A recent study validated the multiple choice examinations generated by LLMs for preclinical medical students. Students in high-stakes testing environments accept only automated grading systems based on the level of trust [30].

Accuracy and transparency of grading assessments using LLM-powered grading tools need to be just, unbiased, and consistent. It was meant to be an attention drawer to the possibility that incomplete or inaccurate information might be provided by LLM, especially in medical education, when used in educational assessments. Indeed, in their analysis of LLM responses to questions related to dermatological education, some of those responses contained substantive errors or omissions that would greatly impact student learning [31]. To establish the validity and reliability of automatic grading systems, it is necessary to apply XAI technologies to evaluate them.

Limitations and Ethical Issues: While automated grading works well, the latter gives rise to ethical concerns relating to students' information privacy and risks of reinforcing humanities' institutional biases. As noted in a study [27], there is growing concern that formatting and writing style unrelated factors will influence more than grading algorithms than the quality of the content. It is vital to integrate XAI methods that will allow us to audit the decision-making flow of the grading classifier and allow instructors to review and modify the model heuristics to prevent such biases. It should be emphasized that LLM-based automated grading systems need to allow teachers and students to expose the logic behind such reasoning on what has brought about a particular grade; this therefore calls for objectivity of the grading process and a deeper analysis of the assessment criteria. Incorporation of XAI into the grading system can avoid the bias and errors that might otherwise go unnoticed while building trust.

The use of LLMs is bound to revolutionize education, from making educational experiences personalized to offering efficient ways of assessment, whether in automated grading or personalized learning.

However, explainability and transparency are going to be the key determinants for the success of these technologies. Ensuring that LLMs are fair, interpretable, and useful to both instructors and students will be quite necessary with XAI technologies. In the long term, the integration of explainability frameworks will be crucial to addressing fairness, bias, and trust problems as long as LLMs remain the leading edge in educational technologies. Further research is needed regarding the role of XAI in educational settings to maximize the benefits of LLMs while mitigating any potential risks. Table 6.3 lists a few use cases in the education domain and the XAI techniques applicable to them.

The efficacy of LLMs in education has been realized in personal learning, grading of students, student feedback, and tutoring systems. LLM analyzes the performance of each student to curate an educational experience that meets their needs and aspirations by providing them precisely with what they need. However, for educators and students to be able to trust these AI-driven systems, the decision-making process needs to be understandable. For example, in grading and assessment, LLMs should not just provide a score but also explain why they have given a particular score, such as how the model has appraised certain aspects of an essay or examination. Similarly, for personalization of learning, students need to understand why particular learning pathways are recommended for them, based on performance data and learning habits. LLMs ensure that decisions taken in education are transparent and that educators and learners understand how these AI tools make a difference. This is achieved by the

**Table 6.3** LLM use cases in education, explainability methods, and challenges

| Use case | Description | Explainability methods | Challenges addressed |
|---|---|---|---|
| Personalized learning | Customized learning via student data | LIME, SHAP | Fairness in suggestions, avoiding bias |
| Grading | Automates essay scoring | Feature attribution, LIME | Fairness in grading |
| Student feed-back | Generates feedback on assignments | NLP explanations, attention | Providing accurate feedback |
| Curriculum dev | Identifies gaps in course content | SHAP, content analysis | Aligning to standards |
| Tutoring systems | Acts as virtual tutor | Attention mechanisms | Engagement, adaptability |
| Performance prediction | Predicts academic performance | SHAP, feature importance | Accuracy versus fairness |

incorporation of explainability techniques like attention mechanisms. In models that use attention, each token in the input sequence is assigned an attention score relative to other tokens. This score determines how much focus the model gives to each token when generating an output. These attention scores can be visualized as heatmaps, where more attention is paid to certain words or phrases. For instance, in a machine translation task, if the model is translating a sentence, the attention mechanism shows which words in the source language are most influential in producing a particular word in the target language.

Fairness within AI-driven assessments, avoidance of biases potentially affecting specific groups of students, and assurance of student data privacy all pose challenges in using LLMs in education. The more interrelated education becomes with AI, the more any future work will need to be oriented toward the development of systems that guarantee transparency, equity, and ethics in data use. Addressing such challenges would imply that LLMs have the potential to revolutionize education with personalized data-driven support, using improved outcomes fairly and with trust. The more interrelated education becomes with AI, the more any future work will need to be oriented toward the development of systems that guarantee transparency, equity, and ethics in data use. Addressing such challenges would imply that LLMs have the potential to revolutionize education with personalized data-driven support, using improved outcomes fairly and with trust.

## 6.5 Explainability in Autonomous Systems

The most critical aspects of self-driving cars are also overly complex. Decisions related to lane changes, accelerations, and braking mandate transparency that is needed for public safety and regulatory acceptance. Integrating LLMs into the self-driving system opens up a grand insight into how the derivation of decisions is handled.

LLMs can significantly contribute to explainability, such as providing insights into decisions like why the car would choose to change lanes or why it would apply brakes at a certain time. Those are not simple predictions but show the rationale behind actions that make it easier for regulators and users to trust the system. On the other hand, an LLM-based lane-change model is also capable of predicting vehicle trajectory, with crystal-clear explanations as to such a decision; perhaps because the flow of traffic is around or the speed of other vehicles is circulating. This added transparency helps a great deal in reassuring regulatory bodies that self-driving cars will be able to meet strict safety standards and reduce public skepticism by making more understandable the process of how decisions are reached.

Another promising application in the domain of autonomous driving involves the integration of LLM with imitation learning. Imitation learning allows for the training of self-driving systems from natural driving on roads. Teaming this with LLMs, the approach enables the system to predict subsequent activities based on previous experiences. Above all, it improves the explainability of the decisions during those scenarios made by the LLM. For example, LLM can explain why the scenario was identified based on the patterns found in this case, the car was forced to make an emergency stop. Only then does the decision become more interpretable by the passengers and regulatory bodies. Besides, LLMs can process multiple modalities of tokens simultaneously for better decisions due to integrated data analysis. These decisions could then be explained, with the use of LLMs, briefly and succinctly such that the end understood why the system acted in such a way.

One of the notable examples of such a developed fusion model can be considered in self-driving systems where multiple sensors combined into one object-level vector modalities form a comprehensive view of the driving environment. This model not only foretells the results but also provides a rationale for each decision. Aggregation from multiple sensors, an LLM-based system enhances accuracy in decision-making while justifying those decisions. Full transparency is availed with it, which ensures self-driving cars operate reliably and safely, giving justifications understandable by humans for increased public trust in the technology.

### *6.5.1 Regulatory Compliance and Public Trust*

Complying with regulatory standards plays an important role in self-driving cars. During the development process, the rules of local and international traffic will also be applied to self-driving systems. However, autonomous systems need not just follow but also describe how they interpret and apply such rules. LLMs can give real-time feedback and alter vehicle behavior based on traffic rules [32]. This realtime analysis not only ensures the vehicle is moving within all legal parameters but is also able to give reasons for why it is doing so. For instance, if an autonomous vehicle speeds up or slows down due to a change in the speed limit, the LLM can explain these changes concerning the particular traffic law at play indeed concerning a particular road sign.

Another key success factor in self-driving technology will be the public's trust in the system. Users need to feel comfortable with self-driving cars, and much of that comfort involves understanding why the system is making certain decisions. It has been shown that users are more trusting of the technology when clear explanations for AI-driven decisions are given to them. LLMs can simulate real-world driving scenarios and explain how the vehicle navigates complex situations. For example, the LLM can describe the decision-making process of an artificial environment in the case of an emergency braking situation, providing that the system is safe and reliable. This transparency engenders public trust and wider reception.

### 6.5.2 Robotics: Ensuring Safe and Predictable Operations

LLMs do not stop there at autonomous driving. Applications in robotics abound, especially in domains where safety and predictability of operation are of paramount importance. For instance, consider an industrial or healthcare setting where the robot has to make traceable decisions. Among the ways LLMs do this is by realtime explanation of complex robot actions. This would involve LLMs, in generating behavior trees that detail actions performed by a robot in sequence. Behavior trees [33] can be employed to explain why a robot has made any particular decision, thereby making the act of the robot more predictable and understandable for the operators.

Such explanations of LLMs' decisions are highly valued in environments like healthcare, manufacturing, and disaster relief, where the robots are in close interaction with humans. In this type of high-consequence environment, predictability and transparency are vital for safety. LLM-assisted generation of BTs lets operators foresee what the robot is going to do; this reduces the chances of accidents or misunderstandings. For example, a robot in a manufacturing plant would decide on its own to move an object this way or that way. The system would point out, through the use of an LLM, that it was based upon a decision concerning weight and shape, but also considering the environment for providing information to operators as to why the robot did what it did.

### *6.5.3 Enhancing Collaboration and Predictability in Robotics*

While safety is an enabling ingredient for collaboration, predictability in operation is another. The prospect of autonomous driving robots cooperating with humans would increase considerably through the deployment of LLMs with multimodal capabilities. A possible example can be explicitly explaining the decisions and actions that robots and workers share nearby, say, in warehouses or medical facilities. More efficient operations, which are characterized by predictable and understandable robot behavior, would be promised only with the improved collaboration between the two agents.

## 6.6 Challenges and Future Work

A few challenges and scope for future work are highlighted below.

- **Regulatory Compliance**: Application and wide acceptance of the autonomous system depend on consideration and adherence to both national and international regulations. LLMs can be used to either provide feedback in real-time or adjust the behavior of the vehicle according to traffic laws [32]. However, how these systems are going to explain the interpretation and compliance with the regulators remains to be a challenge. The regulatory bodies require those systems to be transparent so that they would go for a probe of decisions of AI-driven vehicles. This, therefore, contributes to the fragility of the results of LLMs, whose complexity normally doesn't allow for readable and understandable reasoning that can be presented to any regulatory body for review [32].
- **Public Trust**: Building public trust in the systems remains another big challenge. As much as LLMs work on explainability, the general public still finds a way to distrust AI systems, especially in situations with high-stakes decision-making like self-driving cars. Users trust a system if they can understand how the latter made decisions. LLMs have a long way to go for the former situation to materialize where LLMs also needed to become more transparent about how they make their decisions in real time.
- **Adaptability to dynamic environments**: In general, robotics requires adaptability to changing environments. Robots acting in dynamically changing environments such as warehouses, disaster areas, etc., will have to make decisions that are safe and explainable. LLMs may indeed generate complex behavior trees for explaining robotic actions, but it is still a considerable technical challenge to ensure these actions are predictable and adaptive to the changing environmental dynamics [34].

Several areas of future work could enhance the performance and explainability of LLMs in autonomous systems:

- **Improving Model Transparency**: Future research should be focused on designing hybrid models that merge the ability of LLMs to predict with the explainability of simpler models. By constructing such systems, which are capable of delivering explanations in real time, researchers can minimize the"black box" aspect of LLMs and, consequently, make their decisions more understandable to users and regulators. Another promising gap-reducing strategy involves developing visualization tools that decompose complex decisions into more understandable components.
- **Improving HumanRobot Collaboration**: Work in robotics in the future should strengthen collaboration between robots and humans in shared environments. Multimodal capabilities of LLMs so that they can process both visual and linguistic inputs could make robots even more adept at operating in dynamic environments while providing real-time explanations for their actions [34]. Improving robot explainability will not only help make operations safer but will also enhance the efficiency of human–robot collaboration.

In autonomous systems, such as self-driving cars and robots, LLMs also play an important role in enhancing both the decision-making process itself and explainability. For example, self-driving cars should make quick decisions, such as changing lanes or performing emergency braking. These decisions can be lifesaving but also need to be explainable to regulators, passengers, and manufacturers. LLMs let autonomous systems explain these actions in real time and give transparency into how the decisions were made based on the data around them, such as where the traffic was or how pedestrians behaved. In robotics, LLMs can explain complex actions in dynamic environments, like why a robot chose this particular path in the manufacturing setting or how it interpreted sensory inputs to avoid obstacles. This predictability and transparency are crucial for guaranteeing that robots will be able to work safely with humans, especially in high-stakes industries like healthcare and logistics. The remaining key challenges include bias, interpretability, and adaptability in dynamic environments.

Concretely, decisions made by autonomous systems should be transparent and understandable to humans, while their actions should be treated fairly under all environmental conditions. In this regard, future work on multimodal fusion models and explanation techniques is desirable to overcome the challenges mentioned above since these enable the safe and reliable operation of autonomous systems. In other autonomous systems, such as self-driving cars and robotics, LLMs also play an important role in enhancing both the decision-making process itself and explainability. For example, self-driving cars should make quick decisions, such as changing lanes or performing emergency braking. These decisions can be lifesaving but also need to be explainable to regulators, passengers, and manufacturers. LLMs let autonomous systems explain these actions in real time and give transparency into how the decisions were made based on the data around them, such as where the traffic was or how pedestrians behaved.

In robotics, LLMs can explain complex actions in dynamic environments, like why a robot chose this particular path in the manufacturing setting or how it interpreted sensory inputs to avoid obstacles. This predictability and transparency are crucial for guaranteeing that robots will be able to work safely with humans, especially in high-stakes industries like healthcare and logistics. The remaining key challenges include bias, interpretability, and adaptability in dynamic environments. Concretely, decisions made by autonomous systems should be transparent and understandable to humans, while their actions should be treated fairly under all environmental conditions. In this regard, future work on multimodal fusion models and explanation techniques is desirable to overcome the challenges mentioned above since these enable the safe and reliable operation of autonomous systems. Table 6.4 elucidates the use of XAI techniques in robotics and self-driving cars.

XAI improves the possibility of revolutionizing decision-making processes to be much quicker, more accurate, and larger in scale across all such industries as healthcare, finance, education, and autonomous systems. The full potential of LLMs can be tapped only when the models that drive these are explainable and transparent. As the LLMs continue to increase in power, the urgency for having robust explainability techniques will be much more urgent. Techniques such as SHAP, LIME, and NLP-based explanations are only a beginning. These tools will, in turn, facilitate professionals, regulators, and end users to understand and trust the decisions that are driven by AI, for the responsible and ethical use of LLMs.

In the future, bias, interpretability, and privacy will be significant challenges for all LLM application areas. Fair and transparent AI systems will need to be constantly refined both in terms of models and datasets they are trained on. Besides, it is very important to develop certain standards of explainability and ethical conduct of AI in cooperation with AI developers, industry professionals, and regulatory bodies. In other words, LLMs hold the keys to a potential industry transformation by empowering better decision-making, automation of processes, and personalization. But this integration is successful only when they are transparent, nondiscriminatory, and explainable so that stakeholders in healthcare, finance, education, and autonomous systems can have confidence and benefit from this evolution.

**Table 6.4** LLM use cases in self-driving cars and robotics, explainability methods, and challenges

| Use Case | Description | Explainability methods | Challenges addressed |
|---|---|---|---|
| Lane change (Cars) | Explains lane change via sensor and safety data | SHAP, LIME | Multisensor data, safety |
| Emergency braking (Cars) | Explains real-time braking events | Attention, SHAP | Real-time safety versus speed |
| Path planning (Robots) | Explains path changes in dynamic scenes | Behavior trees, attention | Safety in changing environments |
| Human–robot collaboration | Explains robot actions in shared spaces | Attention mechanisms | Transparent and safe interaction |
| Compliance (Cars) | Explains decisions per legal standards | SHAP, legal XAI | Legal rule alignment |
| Predictive maint. (Robots) | Explains maintenance needs | Feature attribution, analytics | Avoid false positives, trust |

## 6.7 Conclusion

While LLMs are in development and acting in several other industries, the integration of LLMs into healthcare, finance, education, and autonomous systems holds immense potential. These deeper AI systems possess unparalleled capabilities in terms of processing large volumes of information, making predictions, automating tasks, and delivering key insights across several domains. At the same time as this development, however, explainability, fairness, and transparency have become an urgent concern. As much as the LLM would be integrated into high-stakes fields, it is important to understand their decision-making process for widespread adoption and building trust in their ethical application. One research direction that should be closely watched is mechanistic interpretability.

## References

1. De Schutter L, De Cremer D (2024) How counterfactual fairness modelling in algorithms can promote ethical decision-making. Int J Hum Comput Interact 40(1):33–44
2. Golgoon A, Filom K, Kannan AR (2024) Mechanistic interpretability of large language models with applications to the financial services industry. arXiv:2407.11215
3. Pendyala VS, Hall CE (2024) Explaining misinformation detection using large language models. Electronics 13(9):1673
4. Pendyala VS, Chintalapati A (2024) Using multimodal foundation models for detecting fake images on the internet with explanations. Futur Internet 16(12):432
5. Lucas HC, Upperman JS (2024) A systematic review of large language models and their implications in medical education
6. Freyer O (2024) Data privacy and regulatory challenges in AI-driven healthcare systems. AI Law and Ethics 7:78–90
7. Pendyala V, Kim H (2024) Assessing the reliability of machine learning models applied to the mental health domain using explainable AI. Electronics 13(6):1025
8. Liu J, Wang C, Liu S (2024) Applications of large language models in clinical practice: path, challenges, and future perspectives
9. Omiye JA, Gui H, Rezaei SJ, Zou J, Daneshjou R (2024) Large language models in medicine: the potentials and pitfalls. Ann Intern Med 177(2):210–220
10. Oberste L, Heinzl A (2023) User-centric explainability in healthcare: a knowledge-level perspective of informed machine learning
11. Omiye JA, Gui H, Rezaei SJ, Zou J, Daneshjou R (2023) Large language models in medicine: the potentials and pitfalls. arXiv.org:abs/2309.00087
12. Subramanian A, Yang Z, Azimi I, Rahmani AM (2024) Graph-augmented LLMs for personalized health insights: a case study in sleep analysis
13. Cohen-Setton J, Bulusu K, Dry J, Sidders B (2024) Explainable AI: graph machine learning for response prediction and biomarker discovery. Can Res 84(6 Supplement):902–902
14. Freyer O, Wiest IC, Kather JN, Gilbert S (2024) A future role for health applications of large language models depends on regulators enforcing safety standards. Lancet Digit Health 6(9):e662–e672
15. Nolin-Lapalme A, The´riault-Lauzier P, Corbin D, Tastet O, Sharma A, Hussin J, Kadoury S, Jiang R, Krahn AD, Gallo R, Avram R (2024) Maximizing large language model utility in cardiovascular care: a practical guide. Can J Cardiol

16. Huang G, Li Y, Jameel S, Long Y, Papanastasiou G (2024) From explainable to interpretable deep learning for natural language processing in healthcare: how far from reality? Comput Struct Biotechnol J 24:362–373
17. Ali S, Akhlaq F, Imran A, Kastrati Z, Daudpota SM, Moosa M (2023) The enlightening role of explainable artificial intelligence in medical and healthcare domains: a systematic literature review
18. Cheng F, Liu D, Du F, Lin Y, Zytek A, Li H, Qu H, Veeramachaneni K (2021) Vbridge: connecting the dots between features and data to explain healthcare models. IEEE Trans Vis Comput Graph 28(1):378–388
19. Omiye JA (2024) The interpretability enigma in healthcare AI. J Healthc AI 12:45–56
20. Nolin-Lapalme A (2024) Human-AI collaboration: integrating LLMs in healthcare decision-making. J AI Healthc Pract 5:21–33
21. Explainable AI-driven iomt fusion: Unravelling techniques, opportunities, and challenges with explainable AI in healthcare (2024)
22. Liu X-Y, Wang G, Yang H, Zha D (2023) Fingpt: democratizing internet-scale data for financial large language models. In: NeurIPS 2023 workshop on instruction tuning and instruction following
23. Xing F (2024) Designing heterogeneous LLM agents for financial sentiment analysis. ACM Trans Manage Inf Syst. Just Accepted
24. Deng X, Bashlovkina V, Han F, Baumgartner S, Bendersky M (2022) What do LLMs know about financial markets? A case study on reddit market sentiment analysis
25. Head CB, Jasper P, McConnachie M, Raftree L, Higdon G (2023) Large language model applications for evaluation: opportunities and ethical implications. New Dir Eval 2023:33–46
26. Suresh S, Misra SM (2024) Large language models in pediatric education: current uses and future potential. Pediatrics
27. Extance A (2023) Chatgpt has entered the classroom: how LLMs could transform education
28. Pirzado FA, Ahmed A, Mendoza-Urdiales RA, Terashima-Marin H (2023) Navigating the pitfalls: analyzing the behavior of LLMs as a coding assistant for computer science students—a systematic review of the literature. IEEE Access 1–1
29. Condor A, Pardos Z (2024) Explainable automatic grading with neural additive models. In: International conference on artificial intelligence in education, pp 18–31. Springer
30. Berman J, McCoy L, Camarata T (2024) LLM-generated multiple choice practice quizzes for pre-clinical medical students; use and validity. Physiology
31. Cai ZR, Chen ML, Kim J, Novoa RA, Barnes LA, Beam A, Linos E (2024) Assessment of correctness, content omission, and risk of harm in large language model responses to dermatology continuing medical education questions. J Investig Derm S0022–202X
32. Thakur DK (2024) Automatic self driving car. Indian Sci J Res Eng Manag
33. Colledanchise M, Ögren P (2018) Behavior trees in robotics and AI: an introduction. CRC Press
34. Lykov A, Tsetserukou D (2023) LLM-brain: AI-driven fast generation of robot behaviour tree based on large language model

# Chapter 7
# Ethical Considerations in Using XAI for Smart Applications, Model Bias, Moral and Regulatory Considerations

**Valeriu Manuel Ionescu, Madalin Ciprian Enescu, and Cristina Claudia Bizon**

**Abstract** The Artificial Intelligence (AI) presence in devices and services that have become ubiquitous to our daily lives creates the need to specify development guidelines in order to protect the user's privacy, rights and to prevent criminal use. As AI models have often become too complex, Explainable Artificial Intelligence (XAI) methods were introduced to allow the understanding of the decision-making process of AI tools. This Chapter investigates aspects related to fairness, trust and privacy of AI implementations, aspects that are often overlooked in XAI method implementation, because they are not critical as interpretability, transparency or robustness, but are becoming important in the light of national and international regulatory considerations. The Chapter creates an overview of existing literature, current and future regulatory space, the advantages and disadvantages of integrating XAI methods and the potential effect on future explainable ethical AI development.

**Keywords** Artificial Intelligence (AI) · Explainable AI (XAI) · Explainability · Artificial intelligence · Ethics · Regulatory considerations · Privacy · Guidelines · Ethical theory

---

V. M. Ionescu (✉)
National University of Science and Technology POLITEHNICA Bucharest, Pitesti University Center, Pitesti, Romania
e-mail: valeriu.ionescu@upb.ro

M. C. Enescu
Doctoral School, Political Sciences National School of Political and Administrative Studies Bucharest, Bucharest, Romania

C. C. Bizon
Doctoral School, Bucharest University of Economic Studies, București, Romania



N. Bizon and B. Appasani (eds.), *Explainable Artificial Intelligence for Trustworthy Decisions in Smart Applications*, https://doi.org/10.1007/978-3-031-97007-8_7

## Abbreviations

| | |
|---|---|
| AI | Artificial Intelligence |
| DARPA | Defense Advanced Research Projects Agency |
| EU | European Union |
| IDE | Integrated Development Environment |
| IT | Information Technology |
| LLM | Large Language Models |
| ML | Machine Learning |
| XAI | Explainable Artificial Intelligence |

## 7.1 Introduction

Artificial Intelligence (AI) tries to simulate human intelligence through software (methods, algorithms, etc.) in order to extend the abilities of human operators or eliminate the human presence in a specific process. AI is present in many domains such as medical diagnosis, search engines, climate predictions and content recommendations. The complexity of algorithms that lead to models used to solve specific tasks has increased in recent years driven especially by perception and cognition tasks [1] at a rate and to a level that exceeds the researcher's ability to understand the workings of the decision-making process [2]. Understanding how a recommendation was obtained by an AI model is critical in increasing the trust in that result but it also helps in model improvement and development.

Explainable Artificial Intelligence (XAI) attempts to improve the trust in AI models by integrating methods that can be queried for information on how a decision is taken. The methods are descriptive and show which features in the data have a significant effect on the AI models decision. Humans, analyzing the information will then be able to tune the model, correct any errors and, from the ethics standpoint, bias.

The development lifecycle of an AI system includes: establishing the problem that needs to be solved with the help of AI; data collection; data preparation; model selection/development; model training and evaluation; model deployment; system monitoring and maintenance (continuous). These main stages are visible for a system without explainability in Fig. 7.1.

Here the *training data* (dataset) is selected to have a good quality, be in sufficient volume, and be representative for the targeted real data where the system will operate. Training is usually verified by humans to remove errors and bias. The *Machine Learning (ML) algorithm* that is designed to solve the problem will be chosen and it will be applied to the dataset resulting in identification of patterns that will become the *ML model*, that makes the predictions. The *Input data* is the real-world data where the AI model will operate. For a system integrating explainability, the cycle is presented in Fig. 7.2.

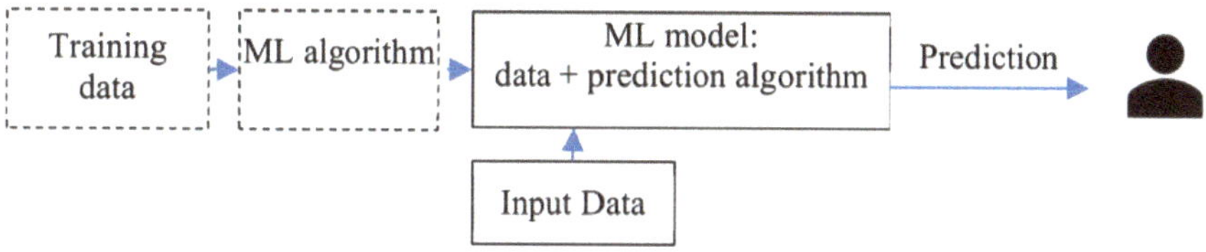


**Fig. 7.1** Operation of an AI system without explainability

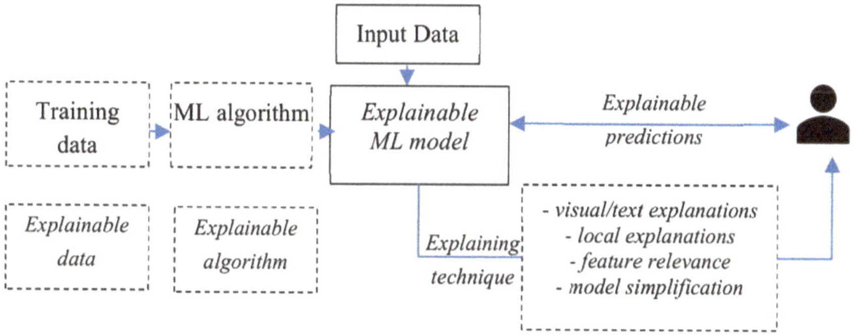


**Fig. 7.2** Operation of an AI system with explainability

Comparing the development of an AI model that does not incorporate explainability methods with one that incorporates them, it is visible there are multiple areas that can integrate explainability:

- *explainable data*: training data used for a model needs to be monitored in order to avoid bias, correlation imbalance and copyright problems. Data needs to be relevant, complete, balanced and accurate.
- *explainable algorithms*: for this the layers and thresholds used for a prediction need to be explained;
- *explainable predictions*: here the features and their weights used in a prediction should be explained. Here it is important the *explainable technique* used, because the user needs to understand the information that is offered by the implemented methods. Too much or too little information are both detrimental and they depend on the level of user understanding.

There two moments in the lifecycle of AI model development when these methods can be introduced:

- ante-hoc: that is explaining how a model operates in order to get a specific result. It requires integrating the methods at the model design stage and during training. This can be very difficult to do and it is difficult to attain good accuracy [3];
- post-hoc: in this case the methods are added after the model was deployed. This will result in a model that is not as transparent as an ante-hoc implementation, but

the model performance is not affected. The danger is that these methods are not very reliable and can be fooled [4] to make biased classifiers undetectable.

The necessity of explainability integration in AI models generated much interest in recent years. One possible way to investigate the level of interest in a topic is by using Google Trends, a service offered by Google that can use the search parameters to identify the interest level for a topic at local or global level [5] and has been used in many papers ranging from research trends analysis [6] to COVID cases prediction [7].

By using Google Trends, we can identify a growing interest in recent years for XAI, as a part of the artificial intelligence domain. The algorithm used by Google was changed in 2016 and 2022 to better collect the data and it can be downloaded as CSV file for further user investigation. In this representation, the interest level is relative to the all-time highest existing interest (which represents 100%). By analyzing the period 2004–2024 (all available Google Trends period) we have noted extreme variations in results before 2013, therefore decided to focus the research to recent years: 2016–2024. As an observation, some investigations are hard to make as certain acronyms (such as XAI) may give results from multiple meanings (such as the xAI [8] which is also in the domain of Large Language Models—LLM) and therefore affect the data validity. This is why for this Chapter we have used the term "Explainable AI" for analysis (Fig. 7.3).

The results show a growing interest starting 2016. This can be linked to the research made by Defense Advanced Research Projects Agency (DARPA) in 2015, at the request of the United States Department of Defense (DoD) [4], where multiple approaches for model design were investigated: interpretable models, integrating explainability from the design stage; deep explanation models where the decisions will provide explanations; finally, model induction where by using different techniques (such as LIME—Local Interpretable Model-Agnostic Explanations) an approximation model will be created based on the analysis of the original black box model. The study also found that it is difficult to measure the effectiveness of an explanation. By using the Glimpse Google Trends supercharged extension [9] on the

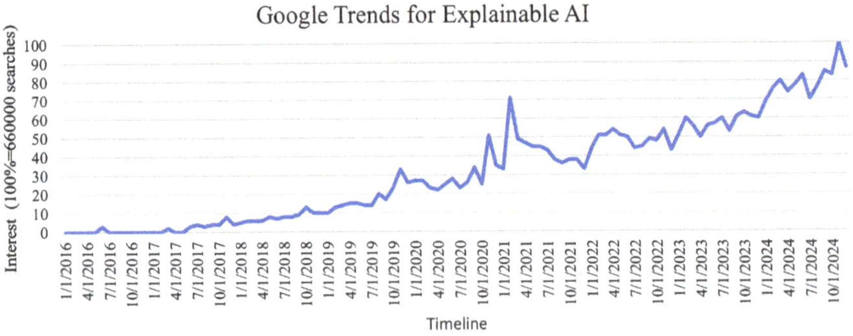


**Fig. 7.3** Search results in google trends for explainable AI for the interval 2016–2024

topics of Artificial Intelligence and Explainable AI, we investigated the interest in the Google search engine and found an increase of searches for both: 1.7 million (2020) to 4.4 million (2025) for Artificial Intelligence and 16000 (2020) to 66000 (2025) for Explainable AI).

It is visible the difference in the interest for the two topics, with the general term of Artificial Intelligence being nearly 67 times more searched at the beginning of 2025, but Explainable AI had a 44% yearly increase in 2024, while interest for Artificial Intelligence did not increase in 2024 compared to 2023, as seen in Fig. 7.4. For the Explainable AI, the most searched terms were "Explainable AI course" and "Explainable AI methods". The interest for Explainable AI is growing outside researchers and developers, as the public that is targeted by these technologies is interested to know that the data is not used in a way that would affect their freedom and privacy.

This chapter novelty includes: a literature overview at the end of 2024 on the topic of Ethical AI models; an investigation of the tools and frameworks used for Ethical AI models; and the presentation of the current national and international regulatory status for this domain, especially of the recently adopted final version of AI Act by the European Union.

The reminder of the Chapter is organized as follows: Sect. 7.2 presents an overview of literature targeting the constantly evolving domain of ethics, its integration in AI

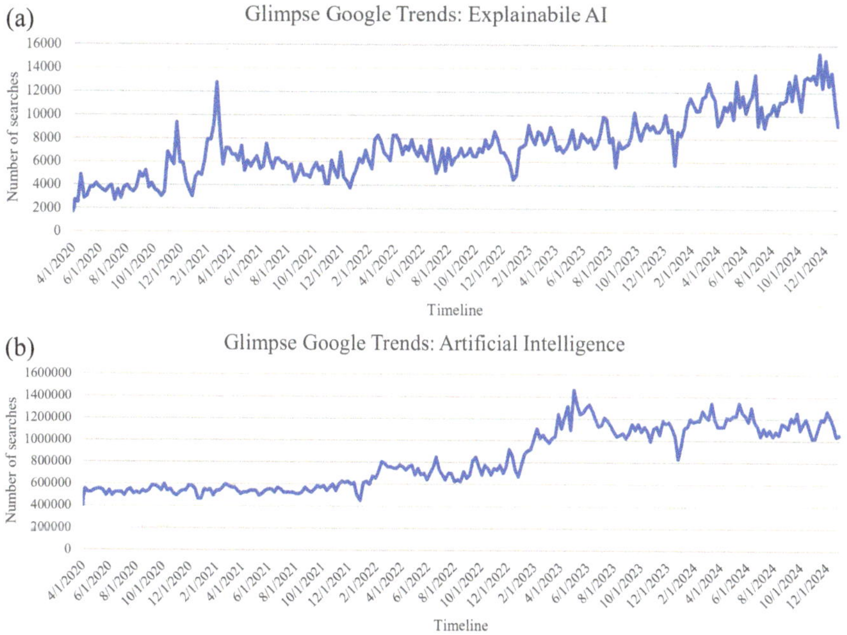


**Fig. 7.4** Search results in glimpse google trends supercharged extension for artificial Intelligence (**a**) compared to explainable AI (**b**) show an increase for the period 2020–2024

models (along with problems related to incorrect selection of the AI model datasets). In Sect. 7.3 the investigation follows how explainability is integrated in ethical AI models. Section 7.4 presents some of the frameworks, programming languages and tools used for monitoring and assessing the created explainable ethical models, a step critical in AI prediction success and bias avoidance. Section 7.4 investigates the regulatory documents that exists today for AI ethics, their limitations, the need for explainability to comply with law requirements. The final section outlines future perspectives and conclusions of this research.

## 7.2 Ethics in the Domain of AI

### *Ethics*

Ethics is the branch of philosophy that deals with questions about what is morally right and wrong, good and bad, fair and unfair. It involves the study of values and principles that guide human conduct. Ethics is essential for individuals and society harmony as it helps us determine how we should act in various situations and what kind of people we should strive to be. It is important to note that, as society is evolving, so are the ethics concepts, as it was outlined in many important historical papers.

Even if ethics was studied from antiquity, we will reduce our investigation to modern times. An important document was written by Charles Darwin in 1871 [10] and views ethics as an evolving complex combination of social instincts (developed according to the theory of evolution), self-introspection about past actions and influence of the community where the person activates. The modified concept was presented in 1975 [11] when Edward Wilson presented that social behaviors are rooted in genetics and have evolved through natural selection. Many other ways to view ethics origins have been presented (be it by philosophers or by biologists) and there is no clear agreement on the causes of ethics evolution and what are the definitive universal standards on what is ethical or not.

The only convergent point of most theories is that ethics is an evolving concept along with human society evolution [12] because it is linked to human evolution and the evolution of our intellectual capacity. The ethics are linked to what humans might want to have or what they might want to be in a particular moment, and this has certainly evolved not only long-term (humanity history) but in short term as well (decades, in the period that machine learning was developed). This means that a machine learning algorithm in the domain of ethics that was trained a few years ago might not reflect well the ethics of humanity today. This also makes the case for introducing explainability in the model as only by understanding how it makes decisions it is possible to verify if it complies with current ethics standards and make future changes to comply with evolving regulations.

When connecting the domain of Ethics with IT, the classical notions of advantage and disadvantage toward a population can introduce novel situations. For example, it was observed [13] that AI algorithms can identify long term ethical goals for a population that, while true, might not coincide with short term goals in ethics that the humans are usually accustomed and cause mistrust and opposition. Another aspect is represented by level of necessary inter-disciplinary training required by the specialists that will be involved in the development of ethical AI systems.

***Certifications and standards***

Applying ethics in AI model development and verifying the implementation is the task of a new brand of specialist that needs skills in AI, ethics and governance. Also, systems need to be certified for offering a certain level of model control for ethics.

For products, services or systems, IEEE proposed a certification (IEEE AI Ethics Certification—CertifAIEd) [14] that allows them to demonstrate being built with respect to the freedoms of individuals and communities. It is organized as a process (represented in Fig. 7.5) with multiple steps that allows authorized IEEE evaluators to understand the product's goal, establish the ethical risk profile for the product and establish the criteria for the final evaluation. Then, evidence is collected that the established criteria for ethics was integrated in the product. Finally, after proof analysis and the criteria passed, an independent certifier will verify the result and issue the certificate that the product met the Ethical criteria.

One of the critical steps in the process above is the Assessment stage, where the specialist needs to determine if a product meets the criteria. If the developers of an AI product do not understand themselves how the AI reached a specific conclusion (due to the complexity and the number of parameters), the assessment specialist will not be able to complete its mission. Therefore, explainability integration can be a solution that allows understanding the decision mechanism factors without exposing the complete implementation.

IEEE Computational Intelligence Society/ Standards Committee (CIS/SC) [15] was established to define the standards for classifying explainability in AI as partially of fully explainable (along with methods and algorithms) and to improve interoperability between systems, for instance exporting an AI solution from one system to another. The ISO/IEC/IEEE 24748-7000 standard [16] that describes how processes can integrate ethics in system design was only adopted in 2022 and has limited availability.

For specialists operating in this area there are many courses and certifications available, such as AI Quality Certification Program (AIQCP) by TÜV SÜD program certifies the quality and ethical standards of AI systems and products across various industries [17] or the course offered by CertNexus for people working in the AI ethics domain [18].

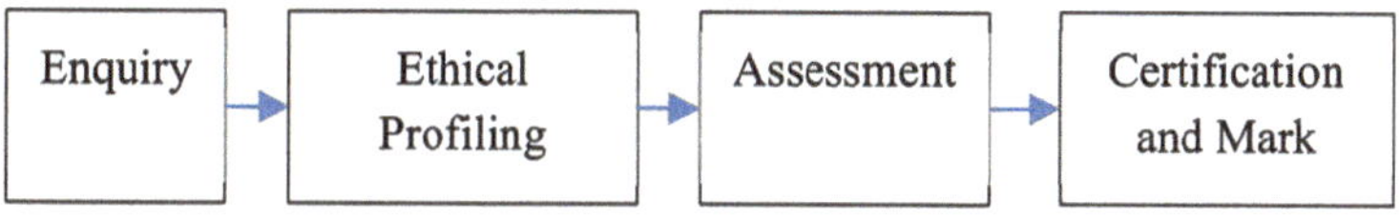


**Fig. 7.5** IEEE CertifAIEd certification steps

***Input data for AI models***

An AI algorithm is a set of rules designed to solve a specific problem. Algorithms are applied on datasets in order to create an AI model that will identify patterns and relationships in the data from the dataset. These will be the basis of the predictions made by the AI model, as seen in Fig. 7.6. If the dataset is too small, not relevant, inaccurate or incomplete, the model will infer biased results.

Today's internet has become flooded with AI generated content, which is—in many cases—impossible to distinguish from human generated content. This is a problem as AI algorithms need to know which data present on the internet was generated by other AI tools in order to avoid it, as they need human input as an intake in order to extract relevant information. Collecting data generated by AI algorithms will initiate a self-feeding loop that will bias the generated results toward directions that are difficult to predict. That is one of the causes why some AI algorithms (generative AI—such as ChatGPT) have only been using data before the emergence of LLM tools for content creation (in the case of ChatGPT, data from before September 2021). These are now opening up to new data inputs that are unchecked and often created by various AI tools for content generation [19].

The approach of including new data, while having the advantage of offering recent information, has the drawback of amplifying the mistakes of the model, and in time (longer or shorter depending on how much synthetic data was inserted) the model may lose substance [20]. The data selection is critical and for some domains there are hardware/physical references that can be used. For the domain of ethics however there is no clear ground truth data therefore developers need to integrate explainability in the model in order to understand how the model is affected by the new data, gets to a specific conclusion and how to correct it [21].

The term "model collapse" is discussed in [22] and occurs when AI models are trained on data generated by other AI models. This recursive training leads to a rapid degradation in the quality of the outputs, resulting in nonsensical responses because the models can no longer identify the correct learning task (as seen in Fig. 7.7). The authors show that this phenomenon can occur in LLMs, in variational autoencoders (VAEs) and Gaussian mixture models (GMMs). For the LLM, after 10 epochs, the trained model will only keep 10% of the original training dataset if not controlled and the generated output will vary significantly. The conclusion is that in the future, content created by humas (usually in the education domain) and pre-2023 stores of data will need to be preserved and placed in a special dataset and used to control the LLM generation and avoid model collapse. Datasets after 2023 will need to be curated in order to be usable.

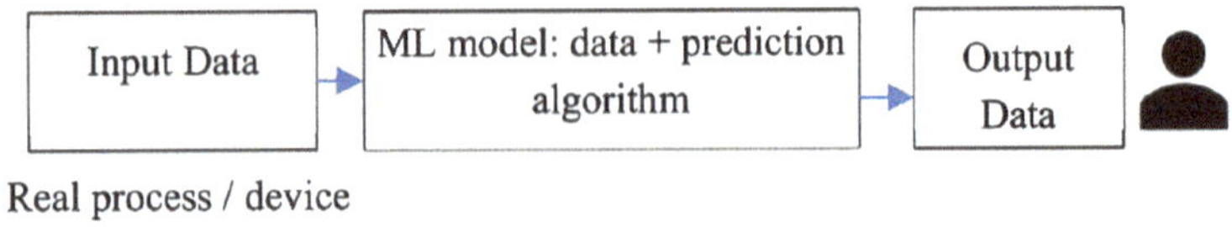


**Fig. 7.6** Input data taken from humans is cured and creates no feedback loop

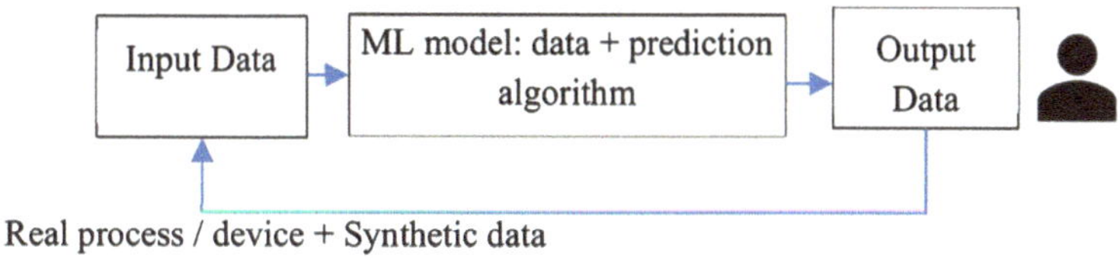


**Fig. 7.7** Input data that includes AI tools output will rapidly generate bias and lead to model collapse

The same is observed in the paper [23] where each version of the created generative model is trained with a dataset composed of elements generated with the previous version of the generative AI model. The results are evaluated by a classifier trained with the original dataset. Results show that by the fifth-generation loop, the results exhibit significant degradation. In regard to the ethics aspect, the authors speculate that bias in the generated content will increase quickly over time if the models are not guided.

The papers [24, 25] investigate the problem of biased datasets used in algorithm training and testing. The generated results will be biased as observed with the gender bias outlined in The Guardian publication for texts, occupations, characteristics and physical terms. Similar aspects were observed in the ImageNet dataset for Computer Vision. The authors propose a dataset audit for content curation. They also warn about the ethical challenges of using low paid human labor for labeling and validation the content [26].

Some authors make the case of training AI in descriptive ethics using the Commonsense Norm Bank (a collection of ethical judgements that is checked and free from AI generated content [27]) as a reference—a form of ground truth—for AI to use. The effect on an experiment they performed gave good results an AI named Delphi that reached 92.1% accuracy on moral judgments in an estimate by humans compared to a lower score obtained by GPT-3 [28].

***Related literature for ethics role in AI***

Implementing AI ethics is crucial across various domains in order to ensure that AI technologies are used responsibly and beneficially.

Some key areas where AI ethics play a significant role are: healthcare (ensuring patient privacy, fairness in treatment recommendations, and transparency in diagnostic tools), finance (preventing biases in credit scoring, loan approvals, and fraud detection systems), law enforcement and security (avoiding discriminatory practices in surveillance, predictive policing, and ensuring accountability), education (protecting student data, ensuring equitable access to educational resources, and transparency in AI-driven assessments), industry (fairness in hiring processes, avoiding biases in performance evaluations, and protecting employee data), transportation (safety and reliability of autonomous vehicles, and ethical considerations in decision-making algorithms), social media and communication (addressing misinformation, ensuring privacy, and preventing manipulation through AI-driven content), environment (promoting sustainable practices and minimizing the environmental impact of AI technologies), entertainment (ensuring fair representation and avoiding harmful stereotypes in AI-generated content) and government (ensuring transparency, accountability, and fairness in AI applications used for public administration and services).

The need for integrating ethics in AI models is reflected in many research papers.

Paper [29] identifies the possibility that in the near future the AI algorithms will be able to fill in social roles but will be endowed with superior intelligence and abilities when compared to a human. These algorithms will be executed in unpredictable contexts, therefore will require transparency in order to get a sense of predictability for their execution in short and long term. Finally, the ethical assessment of these systems will have to be different because of the differences between humans and algoritms.

The paper [30] finds that universal ethical principles which can be applied to specific situations cannot defined. Therefore, a process is proposed (based on two frameworks) that can be followed in the design of Ethical AI, where each design team can answer specific questions related to that implementation, leaving a log of the decisions made and the logic behind the decision. While not totally flawless (because it is limited by the knowledge of the team implementing the design) following these steps can give an AI design that can be improved, it is transparent and can be trusted. The paper also makes a case study in digital mental health applications and makes the case for a design team that includes not only programmers, designers, and managers but also other stakeholders related to the target audience, including the potential users.

AI ethics in the domain of lawmaking and practice is investigated in [31], based on a survey with participants from multiple countries (mainly Europe and Asia). The paper identifies that out of the six AI ethics principles (transparency, privacy, accountability, fairness, autonomy and explainability) the main AI ethics principles are transparency, accountability, and privacy while the lack of ethical knowledge, legal frameworks and monitoring bodies are the main challenges.

The paper [32] is an overview paper that makes a global analysis of the current corpus of principles and guidelines on ethical AI. The ethical principles found are: transparency, justice and fairness, non-maleficence, responsibility, and privacy with the main being transparency. The results show that the main divergences target the way the principles are interpreted and implemented. The study also shows that the regions that lead in this domain are Europe and North America.

The views of people in social media on AI ethics with accent on the Reddit social network (and AI related topics of discussions) were investigated [33] and it was found that the level of anonymity gives users the sense of security to express freely their opinions. The themes of discussion are gender bias, regulation and patent laws, spreading disinformation, the generation of fake content, misuse of personal data and the impact on crime. The author makes the case that new topics of discussion outside of those identified by scientists were found and should be the target for further research in order avoid speculation from ill intended individuals.

In [34] investigates the creation of an ethical AI framework in the context of GDPR (General Data Protection Regulation) where, even if AI is not explicitly mentioned, many of its content can be applied to the way AI interacts with people's data. One of the challenges in developing ethical AI is the balance between innovation and people's rights. The authors find that governance is necessary to achieve this.

Paper [35] analyses the domain of rescue robots where two ethical challenges are found: the lack of AI design guidelines and the lack of identification of relevant values for a specific implementation. For example, some guidelines are critical for a design while being less important for a different design. The solution proposed is to implement such an indicator for design assessment for rescue robots with the help of various stakeholders, in an iterative design process that continuously incorporates ethics related feedback.

A bibliometric perspective is created in [36] for 20 years (2004–2023) with SCOPUS as the main data source and VOSviewer as the data aggregator. The authors identify a 3-phase AI Ethics development, namely incubation, making AI human-like machines, and making AI human-centric machines. The problem detected is that we have already started building human centric machines without having completed the initial phases of defining the rules for prohibiting unethical development of machine augmented non-humans and the audit tools for such development.

Paper [37] investigates the development of AI ethical systems for policing, law enforcement and judicial proceedings with accent on predictive policing that tries to predict future criminal behavior based on the historic data (and is highly prone to bias based on race or gender). As AI is able to identify patterns in datasets that are not visible to humans. The paper proposes the concepts of "Models of Threat" (that tries to identify where a thread will appear and who be commit the act) and an "Ethics of Care" (that tries to identify actions that will lessen or prevent the occurrence of a crime and make law enforcement people aware of how predictions are made) and makes a case that the second is better because it prevents threats while increasing transparency.

Gardner [38] focuses on AI ethics in education and analyses the case from UK where the university entry results from 2020 were determined by a biased algorithm that was only challenged because the individuals had the means to challenge the results. This led the authors state that responsibility, recourse, and redress should enable people to exercise their rights in front of an AI automated decision system by allowing them to appeal and change an unfair decision, and even receive reparations for this. Again, GDPR is specified one of the few pieces of existing regulation that addresses this aspect.

In [39] is an overview paper that identifies transparency, privacy, accountability and fairness as the most common AI ethics principles. Lack of ethical knowledge, vague or conflicting principles are the main challenges of implementations.

Placing AI in smart healthcare products in comparison with traditional ethics is analyzed in [40]. AI in healthcare misuse rages from over trust to mistrust and the "white box" approach is recommended to avoid such situations and biased decisions. Also, because of the speed that AI tools are being developed, not every aspect of an implementation is analyzed and vetted, therefore making hasty implementations is a detriment in the long term as they will decrease the trust in such systems.

The paper [41] investigates AI use in achieving the 17 sustainable goals established by the United Nations in 2015 [42]. Even if the goal is to improve systems, a bad implementation of AI may lead to bad results. The applications identified are $CO_2$ emission monitoring, infection tracking and wildlife GPS monitoring.

The authors of paper [43] propose that AI technologies need to be studied by using o holistic approach. Topics like technological singularity (the moment when AI will exceed human control), impact of automation on economy and employment, privacy considerations and human—AI relation are discussed, but also novel ones, like raising the issue of robot rights with the example of 2015 Boston Dynamics developers kicking a robot dog to show its stability. In this case, the symbolism raised ethical question. References to science fiction authors are used as they have envisioned a human—AI world long before the technology was ready.

In education, many research papers and books are placed behind a paywall and not all researchers have the resources to buy access to these repositories. However, many of these documents are part of the dataset that AI tools have (such as GPT4, Claude or Mistral) therefore, as observed in 2013 in [44], they can be interrogated by anyone for free, exposing the paper contents with various degrees of success [45]. This raises the question of ethics both for the persons obtaining the data but mainly for the smart applications that have access to the documents and how they distribute that information. Some authors make the case of adding metadata to research papers [46] in order to be able to determine the impact of AI tools in academic writing. This allows multiple purposes, for example to determine where AI was used (possibly to remove this content from a dataset used for training AI to prevent model collapse) and if greater AI use in the creation process leads to greater popularity for a document.

The paper [47] outlines the problems like: the use of autonomous weapons, where hackers can direct weapons or smart vehicles in attacks; the fact that AI can find intelligent ways to discriminate and the use of AI in elections; the challenges of implementing ethics in the context of General Data Protection Regulation (GDPR) legislation and the understanding by legislators of the ethical implementation of AI. All these lead to the conclusion that AI opaqueness is the main concern and that there are initiatives like Global Initiative for Ethical Considerations in AI and Autonomous Systems, Algorithmic Transparency and Accountability to implement algorithmic accountability and verifiability.

Paper [48] shows that, as AI systems become more present in everyday life, various failures will highlight ethical issues in these systems. The authors investigate the current state of industrial practice in AI ethics by presenting the results of a survey conducted with over 200 software companies that explores whether high-level guidelines and tools for managing AI ethics have been adopted by the industry and examines the extent to which these guidelines are implemented in practice. The findings reveal a gap between the theoretical frameworks for AI ethics and their practical application in the industry, suggesting a need for more effective integration of ethical principles in AI development processes.

In conclusion, it is a consensus that ethics must be implemented in the AI especially for models that interact directly with humans. The main ethical principles have been established, but the method of their integration in AI models is under debate and it must take into consideration the differences between humans and machines. The researchers also agree that the implementation of AI systems is ahead of the governance regulatory context.

## 7.3 Literature Review of Explainability for Implementing AI Ethics

Explainable AI (XAI) aims to make systems transparent, comprehensible, and accountable by integrating the means to understand the process of decision making. As presented in the previously analyzed research, transparency is one of the characteristics most needed in AI in order to increase the user's trust and analyses the way decisions are made, and in particular for the domain of ethics. Therefore, ethics implementation is often not the purpose of XAI but a side product of the stated goals. XAI needs to integrate explanations that are not misleading or biased and has to handle the problem of competing values: explanations must target human preferences not abstract best cases and the complexity needs to be between too complex and too shallow. In this section we will focus on papers that target ethical implementations of Explainable AI.

Paper [49] investigates how explainability was proposed for implementing ethics for AI. It is a literature overview paper which finds that XAI development and implementation depends on the philosophical theories regarding the meaning of ethical behavior: deontology (where designing an ethical AI should be done with only good will and truth in mind, regardless of consequences), consequentialism (where the evaluation of the implementation will be done based on the consequences, impact on people and risks), and virtue ethics (where the AI will try to promote virtues by offering explanations when needed and in manner shaped to the user level). Finally, the authors find that the papers that come from interdisciplinary backgrounds provide more depth on the subject than pure technical papers.

There is an IEEE Intelligent Systems volume [50] that investigates the implementation and operationalization of ethical and legal frameworks for AI in practice, in articles like [51], where top-down, bottom-up and hybrid approaches to designing moral AI are analyzed, and [52] where a real-world scenario of the development process of a financial service chatbot outlines that problems appear in all stages of development and model-only planning is not sufficient for a functional XAI enabled final product.

Paper [53] analyses the use of AI in medicine to offer a second opinion and the danger of false confirmation. By using explainability however, the algorithm decision can be understood by the physician and a good decision can be taken. In the same domain, the role of XAI is investigated in [54] by using generative tools (DALL-E 3,

Firefly 2, Midjourney 5.2 and Stable Diffusion 2.1) for training students and detecting potential ethics problems. Some of the found problems are: copyright and privacy issues are an issue; as these tools generate content with incorrect anatomical structure (called AI hallucinations), it is unethical for professors to recommend them; all tools were found to amplify gender and ethnicity stereotypes for occupation, physical appearance and roles; finally, attempts to mitigate the observed problems by using prompts and negative feedback were successful only if precise wording was used. As in other papers, it is observed that the lack of regulation affects the development of algorithms and implementations.

The relation between stakeholders (and why they want explainability) and the technical dimension of explainability is investigated in [55]. It is another paper with an applied component that places explainability methods on a healthcare application.

Bertrand et al. [56] presents that explainability assisted decision making can be influenced by cognitive biases by designing incorrect XAI methods and incorrectly evaluating them. A correct XAI implementation can reduce cognitive biases and an incorrect design can increase them. This outlines the need for correct evaluation of the information given by explanation of methods.

An experimental design of a supply chain decision making process where XAI methods are present was created in [57]. The tests performed show that the increased transparency improves the system resistance to cyber-attacks. The database used for acquiring the population sample (individuals that work in the domain of risk and supply chain management) was the Amazon Mechanical Turk, which is a popular database due to its size and coverage of multiple countries. This is one of the few papers that have an implementation made. Results also showed that creating an algorithm that is transparent increases trust, makes people with less knowledge understand the decision process and can help detect decision errors.

Speith et al. [58] presents the novel concept of understanding AI by linking explanations to abilities as outlined in cognitive psychology, grouped as: recognizing, assessing, predicting, intervening, explaining, and designing. In this way it is easier to understand and quantify in studies because, for example, a person with a higher ability is linked to a higher degree of understanding, therefore the explainability can be scaled according the ability level.

Adamson [59] examines the accuracy of post-hoc explaining models (inductive explanation—why the system has reached an answer), a technique favored method by DARPA agency. The conclusion is that contrary to pre-hoc (that shows how the system has reached an answer but can be understood by only few people that have the knowledge), post-hoc is easier to understand but unsatisfactory as explanations can be too vague and unclear, contributing in fact to the mistrust in the system even if it integrates explainability.

Farrow [60] investigates how XAI increases ethical practice for several educational stakeholders (for example the tools and applications that are used in the educational process: assessment automation, class administration, tutorial generation and student assistance for homework– like ChatGPT). For the investigation Google scholar and Elicit datasets were used. Conclusions show that explainability in education is necessary for teachers in order to propose to the students the best tools and avoid recommending those that can potentially introduce biases and reduce the learning incentive.

Brand and Nannini [61] observes that in a world where AI becomes ubiquitous, the right to implement transparency in algorithms becomes a moral duty and that various stakeholders should work together to provide meaningful and cohesive ethical framework for development.

Vainio-Pekka et al. [62] creates a literature review on the topic of ethics in AI targeting multiple databases (IEEE Xplore, ACM Digital Library, Scopus, ProQuest and Web of Science) and acknowledges the broadness of the field (with significant research increase since 2018) while pointing the gaps in research. One observed tendency is to shift from theoretical and philosophical approaches to practical implementations/frameworks, that target specific application issues. For those implementations, some researchers have found that the main purpose of XAI is not to explain the algorithm decisions to the stakeholders that are affected by them (end users, regulators, managers) but to the engineers that work on the system [63]. End users are not the main target because of lack of experts to evaluate the explanations and the lack of computing power to show the explanations in real time.

By analyzing ChatGPT design and history, the overview paper [64] identified ethical considerations regarding intellectual property rights, privacy issues regarding personal data misuse, lack of accountability, misinformation, bias, discrimination and lack of transparency.

Another overview paper [65] finds the following ways to implement fairness: preprocessing fairness (focused on reducing bias in the data before training); in-process (by modifying the learning algorithm); post process (to align the output to the fairness goals); regularization fairness (applied during training stage, to remove the influence of sensitive attributes); finally, counterfactual fairness (by changing some attributes while keeping the rest unchanged to make the algorithm understand bias).

In the overview paper [66] starting from the analysis of well-known AI ethics related incidents (originating from biases to dataset, algorithm and model), the authors organize the results in three categories: types of bias (measurement, labeling, sampling, aggregation, confirmation, temporal feature selection and automation), consequences of data bias and unfairness (discrimination, misinformation and legal implications) and proposed solutions (pre-processing training data, algorithm fairness, Explainable AI and interdisciplinary collaboration). The authors also observe the challenges of working with real time data as trying to minimize delays on limited performance hardware may lead to lack of useful insights in decision making (for example, explainability).

Book [67] offers examples, in R (a language and environment for statistical computing and graphics) using different packages like *iml*. The datasets used are bicycle rental from the Capital-Bikeshare company, YouTube spam comments and risk factors for cervical cancer from UCI Machine Learning repository. Interpretable models, model-agnostic methods, example-based explanations are discussed and tested for algorithms implemented for these datasets. The conclusions and future predictions are that in the near future all data will become digitized and because of this all the tasks that are automatable will become automated, no matter the complexity, and imprecise specification of goals will forever interfere with the automation (and this is the place where all the human specialist effort will be needed). In regard to interpretability, model agnostic methods will be the primary tools, and programs and systems will be programmed such that they will be able to explain themselves.

In the whitepaper [68] from Fujitsu, the gap between principles and practice is acknowledged and the proposed solution is to focus on the relation between stakeholders and AI, where most of the problems are observed. As described in other papers, an incidents database was created (with the help of the Partnership on AI non-profit organization) in order to identify the developed AI ethics impact. In regard to regulatory issues, Europe is identified as the leading player in the field, with the Artificial Intelligence Act. For ethical AI development, algorithm fairness can be achieved in multiple ways such as: using algorithms that equalize the probability of the potential outputs (for example in the case of race or gender), using algorithms that equalize the probability of AI output being adopted in correct data and using XAI to understand the decisions of the algorithm. For ethical AI impact assessment, Fujitsu proposes a sequence that was tested to correctly identify potential problems:

- An ethics model is created based on the established ethical requirements and definitions with links to the incident database;
- By observing the ethics model interactions with the AI system (based on the interaction between the users and the AI), the list of potential dangers is created.
- The problems are then analyzed and the algorithm is corrected to avoid their appearance.

Finally, in [69] the authors focus on translating theory to practice and for this they propose a three-stage approach: a theoretical initial step to define the principles; a secondary step with the practical implementation of AI ethics; the final step where the algorithm is tested in a real situation, where feedback is gathered to detect problems and generate feedback for further refinement. The problems today are the excessive focus on theory and the lack of toolkits that can be used to develop practical applications.

## 7.4 Development Tools and Evaluation of AI Ethics

***Development tools***

The lifecycle of machine learning systems includes multiple stages as described in the first section of this chapter. It is a complex development chain and this section investigates the tools that are available for developers. There are multiple platforms that can assist in the process of development with some encompassing all aspects of the development chain, while other are specialized tools.

One of the platforms is Google's Vertex AI platform that covers tools to train, deploy, and monitor models and AI application and offers details on model evaluation metrics and feature attribution, developers being able to predict the effect of a feature on the model predictions [70]. This platform is available in Google Cloud and Python is the development language of choice. The users are not charged unless they use paid resources. New Google Cloud users receive $300 in free credits to use on cloud services and free usage of over 20 products, up to set monthly limits [71].

IBM Watson Studio is a different platform, from IBM, that can also be used for all stages of development, from data preparation to introducing explainability and transparency. It integrates well with IBM cloud and other tools from IBM [72]. There is a Lite plan of Watson that does not incur charges in order to try tutorials and services. Even if libraries are offered for multiple programming languages, again tutorials are offered in Python and Jupyter notebook computing platform.

Amazon SageMaker is a managed service offered from Amazon, an important player in the cloud business, building, training, and deploying machine learning models that integrates with Amazon Web Services (AWS) services. As is the case with the other services, there is no cost until some functionality is used. There is a free tier available and educators (with AWS Educate) receive free credits to use resources for learning and teaching [73].

Azure Machine Learning offers the same options from Microsoft, integrating with the development environments offered by this company and the Azure cloud services. New users receive $200 credits to use in 30 days but there are 25 services free for 12 months and others that are always free. The programming language of tutorials is Python and the development IDE is Visual Studio Code [74].

Nvidia NeMo Framework [75] is designed for building, customizing, and deploying AI models and includes multiple tools for managing the stages of the life-cycle development. Nvidia is not one of the large cloud providers but is relevant to AI systems because it develops specialized hardware that offers the computing power necessary for AI operation and it also has partnerships that make its services available on other platforms of major cloud service providers (CSPs) like AWS, Google Cloud, and Microsoft Azure. The NVIDIA NeMo Framework is primarily developed using Python and it leverages the PyTorch deep learning library for building and training models. C and C++ are also used in AI development for their CUDA platform.

Apple offers a variety of services and tools for AI development, focusing on on-device machine learning and AI capabilities such as Create ML and Core ML [76]. Apple doesn't have a direct equivalent to end-to-end development platforms like Google Vertex AI or Amazon SageMaker and they focus mainly on the development for their platform. On the explainability aspect, in 2024 Apple has acquired Dawin AI, that offers a platform for XAI in the domain of visual inspection of processing lines [77] and in software it provides frameworks (for example ResearchKit and CareKit [78]) for developing explainable AI methods in the medical domain, and Swift as a development language. Libraries like Tensorflow an PyTorch are also available.

These are the tools offered by the most important companies in AI development and the conclusion is that most are well integrated in the cloud provider's ecosystem, making developers partially locked into that ecosystem if they have chosen at one point to develop AI systems using a specific set of tools. In general, Python is the development language used, even if libraries and frameworks are offered for other programming languages. Most development tools are now offered for free in order to attract new developers.

***Monitoring and evaluating ethics***

One aspect that is relevant to ethics is the model monitoring and detection of bias. Monitoring tools are frameworks that continuously analyze the implemented models running on real data (which is also monitored because itself can experience drift) that compute the prediction drift of the model relative to the ideal output, indicative of conceptual drift and model degradation. In case of ethical AI implementations, it is usually the detection of bias and fairness change in the model output exceeding set thresholds or the crossing of use case specific validations. Other things that the tools offer: computing the model latency (especially when there is a surge in traffic), visualization dashboards and verifications of AI model availability.

Without monitoring, AI models have diverged often from the intended purpose and ethical boundaries. Some famous cases where this has happened are:

- Grok model collapse due (partially) to the use of AI generated content as training and learning data—intentionally and unintentionally [79]
- Microsoft's Tay Chatbot problem from 2016 where the feature of learning from malicious user feedback turned to a flaw that led the bot to generate offensive comments [80]
- The Correctional Offender Management Profiling for Alternative Sanctions (COMPAS) Risk Assessment Tool used in the United States for determining the probability of recidivism was shown to exhibited racial bias, thus defeating to purpose of impartial decision assisting tool in the criminal justice system.

This is only a part of the development chain, but a very important section, and because of this there are many tools available that aim to prevent this.

A platform used in checking for bias in AI algorithms before they are deployed is Fiddler AI [81]. It allows model monitoring, analysis of model performance and explanations for model predictions using methods like Shapley Values and Integrated Gradients. Integrating observability in some domains (finance, law) will soon become a requirement in order to comply with a country's regulations, therefore making tools like this compulsory. Fiddler AI is focused on the observability, explainability, monitoring of models and is integrated with multiple platforms like Google Vertex AI and Amazon Sagemaker.

NeMo Guardrails [82] is a toolkit offered by Nvidia to control the communication between the application code and the Large Language Model, in order to introduce restrictions on the dialog the LLM will initiate (for example enabling ethics by fact checking and output moderation). The decision to vet the input or response dialog is made by another LLM (that is verified to be ethical and is not accessible to the user) through questions with yes/no answers. It also allows connecting other applications to the LLM. It uses the annoy C++ library with Python bindings in order to allow development in Python.

Captum AI [83] is an open-source library for model developers developed by Facebook AI research, that provides implementations for interpretability methods using algorithms like Integrated Gradients, Saliency Maps and SmoothGrad. It has a widget called Captum Insights for visualizations to understand models easier. It is using Python for development, and is extensible via the API that is documented on the project web page.

XMANAI is a European Project [84] aiming to implement explainability to models used in the manufacturing sector. It uses post hoc techniques like LIME and SHARP to implement explainability. In all application cases presented, one of the ethical principles was the trust of the users in the AI decisions through the implementation of transparency (for example understanding the meaning/semantics of the command that was given to the AI model for assisting the human and meeting regulatory compliance requirements).

As previously discussed, IBM was one of the initiators of the explainability for AI models, with accent on the black box models. Their open-source toolkit AI Explainability 360 (AIX360) [85] allows multiple algorithms to implement explainability and evaluation metrics. The development is made in Python and the authors aim to make explainability algorithms framework independent, given that their large number (TensorFlow, Keras, PyTorch, etc.) makes the task of developers harder.

InterpretML [86] is an open-source Python library used to implement interpretability in black box AI models, available online on GitHub: https://github.com/interpretml/interpret. The ethics accent in its presentation is placed on fairness, regulatory compliance, and trust in human–machine communication.

Finally moDel Agnostic Language for Exploration and eXplanation (DALEX) developed by Data Lab at the University of Warsaw [87] is an open-source package developed in R (but also available for Python) that contains a set of tools for global and local interpretability, making it easy for developers to explain model predictions and performance.

There are other tools for monitoring available (like Evidently AI or Qwak, OpenXAI), but we have focused on those that also offer tools for explaining predictions besides the monitoring of models. We should also mention in this section the multitude of toolkits available for data analysis and explainability implementation (for example the open source *scikit-learn*) which offer not only code examples and methods but also datasets for testing.

## 7.5 Regulatory Aspects for AI

The purpose of introducing explainability to an AI model is to increase transparency, robustness, interpretability, fairness, trust and privacy. The importance of each feature differs depending on the stakeholder involved. From the developer's perspective, only transparency, robustness and interpretability are critical to the operation of the system. Fairness, trust and privacy are only relevant to the persons using the model. In order to make companies and developers invest in implementing those three features, national entities need to get involved and adopt legislation to regulate the AI applications [88]. Enforcing the laws on the black box AI models will be done through the development of explainability methods. In this section we will investigate the current recommendation and regulatory context of national and international organizations.

In [89] World Health Organization proposes 40 recommendations for the ethical development of large multi-modal models (LMMs) for specific health-related applications. This AI branch was identified as potentially dangerous due to their imitation of human communication and their rapid integration in many areas concerning the health of populations.

The recommendations that this organization makes are different for governments (that need to create conditions and encourage the ethical development, create the legislative environment, create agencies that can approve applications and audit them, after deployment, and assess the impact on people) and for developers (to investigate developed services with the help of the stakeholders and not limited to their primary task but also secondary outcomes of such applications).

UNESCO (United Nations Educational, Scientific and Cultural Organization) organization that aims to promote peace and security through international cooperation in education, sciences, culture, communication, and information also identified the opportunities of AI use in healthcare, education and task automation. One of the main concerns regarding the use of AI was identified as its ethical use, due to biases in analyzing information especially regarding climate change and human rights. Therefore, UNESCO adopted a recommendation in 2021 [90] that identifies nearly all aspects of the society where AI can have an impact, ranging from education and research, to economy and labor. The organization also created the Global AI Ethics and Governance Observatory, a platform for good practices on the implementation and use of ethics in AI [91]. The mechanism for monitoring is left at the discretion of the governments but should include commissions for continuous ethics monitoring of AI systems and the creation of a repository of complaints to track the concerns and problems that the people have raised. The paper [92] investigates the proposed recommendations and concludes that people education and awareness are also important in order to make everyone (from developers to policymakers) understand the necessity and risks.

In the United States of America, the Congress presented in October 2022 a bill called "Blueprint for an AI Bill of Rights making automated systems work for the American People" [93] concerning guidelines (principles and practices) for AI system that can be followed in order to avoid making them unsafe, ineffective, or biased. A Bill is not a law therefore it does not compel companies to follow it, instead, it is designed to guide the development and use of AI systems. Released by the White House Office of Science and Technology Policy (OSTP), it aims to promote transparency, fairness, and safety in AI technologies that impact the public's rights, opportunities, or access to critical needs. Other AI models are not the subject of this bill. The ethical aspects of this document involve data privacy and discrimination protection and these should be made available for inspection by the AI model through explainability, with valid, clear, timely and understandable responses. Examples why these principles are important are offered from the domains of medicine, law and welfare. Individual USA states have started the process of creating laws, for example California, Colorado and Virginia. The Governor of California for example approved by into law on September 28, 2024, the Assembly Bill 2885 on Artificial Intelligence [94] that targets the growing influence of AI and deepfake technology. The bill defines the term "artificial intelligence" and establishes the basis for investigations of feasibility in creating standards and technologies for state departments to determine the provenance of digital content. On ethical issues, the main accent is placed on containing deep fakes, reducing their impact on socicty, protecting the privacy of individuals and the development of mechanisms to validate the authenticity of a media material (using cryptographic signatures).

European Union is the first entity in the world that passed the AI system regulation proposal (created in 2020) into law on 13 March 2024 (with a minor correction on 16th of April 2024), which become active in August 2024 [95].

Similar to the USA approach, in the beginning of the document, AI is defined and a classification of harmful practices is presented. A new governance architecture is also proposed (European Artificial Intelligence Board) to supervise the application of this law and the creation of a regulatory sandbox to test novel AI systems. After a presentation of academic and stakeholder perspectives are presented, a classification of AI systems is presented (using a risk-based approach) in four categories: unacceptable risk, high risk, transparency risk and minimal risk. Unacceptable risk systems are those that violate fundamental human rights and are prohibited. High risk systems can have an impact on the fundamental rights will need conformity assessment and constant monitoring. Transparency risk systems are those that interact with people and can impersonate humans (deep fakes are specifically named in the document). These systems require the announcement that the output is AI generated (for example through a watermark or direct announcement of the user). For the high risk and transparency risk explainability is necessary, but the way this is implemented is not specified (ante or post hoc). Finally, minimal risk systems are only targeted by the current EU legislation. There are exceptions that are not targeted by this document such as: military, defense, national security systems and models developed for the scientific research and development. Existing systems (which were developed prior to the release of this law) will need to be sandboxed to demonstrate that they pass the set requirements.

## 7.6 Conclusions

AI model explainability is a necessity in the current stage of AI development. Some principles that are behind the explainability target the system performance and potential for development, while others are aimed to the users of the AI tools for protecting the private information and increasing the user's trust. A black box approach to AI development without concern for fairness, trust and privacy can have multiple negative effects and can discourage the broad adoption of an AI model that affects public's rights, opportunities, or access to critical resources or services.

The national governments and international organizations are the only ones that can impose guidelines and rules in this regard. This Chapter analyzed the current regulatory context for the domain of ethics in AI and the novelty compared with the previous papers is the analysis of the adopted AI Act of the European Union, a regulatory first in the world. It is envisioned that, based on this document, similar legislation will be adopted across the world, taking in consideration the regional particularities. The legislation development will take years until all AI systems in the world will become subject to regulation by national or international governance.

The existing frameworks still lack effective enforcement structures and cross border infringement legislation is missing. The lack of speed in establishing regulatory documents it is due to the complexity and novelty of the domain and the fact that, when incorrectly specified, it has the potential to restrict AI development and encourage evading the laws.

Our research showed that there is a growing interest in AI and XAI in recent years as outlined by the large number of related papers presented in the literature research section and the online tools interest estimation tools such as Google Trends. Google Trends confirms the observation that the interest for Explainability in AI is behind the general interest in AI because it is limited to developers but also that the specialists, especially in the domain of ethics, have not agreed on a standard of evaluation for an implementation. This lack of standards in the domain of ethics has two important reasons:

- ethics is constantly evolving and it is highly subjective. An ethical implementation needs to be designed to be updatable in the future then the associated concepts it will change. In this case explainability can offer the mechanisms to make these changes.
- ethics standards vary from region to region so there it is hard to establish a worldwide ethics standard, instead it is easier to define regional standards. Explainability again will offer mechanisms to understand where changes need to be made to the AI model in order to offer different ethical implementation variations depending on the country where the user is located.

As research shows, many researchers have agreed that the AI tools implementations are ahead of standards and regulations, making the intentional and unintentional bias and abuse possible. This generates mistrust in the AI tools for the general public. Explainability implementation is therefore necessary and there are multiple points (and ways) in the development process where it can be introduced.

In regard to the development tools and frameworks the conclusion of our analysis is that these are still at an early stage because of their diversity. This splits the development effort in multiple directions with advantages (various and novel methods are developed) and disadvantages (there is no critical mass of development in a specific direction). The framework creators are multiple and each try to attract the developers and keep them locked in that framework. The tools are offered for free and are in most cases open source. The main development language is Python, with few exceptions due to hardware requirements or platform limitations.

Finally, an aspect specific cu Ethical AI models is the evaluation of the model predictions. This is particularly hard to do because ethical standards are hard to quantify, there is no universal ground truth and ethics evolve in time. Despite of this, there are frameworks created for this purpose and they offer through explanations, to the users that monitor the AI system, the chance to determine if the model predictions have become biased, why that has happened, and give insights on how to correct them.

## References


1. Deng L (2018) Artificial intelligence in the rising wave of deep learning: the historical path and future outlook [Perspectives]. IEEE Signal Process Mag 35(1):180–177. https://doi.org/10.1109/MSP.2017.2762725
2. Liu J et al (2018) Artificial intelligence in the 21st century. IEEE Access 6:34403–34421. https://doi.org/10.1109/ACCESS.2018.2819688
3. Adadi A, Berrada M (2018) Peeking inside the black-box: a survey on explainable artificial intelligence (XAI). IEEE Access 6:52138–52160
4. Slack D, Hilgard S, Jia E, Sameer S, Lakkaraju H (2020) Fooling LIME and SHAP: adversarial attacks on post hoc explanation methods. In: Proceedings of the AAAI/ACM conference on AI, ethics, and society (AIES '20). Association for computing machinery, New York, NY, USA, pp 180–186. https://doi.org/10.1145/3375627.3375830
5. Mavragani A, Ochoa G, Tsagarakis KP (2018) Assessing the methods, tools, and statistical approaches in google trends research: systematic review. J Med Internet Res. 20(11):e270. https://doi.org/10.2196/jmir.9366. PMID: 30401664; PMCID: PMC6246971
6. Jun SP, Yoo HS, Choi S (2017) Ten years of research change using google trends: from the perspective of big data utilizations and applications. Technol Forecast Soc Chang 130. https://doi.org/10.1016/j.techfore.2017.11.009
7. Pan Z, Nguyen HL, Abu-gellban H, Zhang Y (2020) Google trends analysis of COVID-19 pandemic. In: IEEE international conference on big data (Big Data), Atlanta, GA, USA, pp 3438–3446. https://doi.org/10.1109/BigData50022.2020.9377852
8. xAI Corp (2023) Announcing Grok. https://x.ai/
9. Double Point Partners LLC (2023) Glimpse. https://meetglimpse.com/
10. Darwin C (2009) Comparison of the mental powers of man and the lower animal—continued. In: The descent of man and selection in relation to sex. Cambridge Library Collection—Darwin, evolution and genetics. Cambridge University Press, pp 70–106
11. Wilson EO (2000) Sociobiology: the new synthesis, twenty-fifth anniversary edition paperback—deluxe edition, March 4, Belknap Press, ISBN 978-0674002357
12. Wilson C (2007) Evolutionary ethics, handbook of the philosophy of science, pp 219–246, ISBN 9780444515438. https://doi.org/10.1016/B978-044451543-8/50014-9
13. Boddington P (2023) AI ethics, a textbook. ISBN: 9789811993817
14. IEEE Standards Association (2023) AI ethics certification. https://engagestandards.ieee.org/rs/211-FYL-955/images/ai-ethics-certification.pdf
15. IEEE Standards Association (2023) P2976 XAI—eXplainable AI working group. https://sagroups.ieee.org/2976/
16. ISO (2022) ISO/IEC/IEEE 24748-7000:2022(en): systems and software engineering—life cycle management—part 7000: standard model process for addressing ethical concerns during system design. https://www.iso.org/obp/ui/en/#iso:std:84893:en
17. TÜV SÜD (2023) AI quality certification program (AIQCP). https://www.tuvsud.com/en/themes/artificial-intelligence/training/aiqcp
18. CertNexus (2020) Certified ethical emerging technologist. https://certnexus.com/certified-ethical-emerging-technologist-ceet

19. Radford A, Kleinman Z (2023) ChatGPT can now access up to date information. https://www.bbc.com/news/technology-66940771
20. Strickland E (2022) Are you still using real data to train your AI? IEEE Spectrum
21. Panetta K, Bakshi S, Timoszczuk AP (2024) Why we need explainable AI. IEEE Trans. https://transmitter.ieee.org/why-we-need-explainable-ai/
22. Shumailov I et al (2024) AI models collapse when trained on recursively generated data. Nature 631:755–759. https://doi.org/10.1038/s41586-024-07566-y
23. Martínez G et al (2023) Towards understanding the interplay of generative artificial intelligence and the internet. In: International workshop on epistemic uncertainty in artificial intelligence, pp 59–73. Cham: Springer Nature Switzerland
24. Leavy S et al (2020) Mitigating gender bias in machine learning data sets. Bias2020 workshop. https://arxiv.org/abs/2005.06898
25. Prabhu VU, Birhane A (2020) Large image datasets: a pyrrhic win for computer vision? https://arxiv.org/abs/2006.16923
26. Gray ML, Suri S (2019) Ghost work: how to stop silicon valley from building a new global underclass. Paperback, Ed. Houghton Mifflin Harcourt, ISBN-13:978-0358120575
27. Choi CQ (2021) Machines learn good from commonsense norm bank. https://spectrum.ieee.org/ai-ethics-machines-learn-good
28. Jiang L et al (2021) Can machines learn morality? The Delphi experiment. https://arxiv.org/abs/2110.07574
29. Bostrom N, Yudkowsky E (2014) The ethics of artificial intelligence. In: The Cambridge handbook of artificial intelligence, Cambridge University Press. https://doi.org/10.1017/CBO9781139046855
30. Peters D, Vold K, Robinson D, Calvo RA (2020) Responsible AI—two frameworks for ethical design practice. IEEE Trans Technol Soc 1(1):34–47
31. Khan AA et al (2023) AI ethics: an empirical study on the views of practitioners and lawmakers. IEEE Trans Comput Soc Syst 10(6):2971–2984. https://doi.org/10.1109/TCSS.2023.3251729
32. Jobin A, Ienca M, Vayena E (2019) The global landscape of AI ethics guidelines. Nat Mach Intell 1:389–399. https://doi.org/10.1038/s42256-019-0088-2
33. Ocal A (2023) Perceptions of AI ethics on social media. In: 2023 IEEE international symposium on ethics in engineering, science, and technology (ETHICS), West Lafayette, IN, USA, pp 1–1. https://doi.org/10.1109/ETHICS57328.2023.10155069
34. Milossi M, Alexandropoulou-Egyptiadou E, Psannis KE (2021) AI ethics: algorithmic determinism or self-determination? The GPDR Approach. IEEE Access 9:58455–58466. https://doi.org/10.1109/ACCESS.2021.3072782
35. Martin D, Schmidt MW, Hillerbrand R (2023) Implementing AI ethics in the design of AI-assisted rescue robots. In: 2023 IEEE international symposium on ethics in engineering, science, and technology (ETHICS), West Lafayette, IN, USA, pp 1–1. https://doi.org/10.1109/ETHICS57328.2023.10155062
36. Gao DK, Haverly A, Mittal S, Chen J (2023) A bibliometric view of AI ethics development. In: 2023 IEEE Asia-Pacific conference on computer science and data engineering (CSDE), Nadi, Fiji, pp 1–5. https://doi.org/10.1109/CSDE59766.2023.10487710
37. Asaro PM (2019) AI ethics in predictive policing: from models of threat to an ethics of care. IEEE Technol Soc Mag 38(2):40–53. https://doi.org/10.1109/MTS.2019.2915154
38. Gardner A (2022) Responsibility, recourse, and redress: a focus on the three R's of AI ethics. IEEE Technol Soc Mag 41(2):84–89. https://doi.org/10.1109/MTS.2022.3173342
39. Khan AA et al (2022) Ethics of AI: a systematic literature review of principles and challenges. In: Proceedings of the 26th international conference on evaluation and assessment in software engineering (EASE'22). Association for computing machinery, New York, NY, USA, pp 383–392. https://doi.org/10.1145/3530019.3531329
40. Pasricha S (2023) AI ethics in smart healthcare. IEEE Consum Electron Mag 12(4):12–20. https://doi.org/10.1109/MCE.2022.3220001
41. Astobiza AM, Toboso M, Aparicio M, López D (2021) AI ethics for sustainable development goals. IEEE Technol Soc Mag 40(2):66–71. https://doi.org/10.1109/MTS.2021.3056294

42. United Nations (2015) What are the sustainable development goals? https://www.undp.org/sustainable-development-goals
43. Zeng D (2015) AI ethics: science fiction meets technological reality. IEEE Intell Syst 30(3):2–5. https://doi.org/10.1109/MIS.2015.53
44. Neylon C (2013) Architecting the future of research communication: building the models and analytics for an open access future. PLoS Biol 11(10):e1001691. https://doi.org/10.1371/journal.pbio.1001691
45. Kiledjian EN (2023) Use AI to read paywalled content. https://www.kiledjian.com/main/2023/6/16/use-bing-chat-to-read-paywalled-content
46. Conde J et al (2024) Understanding the impact of artificial intelligence in academic writing: metadata to the rescue. Computer 57(1):105–109
47. Sadeghi AR (2017) AI industrial complex: the challenge of AI ethics. IEEE Secur Priv 15(5):3–5. https://doi.org/10.1109/MSP.2017.3681049
48. Vakkuri V, Kemell K-K, Kultanen J, Abrahamsson P (2020) The current state of industrial practice in artificial intelligence ethics. IEEE Software 37(4):50–57. https://doi.org/10.1109/MS.2020.2985621
49. Nannini L, Marchiori Manerba M, Beretta I (2024) Mapping the landscape of ethical considerations in explainable AI research. Ethics Inf Technol 26:44. https://doi.org/10.1007/s10676-024-09773-7
50. Chen F, Zhou J, Holzinger A, Fleischmann K, Stumpf S (2023) Artificial intelligence ethics and trust: from principles to practice. IEEE Intell Syst 38(06):5–8. https://doi.org/10.1109/MIS.2023.3324470
51. Seeamber R, Badea C (2023) If our aim is to build morality into an artificial agent, how might we begin to go about doing so? IEEE Intell Syst 38(6):35–41. https://doi.org/10.1109/MIS.2023.3320875
52. Lu Q et al (2023) Developing responsible chatbots for financial services: a pattern-oriented responsible artificial intelligence engineering approach. IEEE Intell Syst 38(6):42–51. https://doi.org/10.1109/MIS.2023.3320437
53. Rosenbacke R et al (2024) AI and XAI second opinion: the danger of false confirmation in human–AI collaboration. J Med Ethics. https://doi.org/10.1136/jme-2024-110074
54. Currie GM, Hawk KE, Rohren EM (2024) Generative artificial intelligence biases, limitations and risks in nuclear medicine: an argument for appropriate use framework and recommendations. Semin Nucl Med. ISSN 0001-2998. https://doi.org/10.1053/j.semnuclmed.2024.05.005
55. McDermid JA, Jia Y, Porter Z, Habli I (2021) Artificial intelligence explainability: the technical and ethical dimensions. Phil Trans R Soc A 379:20200363. https://doi.org/10.1098/rsta.2020.0363
56. Bertrand A, Belloum R, Eagan JR, Maxwell W (2022) How cognitive biases affect XAI-assisted decision-making: a systematic review. In: Proceedings of the 2022 AAAI/ACM conference on AI, ethics, and society (AIES'22). Association for Computing Machinery, New York, NY, USA, pp 78–91. https://doi.org/10.1145/3514094.3534164
57. Sadeghi K et al (2024) Explainable artificial intelligence and agile decision-making in supply chain cyber resilience. Decis Support Syst 180:114194. ISSN 0167-9236, https://doi.org/10.1016/j.dss.2024.114194
58. Speith T et al (2024) Conceptualizing understanding in explainable artificial intelligence (XAI): an abilities-based approach. Ethics Inf Technol 26:40. https://doi.org/10.1007/s10676-024-09769-3
59. Adamson G (2021) Ethics and the explainable artificial intelligence (XAI) movement. https://doi.org/10.36227/techrxiv.20439192.v1
60. Farrow R (2023) The possibilities and limits of explicable artificial intelligence (XAI) in education: a socio-technical perspective. Learn Media Technol 48:1–14. https://doi.org/10.1080/17439884.2023.2185630
61. Brand JLM, Nannini L (2023) Does explainable AI have moral value? https://arxiv.org/abs/2311.14687

62. Vainio-Pekka H et al (2023) The role of explainable AI in the research field of AI ethics. ACM Trans Interact Intell Syst 13, 4(26):39 pages. https://doi.org/10.1145/3599974
63. Bhatt U et al (2020) Explainable machine learning in deployment. https://arxiv.org/abs/1909.06342
64. Ray PP (2023) ChatGPT: a comprehensive review on background, applications, key challenges, bias, ethics, limitations and future scope. IOTCPS 3:121–154
65. Chen P, Wu L, Wang L (2023) AI fairness in data management and analytics: a review on challenges, methodologies and applications. Appl Sci 13:10258. https://doi.org/10.3390/app131810258
66. Aldoseri A, Al-Khalifa KN, Hamouda AM (2023) Re-thinking data strategy and integration for artificial intelligence: concepts, opportunities, and challenges. Appl Sci 13:7082. https://doi.org/10.3390/app13127082
67. Molnar C (2019) Interpretable machine learning a guide for making black box models explainable. ISBN 9780244768522
68. Fujitsu. AI Ethics Impact Assessment. Whitepaper, web, https://www.fujitsu.com/global/documents/about/research/technology/aiethics/fujitsu-AIethics-whitepaper_en.pdf (2022)
69. Zhou J, Chen F (2023) AI ethics: from principles to practice. AI Soc 38(6):2693–2703. https://doi.org/10.1007/s00146-022-01602-z
70. Google Inc (2024) Train a model using vertex AI and the python SDK. https://cloud.google.com/vertex-ai/docs/tutorials/tabular-bq-prediction
71. Google Inc (2023) Set up a project and a development environment. https://cloud.google.com/vertex-ai/docs/start/cloud-environment
72. IBM (2024) Quick start tutorials. https://www.ibm.com/docs/en/watsonx/saas?topic=started-quick-start-tutorials
73. Amazon (2024) AWS free tier. https://aws.amazon.com/free/
74. Microsoft (2024) Tutorial: train a model in azure machine learning. https://learn.microsoft.com/en-us/azure/machine-learning/tutorial-train-model?view=azureml-api-2
75. Rebedea T et al (2023) NeMo Guardrails: a toolkit for controllable and safe LLM applications with programmable rails. https://arxiv.org/abs/2310.10501
76. Apple Inc (2024) WWDC24 machine learning & AI guide. https://developer.apple.com/news/?id=pby7a6ex
77. Dawin AI (2023) Q&A: DarwinAI brings explainable AI to manufacturing. https://www.waterlooedc.ca/blog/darwin-ai-waterloo-manufacturing
78. Apple Inc (2024) ResearchKit and CareKit. https://www.apple.com/lae/researchkit/
79. Nvidia (2023) NeMo 2.0 overview. https://docs.nvidia.com/nemo-framework/user-guide/latest/overview.html
80. Stokel-Walker C (2023) What Grok's recent OpenAI snafu teaches us about LLM model collapse. https://www.fastcompany.com/90998360/grok-openai-model-collapse
81. Schwartz O (2024) In 2016, Microsoft's racist chatbot revealed the dangers of online conversation. https://spectrum.ieee.org/in-2016-microsofts-racist-chatbot-revealed-the-dangers-of-online-conversation
82. Kenthapadi K, Arsanjani A (2023) Built with google AI: achieve better observability for ML models with fiddler AI. https://cloud.google.com/blog/topics/partners/built-with-google-ai-achieve-better-observability-for-ml-models-with-fiddler-ai
83. Facebook Inc (2024) Captum API reference. https://captum.ai/api/
84. Grandi F et al (2024) A methodology to guide companies in using explainable AI-driven interfaces in manufacturing contexts. Procedia Comput Sci 232(2024):3112–3120. ISSN 1877-0509. https://doi.org/10.1016/j.procs.2024.02.127
85. Arya V et al (2020) One Explanation does not fit all: a toolkit and taxonomy of AI explainability techniques. https://ar5iv.labs.arxiv.org/html/1909.03012
86. The InterpretML Contributors (2023) Getting started. https://interpret.ml/docs/
87. Biecek P (2018) DALEX: explainers for complex predictive models in R. J Mach Learn Res 19(1):3245–3249

88. Huang C, Zhang Z, Mao B, Yao X (2023) An overview of artificial intelligence ethics. IEEE Trans Artif Intell 4(4):799–819. https://doi.org/10.1109/TAI.2022.3194503
89. World Health Organization (2024) Ethics and governance of artificial intelligence for health: guidance on large multi-modal models. ISBN: 978-92-4-008475-9
90. UNESCO (2022) Recommendation on the ethics of artificial intelligence, SHS/BIO/PI/2021/1. https://unesdoc.unesco.org/ark:/48223/pf0000381137
91. UNESCO (2024) Global AI ethics and governance observatory. https://www.unesco.org/ethics-ai/en
92. Ramos G, Squicciarini M, Lamm E (2024) Making AI ethical by design: the UNESCO perspective. Computer 57(2):33–43. https://doi.org/10.1109/MC.2023.3325949
93. The White House (2022) Blueprint for an AI Bill of rights. https://www.whitehouse.gov/wp-content/uploads/2022/10/Blueprint-for-an-AI-Bill-of-Rights.pdf
94. California Legislature (2024) California assembly bill 2885. https://legiscan.com/CA/text/AB2885/id/2971555
95. European Parliament (2024) Artificial intelligence act. https://www.europarl.europa.eu/thinktank/en/document/EPRS_BRI(2021)698792

# Chapter 8
# XAI-Driven Analysis of Emerging Hegemonic Trends in Great Power Development Strategies Incorporating Fuzzy Logic

**Mădălin Ciprian Enescu, Cosmin-George Nicolăescu, Clementina Elena Niță, and Cristina Claudia Bizon**

**Abstract** This certification aims to provide a framework for visualising the developmental status of countries, use fuzzy logic as the basis with intermediate systems such as deep learning neural networks for a more detailed examination of the results. This system is created by collecting and analysing data from the USA, China, Russia, Japan, and Germany, continuously monitoring their advancement through various developmental phases. This study strategy involves using data sources via collection, revealing indices, and constructing a model that executes statistical calculations. This system employs data prediction techniques, including mathematical inference, linear regression, and deep learning neural networks, to assess and predict developmental patterns. Moreover, there is an opportunity to incorporate additional countries into subsequent analyses. Additionally, the research may be extended beyond the participating countries and customised to meet the user's requirements. For the purposes

The original version of the chapter has been revised. A correction to this chapter can be found at https://doi.org/10.1007/978-3-031-97007-8_14

M. C. Enescu (✉)
Doctoral School, Political Sciences National School of Political and Administrative Studies Bucharest, Bucharest, Romania
e-mail: enescumadalin89@yahoo.com

C.-G. Nicolăescu
Doctoral School, National University of Science and Technology POLITEHNICA Bucharest, Bucharest, Romania
e-mail: cosmingeorge470@yahoo.com

C. E. Niță
National University of Science and Technology POLITEHNICA Bucharest, Pitesti University Centre, Pitesti, Romania
e-mail: elena.nita@upb.ro

C. C. Bizon
Doctoral School, Bucharest University of Economic Studies, Bucharest, Romania
e-mail: crystinaclau@yahoo.com



N. Bizon and B. Appasani (eds.), *Explainable Artificial Intelligence for Trustworthy Decisions in Smart Applications*, https://doi.org/10.1007/978-3-031-97007-8_8

of this research, the participants were selected according to their greater potential for advancement in numerous key areas. The choice of fuzzy logic was driven by its efficiency in making decisions marked by uncertainty and its importance in selection and planning strategies, being utilised in this system to manage the uncertainty and ambiguity of data in assessing the development level of the analysed countries. Subsequently, deep learning neural networks were employed for detailed data analysis and for making predictions regarding development trends, thereby providing a clear and comprehensive understanding of the complex processes involved. This system encapsulates various data processing capabilities within the Explainable Artificial Intelligence (XAI) terminology, using various data processing enhancement methods. The process of structuring information in the main system focuses on the use of conditional *if–then* rules, opting for formulations based on descriptive language instead of precise numerical values. This technique reflects the way people think, often using rules that are not precisely determined. The set of rules embedded in the core mechanism, which includes four types of input data and one category of output data, is designed not only to facilitate analytical evaluation according to system standards, whilst also upholding the principles of XAI. This allows the system not only to produce results, but also to provide a clear and transparent understanding of the processes and reasoning behind its decisions. The data of the values located by year were extracted for each country from the specialised documentation and were utilised in the statistics of the candidate countries, this constituting the input information in the analytical process. The study develops a perspective on progress in establishing supremacy for the countries examined and suggests several approaches to categorisation. Areas of trade growth, defence financing, science and technology innovation, and foreign investment are all covered. This study will examine the data obtained from system inputs and develop methods for predicting future values based on existing value ranges. The model under discussion is not exempt from constraints and dangers, which requires a rigorous and balanced analysis, since fluctuations in the estimates made can be affected by different parameters that can lead to varied results. The study is intended to be not only an analytical exercise, but also a robust contribution to understanding the complexity of state development in an ever-changing global landscape.

**Keywords** Linear regression · AI · Intelligent system · Relative distance method · Competition for dominance · Model · Hegemony · Fuzzy logic · Neural network · Analysis

## Abbreviation

| | |
|---|---|
| FLS | Fuzzy Logic system |
| AI | Artificial Intelligence |
| GDP | Gross Domestic Product |
| MSB | Military Spending Budget |

| | |
|---|---|
| RDB | Research and Development Budget |
| FDI | Foreign Investment Budget |
| LOW_GDP | Low Gross Domestic Product |
| MED_GDP | Medium Gross Domestic Product |
| HIGH_GDP | High Gross Domestic Product |
| LOW_MSB | Low Military Spending Budget |
| MED_MSB | Medium Military Spending Budget |
| HIGH_MSB | High Military Spending Budget |
| LOW_RDB | Low Research and Development Budget |
| MED_RDB | Medium Research and Development Budget |
| HIGH_RDB | High Research and Development Budget |
| LOW_FDI | Low Foreign Investment Budget |
| MED_FDI | Medium Foreign Investment Budget |
| HIGH_FDI | High Foreign Investment Budget |
| RNN | Recurrent Neural Networks |
| LSTM | Long Short-Term Memory |
| RMSE | Root Mean Square Error |
| MSE | Mean Squared Error |
| USA | United States of America |
| US | United States |
| CHN | China |
| RUS | Russia |
| JAP | Japan |
| GER | Germany |
| MoM | Mean of Maximum |
| XAI | Explainable Artificial Intelligence |

## 8.1 Introduction

In the contemporary era of technological advancement and creative discovery, the pursuit of progress has emerged as a key element in global interactions, shaping a complicated rivalry between the countries of the world. Scientific advancements and economic development are now seen as crucial elements in gaining and sustaining authority and sway within the global community [1]. From this point of view, the United States of America (USA), the People's Republic of China, the Russian Federation, Japan and Germany are among the most powerful and influential entities globally. The intense rivalry between these countries to gain supremacy in various spheres of progress has become more and more evident [2]. This competition transcends the sphere of economic and military development, extending to the vital field of technological innovation and research, which constitutes the essence of power in the contemporary world [3].

In the past few years, the interactions among the United States of America, the Russian Federation, and the People's Republic of China have grown more tense, largely as a result of ideological and geopolitical divergences. However, the competition for hegemony has expanded to a wider level, including an intense struggle between the countries participating in this study in various development sectors [4]. In this framework, a fuzzy logic system (FLS) was designed with four inputs and one output, intended to evaluate the developmental status of each country in terms of economic growth competition, military progress, scientific and research advancements, and military forces competition. Fuzzy logic plays an essential role within the system due to its ability to manage data uncertainty and ambiguity, facilitating decision-making in a transparent and interpretable manner through the use of *if–then* rules and gradual evaluations, which allows complex processes to be modelled in a form that is readily comprehended and interpreted by users. Consequently, the integration of fuzzy logic into the proposed system contributes to the clear explanation of results and the justification of decisions generated by the model, supporting the goals of XAI to provide transparency and trust in the decision-making process. Additionally, deep learning neural networks enhance the processes resulting from the fuzzy logic system by offering a more detailed and complex analysis of the data. These networks can identify hidden patterns and subtle relationships between variables that fuzzy logic alone cannot detect. Combining these two technologies enables the achievement of more precise and adaptable results, blending the interpretability of fuzzy logic with the processing power and machine learning capabilities of neural networks. This integration enables the system to assess extensive data sets and provide more precise forecasts about the developmental tendencies of the examined nations. Furthermore, it enhances the model's capacity to adjust to alterations in the global environment and to provide precise justifications for the conclusions and recommendations produced, therefore enhancing the transparency and consistency of the research. This paper aims to explore how fuzzy logic can contribute to understanding the complexity of the competition between the United States (US), China, Russia, Japan and Germany, whilst providing significant implications for world order. The primary objective of this research is to provide forecasts and track the evolution of competition among these countries in the next years. Simultaneously, the established system is designed to be versatile, allowing the inclusion of additional countries in the analysis beyond the five initially mentioned. The data utilised in this analysis are meticulously extracted from official sources, to provide a robust basis for the study and ensure high precision in the outcomes acquired [5]. Thus, the system offers a wider and more thorough perspective on the contest among the involved nations, thereby presenting an efficient method to discern patterns and development trajectories in this rivalry. According to our knowledge, no analysis has been conducted on the level of development dominance for countries using a FLS approach to date.

The structure of this research paper is organised in the following manner: subsequent to a thorough yet succinct introduction, Sect. 8.2 is dedicated to examining recent research that aligns with the goals of this study; Sect. 8.3 focuses on presenting the design of the proposed system for assessing the level of development;

Sect. 8.4 details the outcomes achieved following the data processing; The document concludes by outlining prospective directions for development, noting certain limitations and multiple risk factors, along with final observations, all of which are available in Sect. 8.5.

## 8.2 Analysis of Related Works

Artificial intelligence (AI) fulfils a significant role in the development of society and the generation of new knowledge by demonstrating its viability in a variety of sectors, including industry, medicine, and social services. Its algorithms fundamentally alter our behaviour based on our surroundings and enhance our problem-solving capabilities according to the circumstances we encounter.

In theory, Artificial intelligence fundamentally transforms society and is closely linked to the future of mankind. A distinctive area of research is on using AI for outcomes in geopolitical analysis via the elucidation and interpretation of models, known as XAI. In analysing the outcomes, it is crucial to derive conclusions on geopolitical predictions. Thus, XAI performs a significant role in ensuring transparency, understanding and acceptance of AI models in geo-political analysis. Moreover, decision-making shapes and influences the generated predictions.

In [6], the impact of extenics on the natural resources market is examined. It is determined that study will impact areas such as the economy, social viewpoints, and the economic environment. Moreover, emphasis is placed on the responsible management of the environment. Nevertheless, geopolitical developments resulted in the formation of business partnerships. Alongside involvement in AI research and commercial endeavours, an examination of natural resources, especially in pivotal countries, will also be undertaken.

Estrada M.A.R. et al. in [7] utilise different methods for complicated socio-economic problems. Inflation control policies benefit from applied AI assistance.

A multidisciplinary perspective for the challenges, opportunities, practice and policy of AI application is discussed in [8]. This perspective reflects the views of researchers, public experts, industry, and, respectively, the academic environment.

With the development of technology and AI in the US, the authors of [9] state that there may be a possibility of national security problems.

Building on Christopher Freeman's (2002) assertions regarding the impact of technological evolution on global hegemony, the authors of [10] predict that China will achieve dominance in AI through the emergence of major corporations and the introduction of a *corporate innovation system*. Additional research is necessary for the United States and China to maintain their global geopolitical dominance.

In the study from [11], there exists a significant connection between technology and politics, which has ramifications for international, social, and economic connections on a worldwide scale. Technological progress and the role of governments influence development, shaping the trajectory of future governance.

The hazards associated with AI supremacy impact the social and public sectors, respectively, as shown in [12]. The authors categorise these dangers and recommendations into technological, informational and communicational, economic, social, ethical, and regulatory-legal domains. The paper examines a series of AI dominance methods aligned with the actions implemented in the public sector.

Feijoo et al. [13] brings arguments in technology diplomacy regarding the recovery of AI in the improvement of activity in most sectors of activity.

Ecological economic development can be influenced by the development of AI, as the authors of [14] will demonstrate with reference to China. In this sense, a mathematical model is proposed that will analyse the productivity and the impact on the ecological economy. As a result of this research, there is a U-shaped evolution in terms of productivity, through dynamic nonlinear regression. The research considers factors such as energy and pollution, which leads to a non-linear model. The quadratic function makes the correlation between productivity and AI.

From the research [15] it is observed that AI is consistently and exponentially present in contemporary society. To avoid negative problems, it is necessary to utilise XAI. The study indicates that it is not a strategy applicable in a standardised manner, since it pertains to human perspectives. A survey is conducted to gather data for a mixed model, emphasising the preference for explanation and visualisation over lengthy written responses.

The screening process for AI identification in different states is addressed in [16]. AI technology is assessed with regard to both technological and human resource factors. A study was conducted that discovered research, investments, and workforce in AI across a number of countries. A matrix for the implementation of the AI plan was developed, including the following categories: a country capable of using a technology, possessing sufficient staff, demonstrating ambition, and having the potential to be a leader. The study examines the framework of an AI technology governance matrix, emphasising the governance and ethical considerations of AI deployment in the countries under analysis. The research emphasises the interest in AI utilisation and differentiates between countries engaged in discussions regarding AI issues and those recognised for their achievements in this field.

The challenges of AI development in Russia are also examined in the research [17]. In contrast to countries such as the USA or China, in Russia, artificial intelligence is mostly funded by the public sector rather than private enterprises. In Russia, substantial sums were allocated to completely different technologies. The study indicates that the advancement of research in Russia is clearly inferior to that of countries such as the USA and China.

In [18], the influence of artificial intelligence on the military is examined and the need for global initiatives to mitigate its adverse impacts via effective risk management, emphasising the importance of international dialogue is highlighted. The progress of AI is examined in [19], focussing on the corporate sector. The research is conducted on China throughout a decade. These findings were determined subsequent to the examination of the investigation:

- the sustainable development of the corporation—with a higher impact in the social and governmental area;
- it presents effects that shape both internal and external pressure—the positive effects are stopped by the public, and the development of technology and data security offers positive effects;
- the possibility of AI is researched by using the corporate potential as well as those interested in this. Constraints are mitigated by: financing, decreasing expenses, efficient supply, increasing productivity, using resources without waste and minimizing risks.

The legitimisation of AI and the media's interest in this context are examined by Korneeva et al. in [20]. The research concludes that it undermines the legitimacy of AI applications, the lack of widespread adoption of AI as a technology, and the regulations enacted by governmental authorities concerning public access to certain AI applications.

The authors of the research conducted in [21] claimed that AI represents a socio-ecological challenge. The case study focused on Greece, examining social identity and the relationship between the workplace and AI, with outcomes of technologies to be scientifically validated by experts (see summary in Table 8.1).

The specialist literature has a comprehensive study of the competition for global domination across several domains of activity. This document addresses economic growth, defence funding, and progress in technology. The United States of America, the People's Republic of China, and the Russian Federation are three forces that are expanding their geopolitical influence worldwide [4].

The United States, characterised by its substantial economy and many industries such as technology, healthcare, and manufacturing, stands as a worldwide leader in innovation and research and development. The United States allocates an impressive military budget, investing in cutting-edge technology and armed forces, which attracts major international investment owing to its size and stability [22].

China has undergone rapid economic growth, rising to become the world's second-largest economy, focusing predominantly on manufacturing, technological development, and service sectors. It has invested substantially in R&D, focusing on sectors such as 5G technology, artificial intelligence and renewable energies. Moreover, it has significantly increased military spending in recent years and developed advanced capabilities, including cyber and space technologies. Regarding foreign investment, it has become a major international investor, participating in infrastructure initiatives and corporate acquisitions [23].

Russia has a natural resource-oriented economy, including energy and mining. It has a rich history of scientific research, but modern efforts are plagued by various challenges. It maintains a strong military sector, developing advanced technologies in areas such as missiles and cybernetics. From the standpoint of foreign investments, military and geopolitical crises have rendered it an unsafe environment for numerous investors. Nonetheless, it draws investment; yet, there are concerns over the economic environment and political volatility affecting long-term stability [24].

**Table 8.1** Synthesis of the latest trends for XAI and AI, respectively

| Ref | Application | Domain | Solutions/results technology | Advantages/ opportunities |
|---|---|---|---|---|
| [6] | The role of patents in technologies that influence China's natural resources market | AI, geopolitical risk | Using AI to influence the natural resources market | Natural resources are influenced by extenics in relation to the evolution of economic development |
| [7] | Policy design | AI—complicated socio-economic problems | Modelling of inflation control policies | Promoting policy shaping |
| [8] | AI perspective—through the vision of researchers, public experts, industry and, respectively, the academic environment | AI—review | – | – |
| [9] | Security | AI—national security and foreign policy | – | National security issues |
| [10] | – | Prediction—China dominates in AI | Predictive solutions regarding world hegemony—global research | – |
| [11] | Technology influences geopolitics and vice versa | Influence—shaping governments | Technology and geopolitics—mutual influence | – |
| [12] | Measures for the public sector | AI development | Risks in the social and public sector | AI dominance strategies |
| [13] | – | Use of AI | Solutions in most sectors of activity | Development |
| [14] | AI—influences the growth of the economy | Ecological economy | Mathematical model—by non–linear regression | Impact on productivity |
| [15] | AI and XAI currently used | Explanation for the indication of certain directions of use | To provide: explanation and visualisation, without offering long texts | Understanding AI |
| [16] | The AI screening process in different countries | Political and economic | Made an AI strategy implementation matrix | Identification of countries that research AI issues and those that only debate them |

(continued)

**Table 8.1** (continued)

| Ref | Application | Domain | Solutions/results technology | Advantages/ opportunities |
|---|---|---|---|---|
| [17] | – | Competition on the market | Russia is inferior to the USA and China | – |
| [18] | Military | AI—trust achievement | Appropriate management of AI in the military field | Risk control |
| [19] | Study | AI—corporate sustainable | – | Major advantages for the corporation |
| [20] | Media | AI—legitimacy | – | – |
| [21] | Socio-economic study | AI—legislation | – | – |

Japan has an advanced economy centred on technology, automobiles, and electronics, allocating resources to research and development aiming at technological and scientific innovation. It has adopted a more limited perspective on the military sector, emphasising self-defence. Regarding investments, Japan often invests internationally, but less actively than other countries [25]. The German economy is defined by its industrial, engineering, and pharmaceutical sectors, emphasising research and innovation in green technology and life science. Its diminished military presence draws significant investments and establishes it as a prominent industrial and technological hub [26].

The interaction between these countries encompasses both collaboration and rivalry, featuring economic partnership as well as competition in the global market and technological arenas. The military rivalry focusses on strategies to obstruct each other from implementing regulations by influencing the perceptions of other countries [3]. In 2014, the US Department of Defence revised its strategies, employing AI to diminish the possibility of military progression of states and to strive for dominance. In general, the competition between the US, China, Russia, Japan, and Germany is multifaceted, encompassing numerous areas, where AI and FLS facilitate strategic decision-making and provide useful insights, particularly within the defence sector [27].

At the beginning of 2019, to address modern challenges, the Pentagon detailed a plan for the smart application of AI in strategy formulation, safeguarding equipment, and predicting supply systems, thereby seeking to enhance operational effectiveness and minimize unnecessary activities [1]. Alternatively, China is engaged in a worldwide contest with the US for military supremacy, endeavouring to emerge as a global frontrunner in AI innovation. The input data for this research encompasses gross domestic product, defence expenditure, research, and the number of active military personnel, as the modernization of the military is reliant on progress in science and technology [28]. The strategies for modernizing China's military system, as examined in [29], have achieved notable advancement by integrating cutting-edge technologies from the civilian industry. Conversely, the US upholds a tight synergy among the defence industry, academic institutions, and the private sector, with ambiguities highlighted in the assimilation of new energy technologies into military tactics, as indicated in a particular study [30]. China's initiatives to enhance its military capabilities and foster collaboration between the civilian and defence sectors have brought about substantial modifications in US military policy, as detailed in [29]. Moreover, within this context, Russia has made progress in the field of artificial intelligence, with extensive research in the areas of speech recognition, natural language processing and related technologies. Japan is leading the adoption of Internet of Things (IoT) technology, integrating this technology into various aspects of daily life, from smart homes to smart cities [31]. Furthermore, Germany is a leader in implementing the concept of Industry 4.0, which involves the integration of information technology into production processes to create smart and efficient factories [32]. Moreover, it has made significant investments in renewable energy technologies such as solar and wind power, including the development of smart grids for efficient energy management. The existing literature provides a complex and detailed panorama of this competition, highlighting the paradigm shifts and their consequences at the global level [3]. Therefore, the rapid pace of development creates challenges for policy makers in these countries, as mentioned in [4].

## 8.3 Methodological Approach

### *8.3.1 Fuzzy Logic in Smart Systems*

The research approach utilised in this study is rooted in the domain of Soft Computing, which embodies a paradigm for creating adaptable, smart, and resilient systems, adept at handling ambiguity and limited information [33]. This comprehensive approach combines advanced methods such as fuzzy logic, artificial neural networks, probabilistic reasoning, and genetic algorithms.

FLS have been recognised as a fundamental approach for addressing imprecise and uncertain information, in a way influenced by the decision-making process used by humans when confronted with inadequate information. Fuzzy systems use verbal descriptions, making them more accessible than complex mathematical models [34]. FLS utilise approximation reasoning, enabling effective management of uncertainty. It facilitates the representation of control programs through condition-based principles, thereby enhancing the convenience of human operators. Their proficiency in managing ambiguous or unclear data provides them with adaptability across numerous fields. Nevertheless, it is important to acknowledge certain limits, particularly with their implementation in procedures that do not leverage prior human experience [35].

This study incorporates fuzzy logic into hierarchical control systems to enhance thinking and decision-making capabilities. This method is very effective for monitoring and controlling an ongoing flow of data and events. The use of fuzzy logic in the top levels of the control system enhances its capacity to manage macro-decisional levels, therefore facilitating the cohesive integration of imprecise information inside complex processes.

### 8.3.2 Approach Based on Relative Distances

The process of predicting the future developments of the variables under scrutiny in this study, conducted from a temporal standpoint, is addressed through meticulous statistical analysis. As underscored in the specialised literature [36], achieving accurate predictions necessitates the inclusion of external factors in the analysis, and the data quantifying these issues may lack extensive coverage, precision, or completeness. According to the cited source [36], the multi-criteria territorial ranking is accomplished through a variety of methods, including the ranking method, the relative distance method, and the matrix method. To ensure accurate and relevant prediction, the analysis process must integrate significant external factors and use data with varying levels of coverage and precision. This multidimensional approach allows a better understanding and anticipation of the complex dynamics of the studied variables. Consequently, more precise findings and predictions can be attained, even within the uncertainties and constraints of the information being analysed. This is a key component in advanced statistical modelling and analysis. This tool may play a crucial role in data management and interpretation. The relative distance method finds applicability in many fields, such as the statistical analysis of the degree of development of countries, providing a flexible way to analyse and compare different data sets. Through this method, the degree of similarity or difference between two or more phenomena can be determined, taking into account not only the raw values, but also their context and variability. In this research, the focus is on the relative distance method, an approach that proves to be particularly effective in evaluating and forecasting the evolution of variables under uncertainty.

The document employs the relative distance approach ($d_r$—refer to Eq. 8.1), to assess the proximity between territories for ranking purposes. The computation is performed as follows:

1. Computation of the particular territorial indices for each region concerning the maximum distance ($Y_{max}$):
   - the territory with the highest value has a relative distance equal to 100,0%.
   - the remaining values of the initial distance (Yi) are determined using the Eq. (8.1):

$$d_r = I_{i/max} = \frac{Y_i}{Y_{max}} \times 100 \tag{8.1}$$

   where we know that: $I_i$—initial specific index, $I_{max}$—maximum index.

2. For each region, the average distance is computed (utilising the basic geometric mean $\left(\overline{d}\right)$ of relative distances, n—the number of entries examined in the current instance 4) (Eq. 8.2):

$$\overline{d} = \sqrt[n]{\prod_{i=1}^{n} I_{i/max}} = \sqrt[n]{\prod_{i=1}^{n} d_i} \tag{8.2}$$

3. Each regional unit is assigned a ranking based on the average distance from the maximum to the minimum value.

This results in the establishment of a hierarchy among the territories.

## 8.4 Visual Structure of the Proposed Predictive System for Evaluating Development Level

### *8.4.1 Conceptualization and Implementation of the Fuzzy Logic System Design*

The FLS offers a theoretical framework that enables a nation's development system to leverage the ability to accommodate additional uncertain information, thereby improving the outcome by incorporating more data sources [37]. Accordingly, in the setting up of this system, fuzzy logic will be utilised due to its adaptability in various domains, including control systems, artificial intelligence, due to its flexibility and ability to model real-world situations. The system can effectively handle imprecision and uncertainty, providing a more realistic and tailored picture of the dynamics of development and hegemony in the global context. This represents an important step in the modelling of socio-economic and political phenomena, contributing to a better understanding and design of development strategies.

Consequently, for this phase of system design, the identified input and output variables were employed as follows: Gross domestic product (GDP), Military spending budget (MSB), The research and development budget (RDB), The foreign investment budget (FDI) and RESULT.

***Input variables***

Each added input variable has three functions implemented:

- LOW_GDP, MED_GDP and HIGH_GDP, designed for GDP—the most precise assessment of economic performance within the chosen country (see Fig. 8.1);
- LOW_MSB, MED_MSB and HIGH_MSB, intended for MSB—The military budget reflects a country's ability to perceive the possibility of a threat or aggression against it. Moreover, it provides information on the level of funding required to increase security status (see Fig. 8.2);

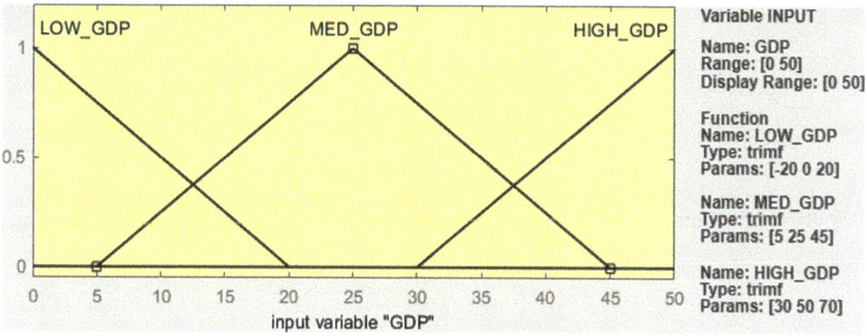


**Fig. 8.1** Graph depicting input functions for the GDP variable

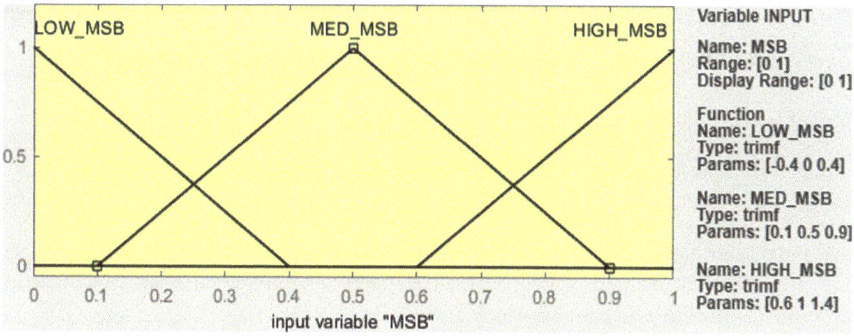


**Fig. 8.2** Graph depicting input functions for the MSB variable

- LOW_RDB, MED_RDB and HIGH_RDB, intended for RDB—The R&D budget is defined as the total expenditure allocated to this category; thus, a substantial value of this variable can elevate a country to a significant competitor on the global stage (see Fig. 8.3);

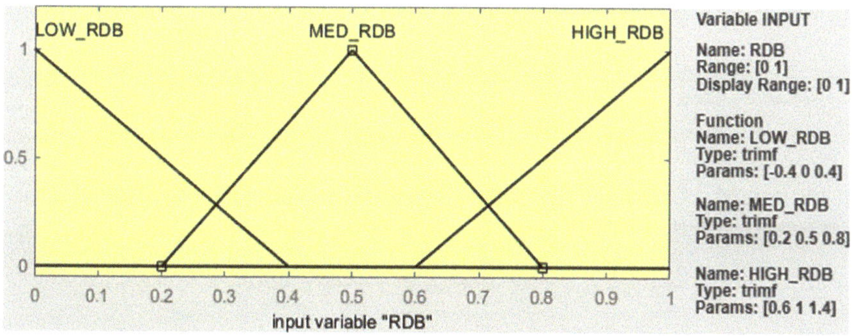


**Fig. 8.3** Graph depicting input functions for the RDB variable

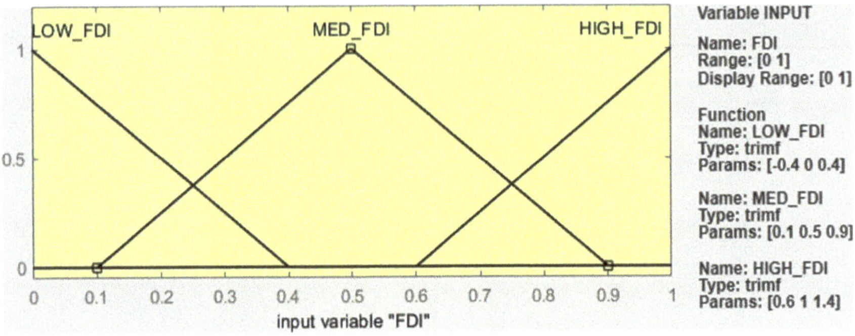


**Fig. 8.4** Graph depicting input functions for the FDI variable

- LOW_FDI, MED_FDI and HIGH_FDI, intended for FDI—This variable encompasses all foreign investments conducted within the country (see Fig. 8.4);

***Output variable***

This variable possesses five functions that align with distinct value ranges:

- VERY_LOW, LOW, MED, HIGH and VERY_HIGH, designed for RESULT—this evaluation will entail categorizing the examined country based on a supremacy score, and the obtained scores can be compared with those of other participating countries in this assessment (see Fig. 8.5).

The set of rules in the envisioned system is a vital component, representing a set of 81 intricate logical rules, fundamental for simulating decisions, based on system inputs. In an effort to optimise clarity and structure, an indexed form has been implemented, aiming to simplify the drafting process, as indicated in the following depiction (refer to Table 8.2).

FLS utilises the Mamdani model due to its intuitive and easy-to-understand features, making it appropriate for applications resembling expert systems such as country development level analysis. The choice of this model integrates rules developed based on the knowledge of human experts. During the validation process,

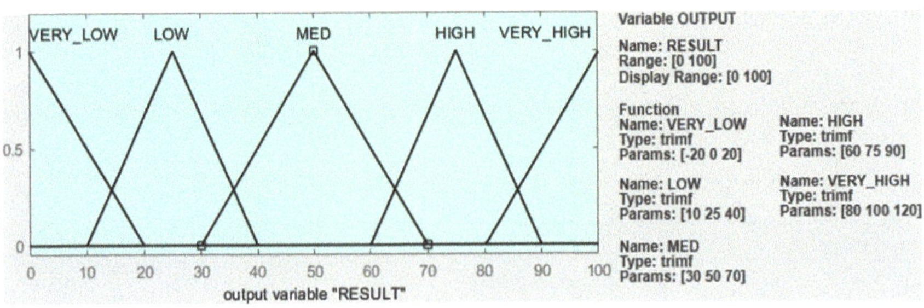


**Fig. 8.5** Graph the functions for the output variable RESULT

**Table 8.2** Rule set of the fuzzy logic system

| No | The rule base | No | The rule base | No | The rule base | No | The rule base | No | The rule base |
|---|---|---|---|---|---|---|---|---|---|
| 1 | 1 1 1 1, 1 (1): 1 | 18 | 1 2 1 1, 1 (1): 1 | 35 | 2 2 1 2, 3 (1): 1 | 52 | 2 3 1 2, 3 (1): 1 | 69 | 2 2 3 3, 5 (1): 1 |
| 2 | 2 2 2 1, 2 (1): 1 | 19 | 1 3 1 1, 1 (1): 1 | 36 | 3 3 1 2, 4 (1): 1 | 53 | 2 1 3 2, 3 (1): 1 | 70 | 3 2 2 3, 5 (1): 1 |
| 3 | 3 3 3 1, 3 (1): 1 | 20 | 2 1 1 1, 1 (1): 1 | 37 | 2 1 2 2, 2 (1): 1 | 54 | 3 1 2 2, 3 (1): 1 | 71 | 2 3 2 3, 5 (1): 1 |
| 4 | 1 2 3 1, 2 (1): 1 | 21 | 3 1 1 1, 1 (1): 1 | 38 | 3 1 3 2, 4 (1): 1 | 55 | 1 1 1 3, 1 (1): 1 | 72 | 1 2 1 3, 3 (1): 1 |
| 5 | 3 2 1 1, 2 (1): 1 | 22 | 1 1 2 1, 1 (1): 1 | 39 | 2 3 3 2, 4 (1): 1 | 56 | 2 2 2 3, 4 (1): 1 | 73 | 1 3 1 3, 3 (1): 1 |
| 6 | 1 2 2 1, 1 (1): 1 | 23 | 1 1 3 1, 1 (1): 1 | 40 | 3 3 2 2, 4 (1): 1 | 57 | 3 3 3 3, 5 (1): 1 | 74 | 2 1 1 3, 3 (1): 1 |
| 7 | 1 3 3 1, 3 (1): 1 | 24 | 1 3 2 1, 2 (1): 1 | 41 | 3 2 3 2, 4 (1): 1 | 58 | 1 2 3 3, 4 (1): 1 | 75 | 3 1 1 3, 3 (1): 1 |
| 8 | 2 2 1 1, 2 (1): 1 | 25 | 2 3 1 1, 2 (1): 1 | 42 | 2 2 3 2, 4 (1): 1 | 59 | 3 2 1 3, 4 (1): 1 | 76 | 1 1 2 3, 3 (1): 1 |
| 9 | 3 3 1 1, 3 (1): 1 | 26 | 2 1 3 1, 2 (1): 1 | 43 | 3 2 2 2, 4 (1): 1 | 60 | 1 2 2 3, 3 (1): 1 | 77 | 1 1 3 3, 3 (1): 1 |
| 10 | 2 1 2 1, 1 (1): 1 | 27 | 3 1 2 1, 2 (1): 1 | 44 | 2 3 2 2, 4 (1): 1 | 61 | 1 3 3 3, 5 (1): 1 | 78 | 1 3 2 3, 4 (1): 1 |
| 11 | 3 1 3 1, 3 (1): 1 | 28 | 1 1 1 2, 1 (1): 1 | 45 | 1 2 1 2, 2 (1): 1 | 62 | 2 2 1 3, 4 (1): 1 | 79 | 2 3 1 3, 4 (1): 1 |
| 12 | 2 3 3 1, 3 (1): 1 | 29 | 2 2 2 2, 3 (1): 1 | 46 | 1 3 1 2, 2 (1): 1 | 63 | 3 3 1 3, 5 (1): 1 | 80 | 2 1 3 3, 4 (1): 1 |
| 13 | 3 3 2 1, 3 (1): 1 | 30 | 3 3 3 2, 4 (1): 1 | 47 | 2 1 1 2, 2 (1): 1 | 64 | 2 1 2 3, 3 (1): 1 | 81 | 3 1 2 3, 4 (1): 1 |
| 14 | 3 2 3 1, 3 (1): 1 | 31 | 1 2 3 2, 3 (1): 1 | 48 | 3 1 1 2, 2 (1): 1 | 65 | 3 1 3 3, 5 (1): 1 | – | – |
| 15 | 2 2 3 1, 3 (1): 1 | 32 | 3 2 1 2, 3 (1): 1 | 49 | 1 1 2 2, 2 (1): 1 | 66 | 2 3 3 3, 5 (1): 1 | – | – |
| 16 | 3 2 2 1, 3 (1): 1 | 33 | 1 2 2 2, 2 (1): 1 | 50 | 1 1 3 2, 2 (1): 1 | 67 | 3 3 2 3, 5 (1): 1 | – | – |
| 17 | 2 3 2 1, 3 (1): 1 | 34 | 1 3 3 2, 4 (1): 1 | 51 | 1 3 2 2, 3 (1): 1 | 68 | 3 2 3 3, 5 (1): 1 | – | – |

the Centroid, Bisector, and Mean of Maximum (MoM) methods were employed to ascertain the values, enhancing the accuracy and reliability of the analysis (refer to Table 8.3).

The values displayed in a unique format indicate the response of the logical fuzzy system to various examined data. This method offers enhanced adaptability and precision in analysing intricate economic data. The utilisation of the Mamdani

**Table 8.3** Values derived during the defuzzification process

| No | Input variables | | | | Defuzzification | | |
|---|---|---|---|---|---|---|---|
| | | | | | Output variable | | |
| | GDP | MSB | RDB | FDI | Result | | |
| | | | | | Centroid | Bisector | MOM |
| 1 | 5 | 0.10 | 0.10 | 0.10 | 6.72 | 6 | 2.5 |
| 2 | 8 | 0.15 | 0.10 | 0.20 | 23.8 | 16 | 5 |
| 3 | 10 | 0.20 | 0.20 | 0.10 | 15.6 | 13 | 5 |
| 4 | 12 | 0.25 | 0.20 | 0.20 | 30.4 | 29 | 6 |
| 5 | 15 | 0.30 | 0.30 | 0.30 | 35.5 | 38 | 50 |
| 6 | 20 | 0.40 | 0.40 | 0.40 | 50 | 50 | 50 |
| 7 | 25 | 0.50 | 0.50 | 0.50 | 50 | 50 | 50 |
| 8 | 30 | 0.60 | 0.60 | 0.60 | 50 | 50 | 50 |
| 9 | 35 | 0.70 | 0.70 | 0.70 | 64.5 | 62 | 50 |
| 10 | 40 | 0.80 | 0.80 | 0.80 | 84.4 | 87 | 95 |
| 11 | 45 | 0.90 | 0.90 | 0.90 | 93.3 | 94 | 97.5 |
| 12 | 50 | 1.00 | 1.00 | 1.00 | 93.7 | 94 | 100 |

inference engine alongside the Centroid, Bisector, and MoM approaches for defuzzification is advantageous, since they offer an optimal balance between accuracy and computational complexity. The Centroid approach facilitates precise defuzzification by achieving balance among varying degrees of truth. The bisector offers a technique for balancing the impact of minor and major values. The MoM technique is particularly useful in scenarios where maxima are of significant importance. These methods enhance efficiency and precision in intricate systems, proving indispensable for comprehensive economic and social assessments.

The graphs generated correspond to: Centroid method (refer to Fig. 8.6), Bisector method (refer to Fig. 8.7), and MoM method (refer to Fig. 8.8).

The graph depicting the bisector mode is displayed in the following figure.

In the last illustration, the MoM method is used, as can be seen below.

### *8.4.2 Realization of the Implemented Graphic Architecture System*

The purpose of the system is to evaluate development performance for countries such as the US, China, Russia, Japan and Germany. This approach considers input variables such as current GDP, military expenditure, research and development budget, and foreign direct investment.

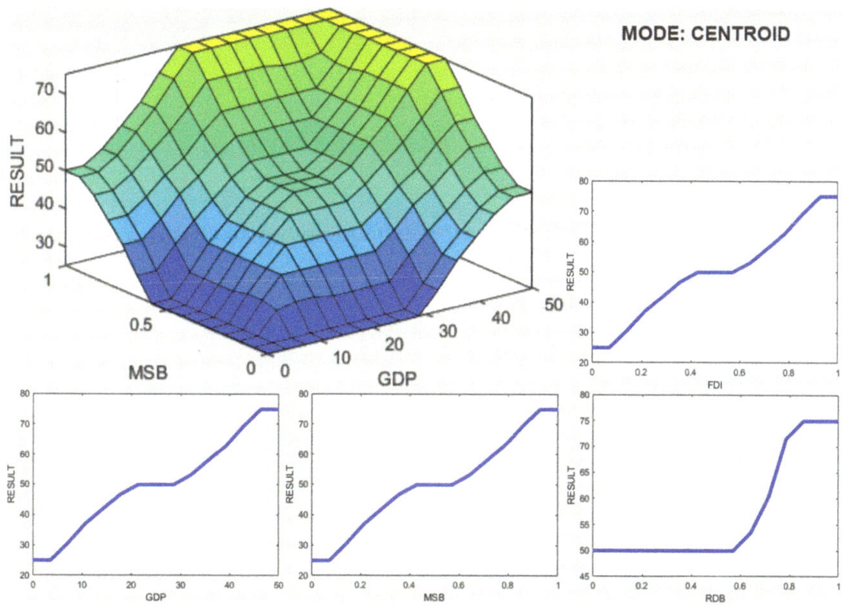


Fig. 8.6 Visual representation of the centroid method

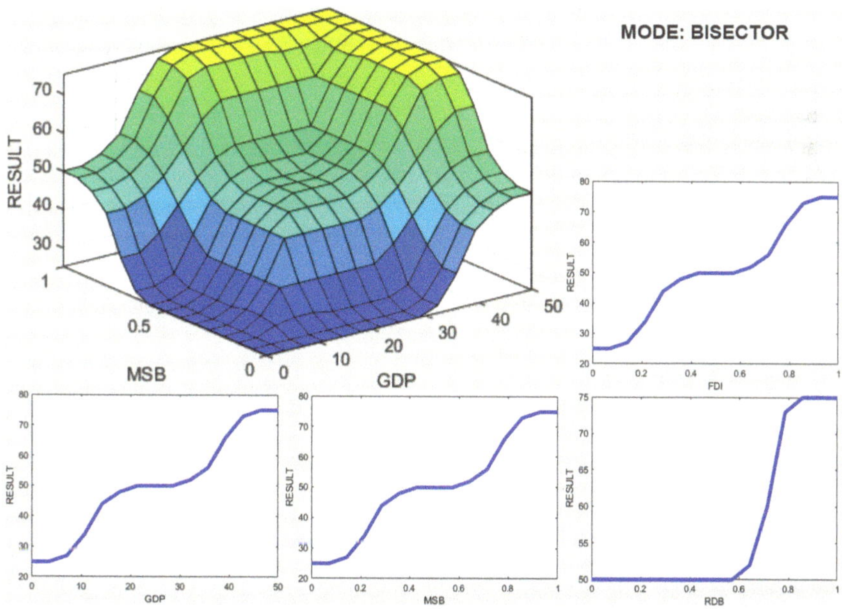


Fig. 8.7 Visual representation of the bisector method

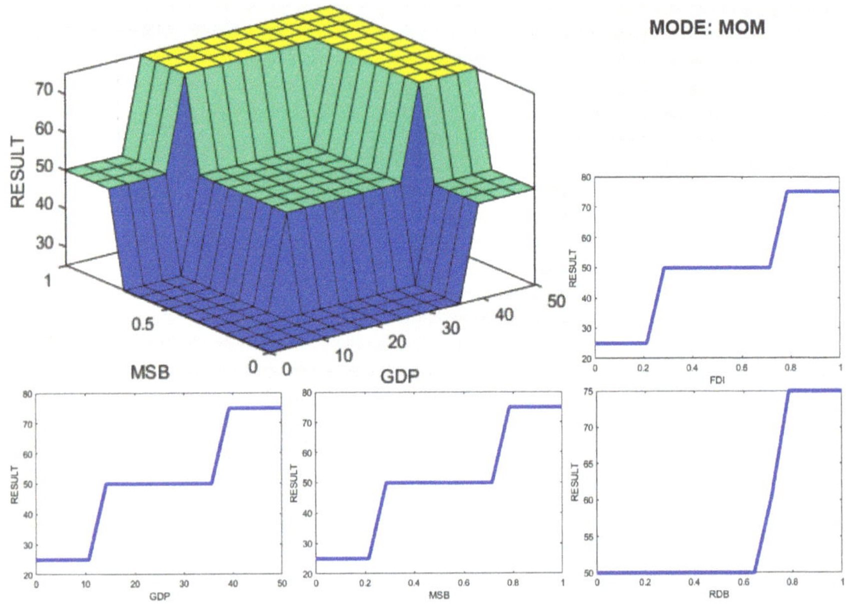


**Fig. 8.8** Visual representation of the MoM method

Five countries have been selected for their substantial development potential across various domains to demonstrate the functionality of the proposed system. The United States of America, People's Republic of China, Russian Federation, Japan and Germany [38] fulfil the requisite criteria for growth and development over the years. The development score for each area of activity, as achieved by each participating country, is assessed and rated in a designated column. For a more extensive study, a chronological time span arranged by years was employed, resulting in the ranking of the input data spanning from 2000 to 2022 [39].

Each year in this ranking is linked to a numerical value recorded in the specialised literature to simulate many scenarios in a realistic setting [22]. The system possesses multiple capabilities and handles different perspectives to encompass multiple scenarios. Another approach is to calculate the arithmetic mean value by summing the values for each variable across the years in the dataset. This may provide a perspective on the overall trend and average performance of each variable over the specified time frame. A simple calculation method was incorporated to forecast the future values for each input variable in the next year. The approach involves utilising the data for each input variable from the preceding two years and employing mathematical inference to estimate the outcomes for a forthcoming year [40, 41]. Hence, utilising this method allows for the projection of future values over an extended time span. This prediction scenario is formulated as a comparison, as the mathematical inference approach assessing the past two years cannot yield highly accurate findings. Therefore, the current system is completed by programs written in the MATLAB programming language, in which the prediction is calculated over a longer interval of years from 2000 to 2022, and the calculation method utilised is linear regression, which provides advanced levels of accuracy.

The overall framework of the system is delineated in stages, with the initial illustration exhibiting the system for the United States of America (see Fig. 8.9).

The second stage illustrates the system for the People's Republic of China (see Fig. 8.10).

The third stage illustrates the system for the Russian Federation (see Fig. 8.11).

The fourth stage illustrates the system for Japan (see Fig. 8.12).

The fifth stage illustrates the system for Germany (see Fig. 8.13).

The final stage in the system is to generate the value indicated below (see Fig. 8.14).

Five system architectures were developed to perform data collection. The system can manually incorporate new data, view the percentage growth of countries for the input variables GDP, MSB, RDB and FDI using the years' values, view the prediction based on mathematical inference for the year 2023 and show the arithmetic mean of the values of the years for each country, being able to easily determine which state experienced a more prosperous development.

The last architecture is the interconnection of all the subsystems, and it displays the responses and visualises the scores obtained. Therefore, the proposed system consists of the following architectures:

1. System architecture for the United States of America;
2. System Architecture for the People's Republic of China;
3. System architecture for the Russian Federation;
4. System architecture for Japan;
5. System architecture for Germany;
6. Interconnection architecture of the subsystems of the candidate countries.

This research intends to enhance the current system, employing a strategy designed to substantially contribute to the forecasting of long-term evolution. The robust linear regression method was employed to generate predictions for the subsequent five years in order to accomplish this objective.

This system development process involved integrating all relevant data for the input variables, using values spanning the years 2000–2022. This data was meticulously collected and organised into five separate Excel files, each corresponding to a distinct variable. This approach ensured a solid and comprehensive basis for the data utilised in our analysis and prediction process.

The analysis of previous data led to the development of a specialized program, which, by implementing the linear regression technique, can anticipate the evolution of the respective variables for the next five years. This program is based on a careful evaluation of trends and relationships identified in historical data, thus providing reliable and relevant predictions for the future period.

The results obtained from this analysis and prediction process are documented and saved in five separate spreadsheet files, ensuring that the generated information is readily accessible and can be utilised for additional investigation or interpretation. This comprehensive and rigorous approach represents a significant step towards the

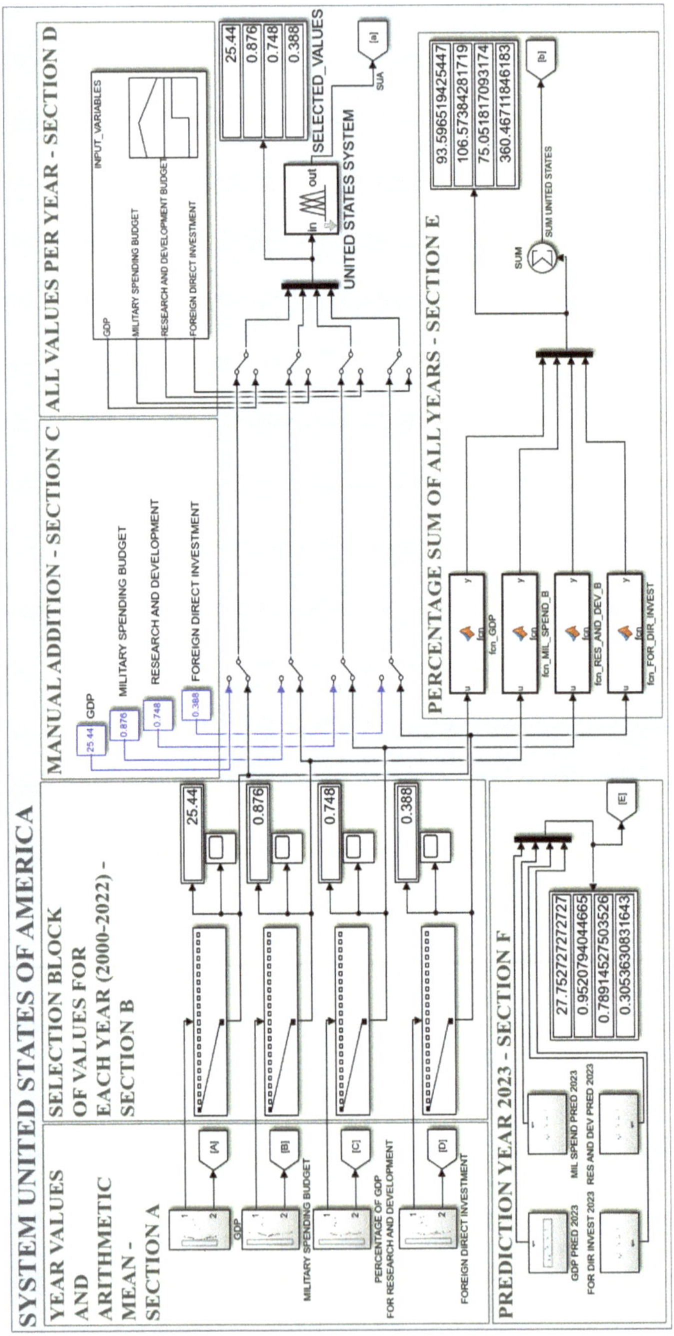


**Fig. 8.9** System for the United States of America

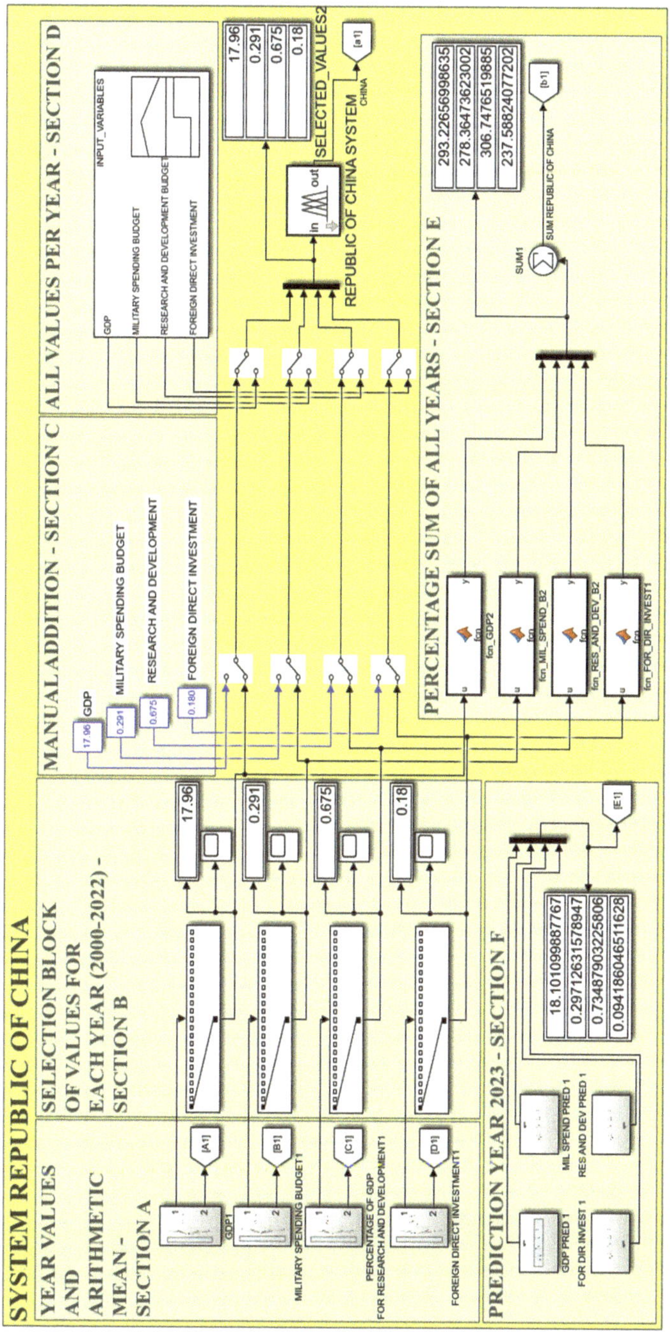


**Fig. 8.10** System for People's Republic of China

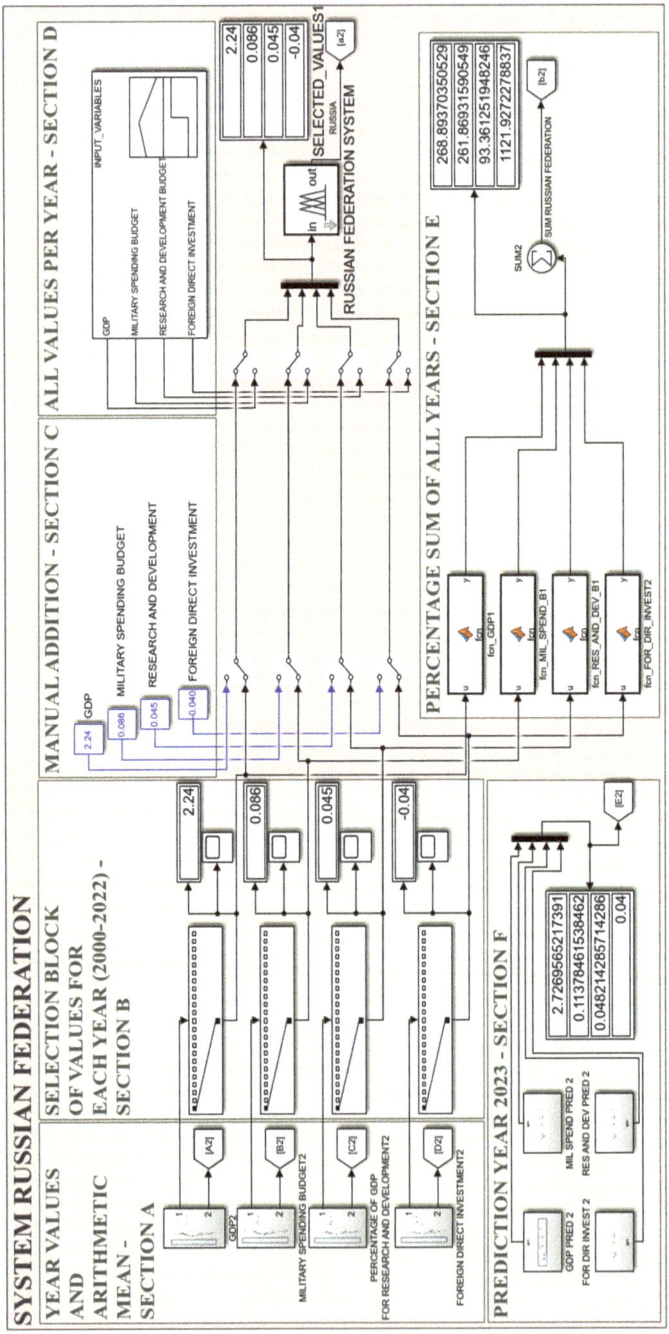


Fig. 8.11 System for the Russian Federation

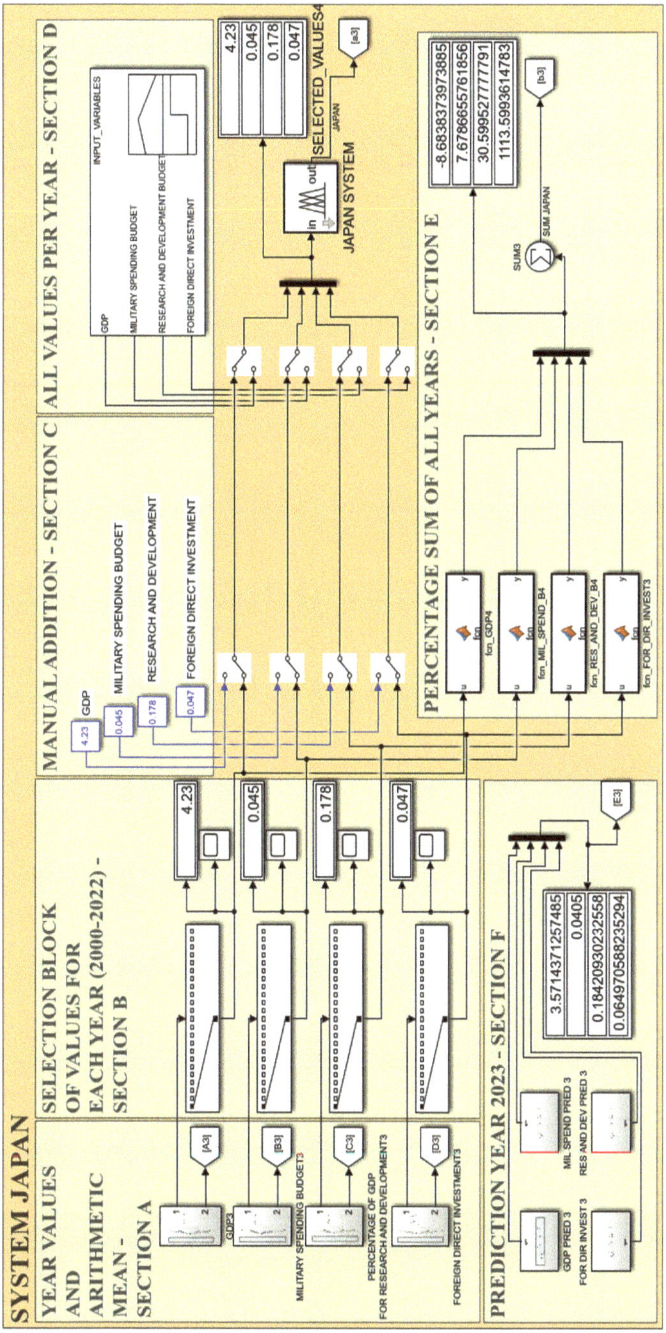


Fig. 8.12 System for Japan

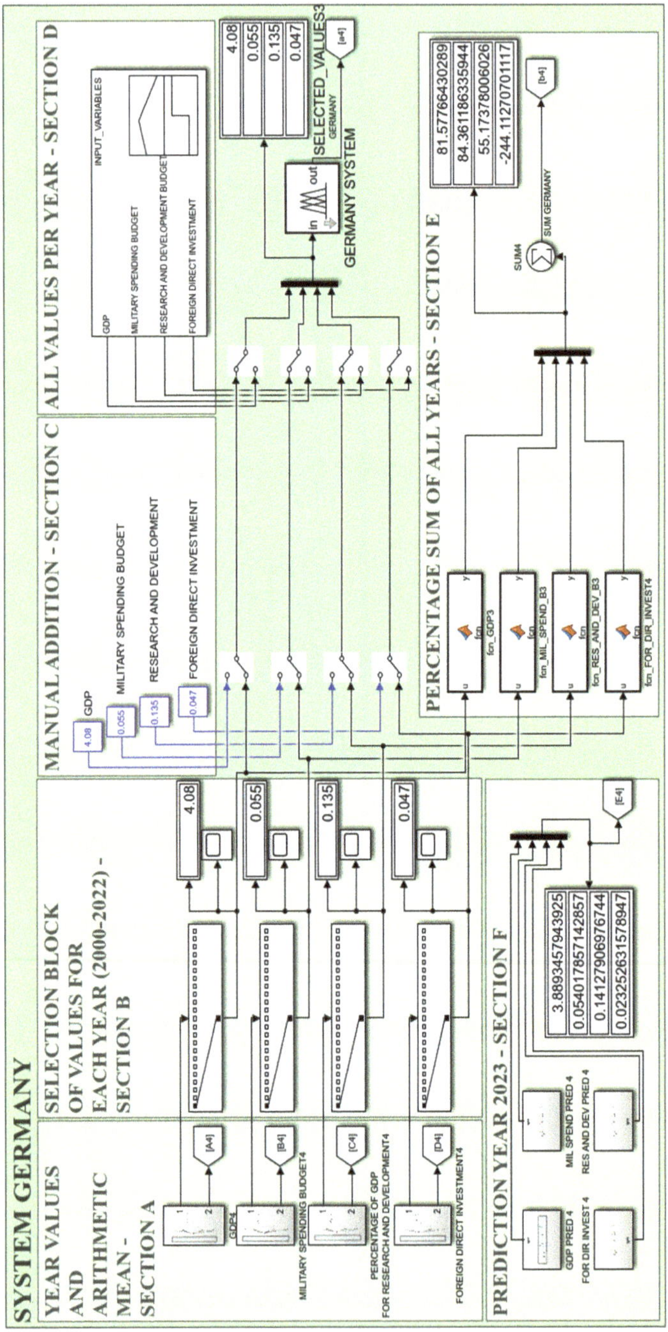


Fig. 8.13 The system for Germany

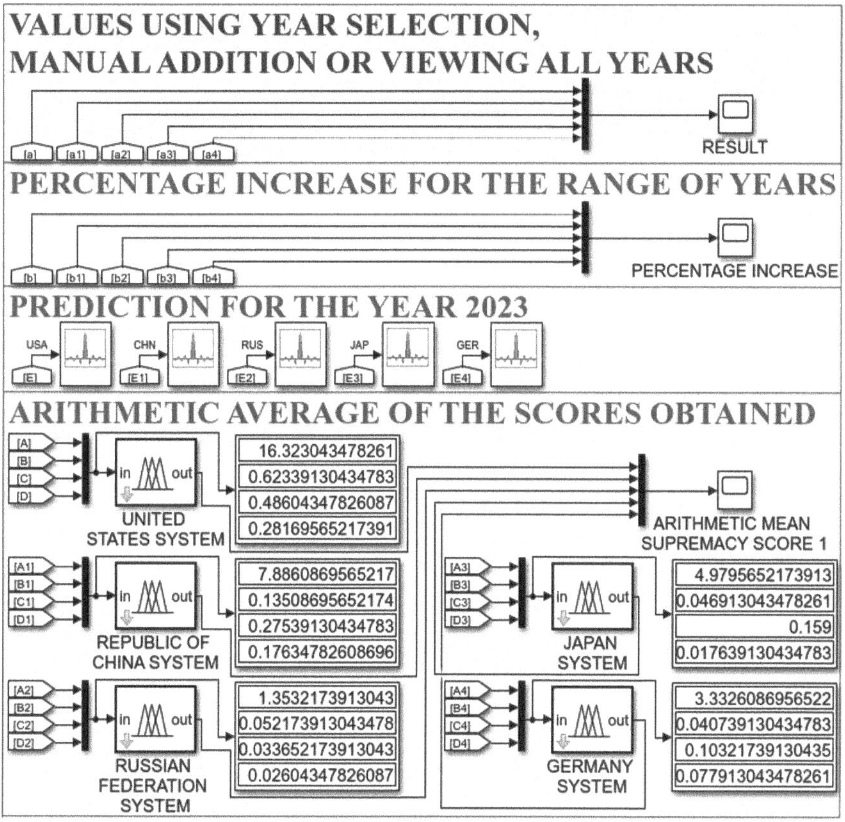


**Fig. 8.14** Results display stage

development of an effective and robust predictive system. A sophisticated automated method was developed to calculate predictions for the next five years in the continued development of the system.

To provide a detailed insight into the implementation, the program code is presented in pseudocode format and can be reviewed in Appendix 8.1. This stage presents a block diagram that illustrates the functioning of the linear regression method in pseudo-code. Consequently, the below graphic illustrates the operation of the proposed algorithm (see Fig. 8.15).

Upon successful completion of the aforementioned linear regression procedure, the resultant prediction data is observable.

The following program will perform the calculation for the relative distances and the geometric mean for the GDP, MSB, RDB and FDI variables of each candidate country the USA, China, Russia, Japan and Germany for the years 2023–2027. For this calculation the equations presented above are utilised (see Eqs. 8.1 and 8.2). In the next step, the program for calculating the relative distances and the geometric mean is presented in the form of a pre-code to simplify the writing of the

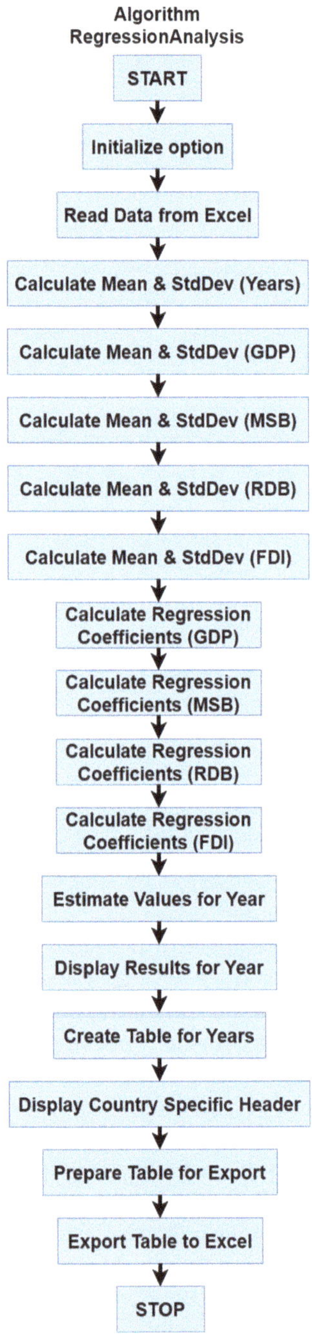


**Fig. 8.15** Block diagram for the regression analysis algorithm

program and can be analysed in Appendix 8.2. This stage presents a block diagram that demonstrates the algorithm's capability for calculating relative distance and geometric mean, expressed in pseudocode. Consequently, the below figure illustrates the functionality of the proposed algorithm (see Fig. 8.16).

Upon concluding the aforementioned step, we will initiate the program for the application of artificial intelligence to verify and recalculate the data spanning the years 2000–2022 of the GDP, MSB, RDB and FDI variables from the tables, for the candidate countries. The developed program implements a prediction approach using recurrent neural networks (RNN) with Long Short-Term Memory (LSTM) specifications. The data from the training is comprised of test and load files from each candidate country. The most recent data is presented first in the training and test data, as the orientation of the lines is reversed. The data is then split into training (70%), validation (15%) and test (15%) sets using the *dividerand* function. Subsequently, the data is normalised to a specified range of 0–1, since this enhances the training of the neural network and mitigates issues associated with divergence or slow convergence. An LSTM network is configured with a sequential input, two LSTM layers with 500 hidden units each, a fully connected state and a regression layer. This architecture was designed to facilitate the management of time sequences. Options for network training are specified, including Adam optimization algorithm, maximum number of training epochs 1000, gradient threshold, initial learning rate, automatic execution environment, training progress display, validation data, batch size and others. The training progress is then illustrated in a graph presented in chapter five. Subsequent to the training stage, the model is employed to generate predictions on the test data. Predictions and test data are subsequently normalised to enhance the interpretation of the results. Finally, several diagrams are generated to evaluate the performance of the model, including the response graph, the residual graph, the prediction and original graph, the cumulative absolute error graph and the distribution of the residuals. These diagrams provide a clear picture of the quality of the model's predictions and can be viewed in the results section. The analysis focuses on interpreting the aforementioned graphs and extracting valuable information from them. The response graph illustrates the correlation between actual and predicted values, offering insight into the accuracy of the model. Nevertheless, the residual plot illustrates the discrepancies between expected and actual values, facilitating the identification of any anomalous patterns or trends in the data. The prediction and original graphic juxtapose the predicted and actual values throughout the whole dataset, offering insight into the model's consistency. The Cumulative Absolute Error plot indicates how errors accumulate as predictions are formulated, providing insight into the model's stability over time. The distribution of residuals ultimately aids in evaluating the normality of errors, a crucial indicator of model adequacy. The subsequent stage presents the deep learning program, based on predictions, in the form of pre-code to facilitate program writing, which can be examined in Appendix 8.3. This section will offer a block diagram that demonstrates the functionality of the deep learning algorithm through pseudocode predictions. Consequently, the below figure illustrates the functionality of the proposed algorithm (see Fig. 8.17).

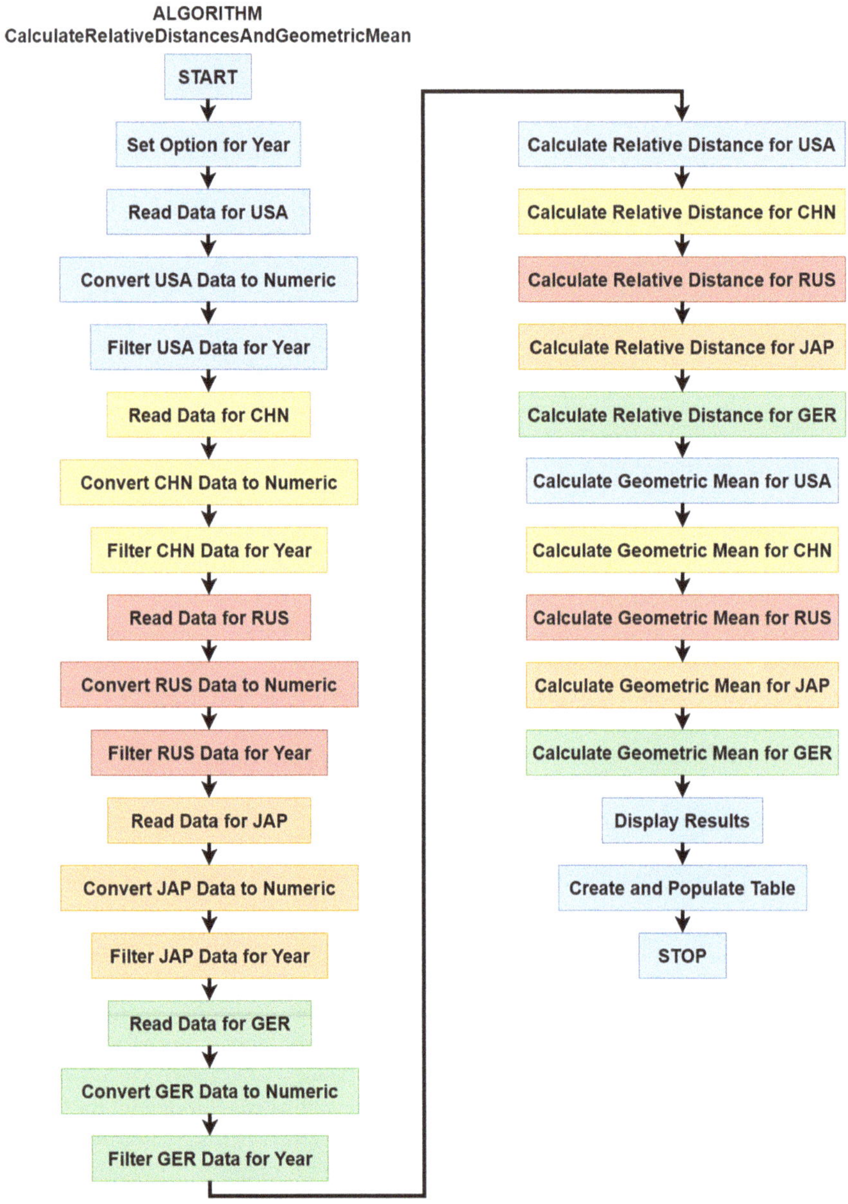


**Fig. 8.16** Block diagram illustrating the algorithm to calculate relative distances and the geometric mean

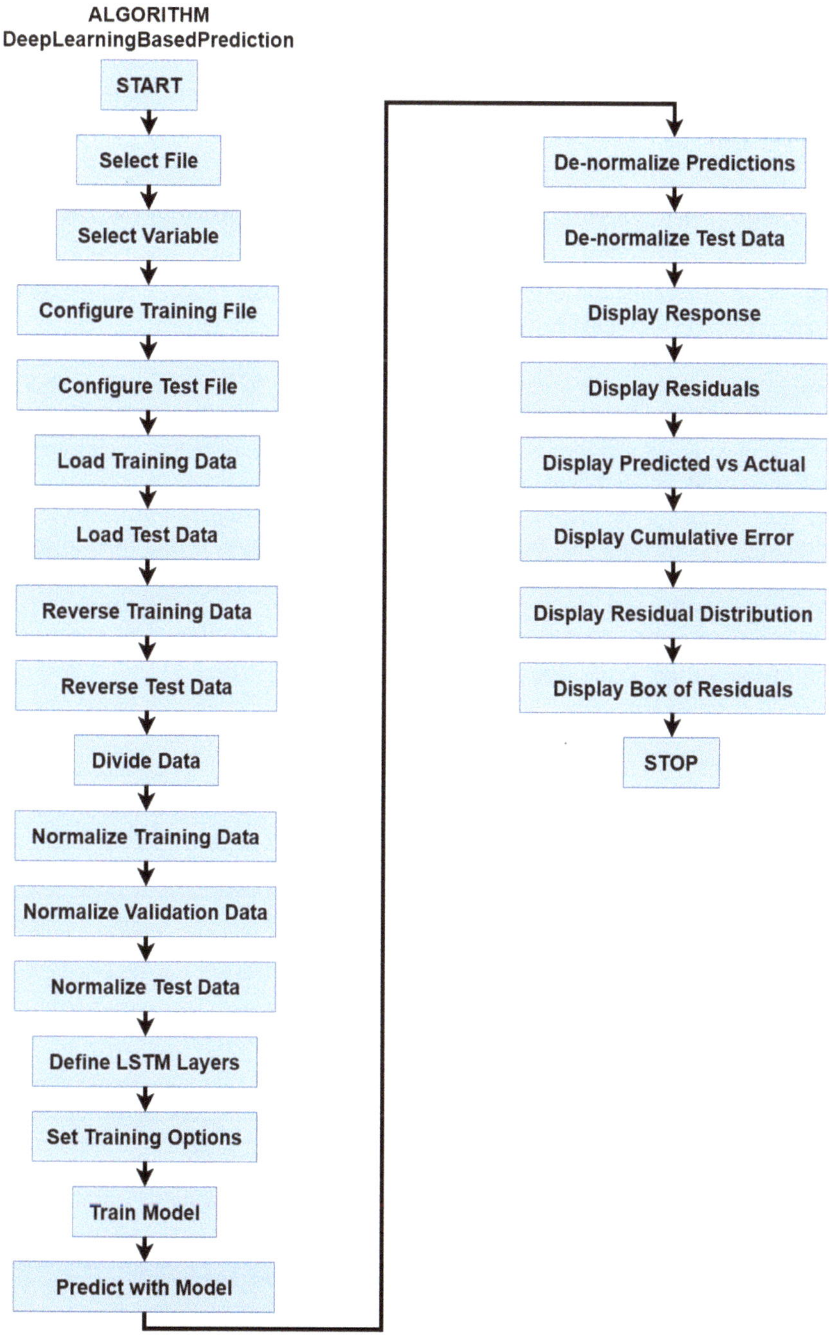


**Fig. 8.17** Block diagram for deep learning based prediction algorithm

## 8.5 Analysis of Outcomes

The methodology for achieving outcomes is based on the careful selection of input data. The data, once entering the designated system, undergoes analysed according to the established rules, and in the final step, the results are produced and illustrated graphically. Diverse testing methods were essential to produce a wide range of output values, so enabling a comprehensive comparison between the two involved nations. Utilising the specified forms to select the required input values within the system enables the determination of the generated response. The inputs employed were derived from the following categories:

Table 8.4 presents the data from the input parameters for the year 2022, selected through the procedure, for each country under research. To facilitate understanding, the numeric figures have been meticulously converted into trillions. This thorough process guarantees a rigorous methodology for data analysis and interpretation, leading to valid and pertinent outcomes.

Upon reviewing the records and implementing the established criteria for data extraction, it is obvious that the USA has a leading position with a superior score for the year 2022. These values were determined by employing the manual data addition mode for each candidate country. The simulation's additional values are documented in Table 8.3. Consequently, the subsequent figure indicates the score achieved according to the specified criteria (refer to Fig. 8.18).

**Table 8.4** Input data for year 2022

| Country | Variables | | | |
|---|---|---|---|---|
| | GDP | MSB | RDB | FDI |
| USA | 25.44 | 0.876 | 0.748 | 0.388 |
| CHN | 17.96 | 0.291 | 0.675 | 0.180 |
| RUS | 2.24 | 0.086 | 0.045 | −0.040 |
| JAP | 4.23 | 0.045 | 0.178 | 0.047 |
| GER | 4.08 | 0.055 | 0.135 | 0.047 |

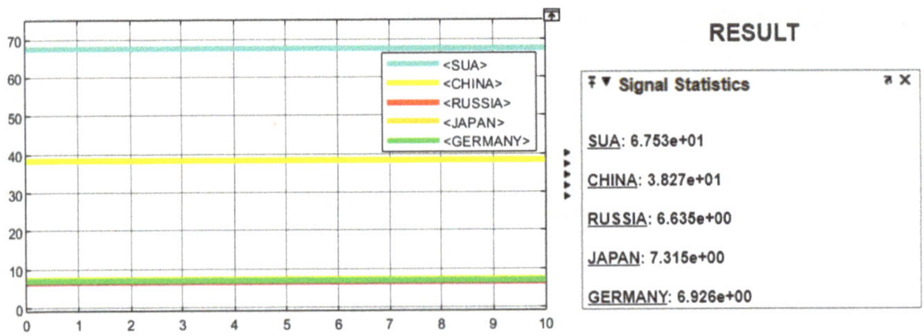


**Fig. 8.18** The rating derived for the year 2022 through manual data entry

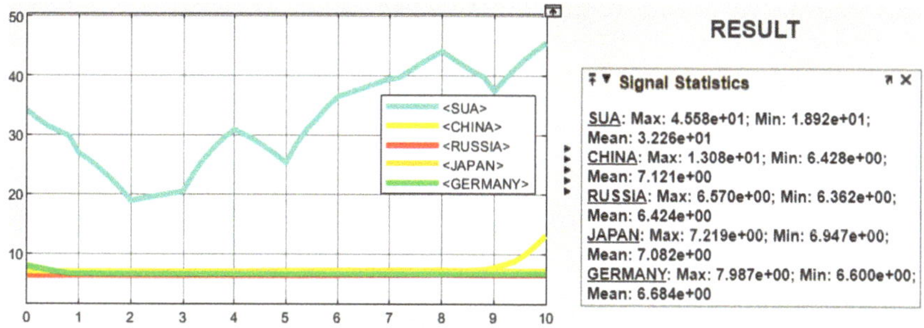


**Fig. 8.19** The rating achieved from examining the span of years

The subsequent chart examines the spectrum of input data from 2000 to 2022 and presents a visualisation of the annual score progression. The highest-ranked country is the USA, which has consistently demonstrated a good trajectory over the years. China, Russia, Japan, and Germany attained a similar outcome. Simultaneously, it is evident that China exhibits a growth trajectory post-2022, which will influence a more favourable position in the future. The data utilised are processed according to the fuzzy logic rule foundation, and the responses are shaped by the rules stated in Table 8.1. The subsequent image depicts the response on the graph for each country throughout the selected range of years and utilising the specified characteristics (refer to Fig. 8.19).

The following stage involves analysing percentage growth within the specified interval from 2000 to 2022 for this category, resulting in a more favourable response compared to other candidates in the analysis. Conversely, states from other positions in the ranking should not be overlooked. Consequently, Japan secured the second position, succeeded by China, the USA, and Germany. The subsequent graph presents the percentage growth trajectory over a specified range of years (refer to Fig. 8.20).

At this stage, the prediction component will be performed to ascertain the values for the year 2023 by mathematical inference derived from the preceding two years, 2022 and 2021. As stated in this research, mathematical inference is not the most

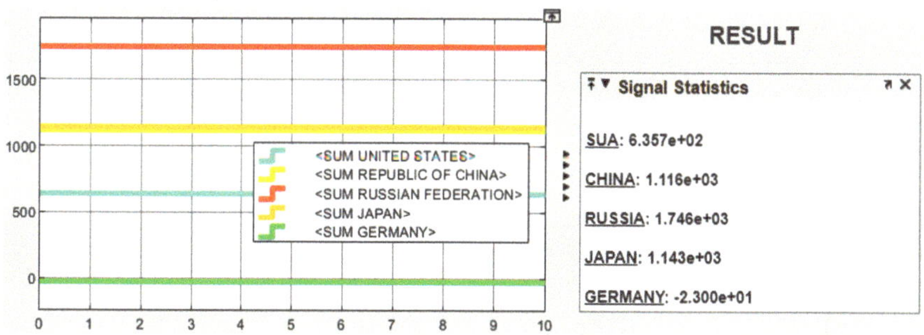


**Fig. 8.20** Growth rate percentage calculated over the span of years

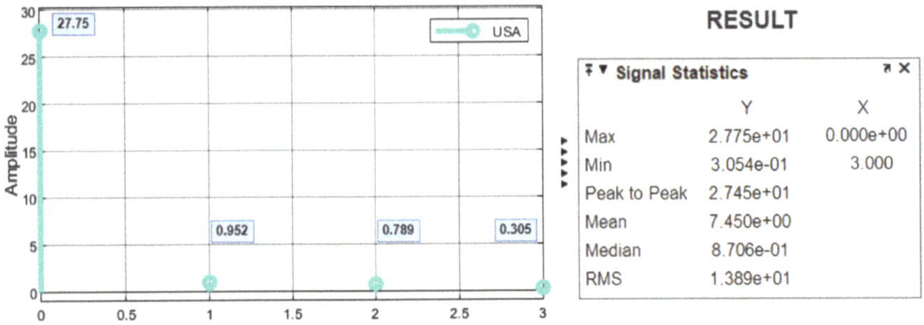


**Fig. 8.21** The rating achieved for forecasting the year 2023 for the United States of America

efficient way for calculating the selected predictions when utilising limited data. This method is employed to evaluate the results with those obtained through linear regression, which achieves significantly greater accuracy. The subsequent statistics illustrate the scores achieved for the 2023 predictions for each candidate country. The initial illustration pertains to the United States of America (refer to Fig. 8.21).

For the prediction of the year 2023 in the context of China, the following illustration is presented (refer to Fig. 8.22).

This graph illustrates the response for this category pertaining to Russian Federation (refer to Fig. 8.23).

The following figure illustrates the response of the prediction values for the year 2023 for Japan (refer to Fig. 8.24).

At the end of this stage of making the prediction for the year 2023 utilising the mathematical differential method, the graph for Germany is illustrated (refer to Fig. 8.25).

The graph below illustrates the method for calculating the arithmetic mean for each reinforcement variable within the fuzzy logic system. Thus, all the data within the interval of years for each individual variable are collected and the arithmetic mean is calculated, and after the values are sent to the four inputs intended for the system variables. The purpose is to illustrate a score of the average values from the range

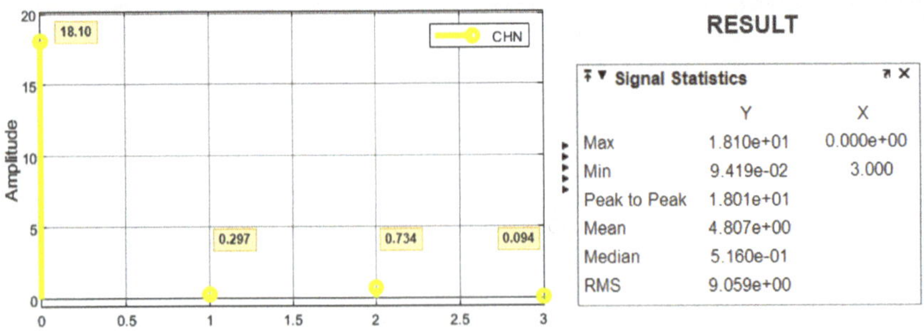


**Fig. 8.22** The rating achieved for forecasting the year 2023 for the People's Republic of China

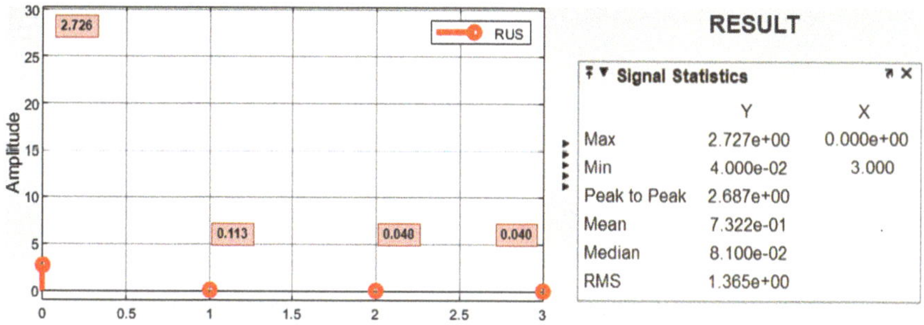


**Fig. 8.23** The rating achieved for forecasting the year 2023 for the Russian Federation

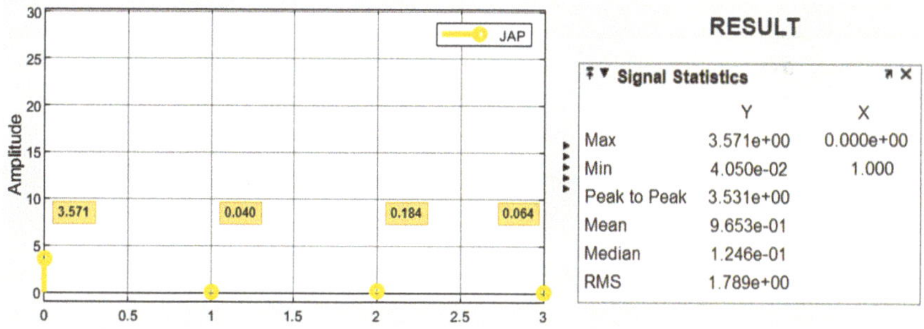


**Fig. 8.24** Prediction score for Japan 2023

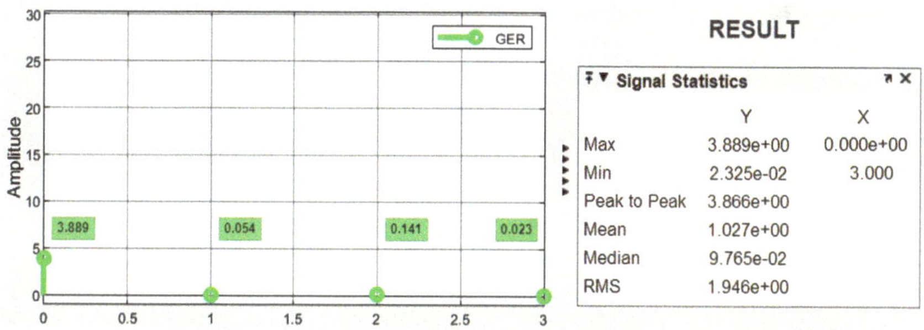


**Fig. 8.25** The rating achieved for forecasting the year 2023 for Germany

of years. The following figure presents the response of the system for the arithmetic mean (refer to Fig. 8.26).

The linear regression calculation method is specified below.

The total of the years within the historical data span is computed (refer to Eq. 8.3):

$$\overline{x} = \frac{a}{Number\_of\_years} \tag{8.3}$$

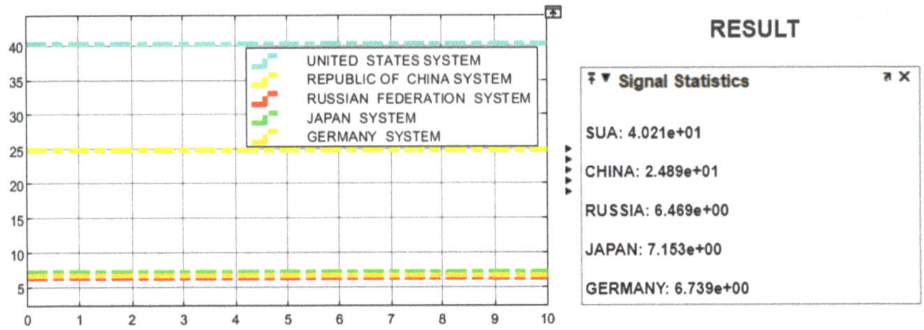


**Fig. 8.26** The rating derived from calculating the cumulative arithmetic average over the series of years

The aggregate of the yearly values within the data history period is determined (refer to Eq. 8.4):

$$\bar{y} = \frac{b}{Number\_of\_values} \tag{8.4}$$

We determine the regression coefficient (refer to Eq. 8.5):

$$\beta_1 = \frac{\sum_{i=1}^{n} (xi - \bar{x})(yi - \bar{y})}{\sum_{i=1}^{n} (xi - \bar{x})^2} \tag{8.5}$$

We compute the intercept $\beta_0$ (refer to Eq. 8.6):

$$\beta_0 = \bar{y} - \beta_1 \cdot \bar{x} \tag{8.6}$$

Prediction (see Eq. 8.7):

$$Value_{Prediction} = \beta_0 + \beta_1 \cdot year \tag{8.7}$$

To assess the system's performance thoroughly, a computation method utilising the linear regression formula was implemented over a long period, with the analysis covering the next five years. The program was utilised to generate the prediction component, which was then subjected to linear regression to produce future values in accordance with the previously mentioned formulas. The first illustration is the display of the prediction values for the United States of America (refer to Fig. 8.27).

The subsequent step presents the values acquired for the People's Republic of China (refer to Fig. 8.28).

This new stage highlights the values of the Russian Federation (refer to Fig. 8.29).

The values acquired for Japan are presented at this stage (refer to Fig. 8.30).

UNITED STATES OF AMERICA

| | GDP | MSB | RDB | FDI |
|---|---|---|---|---|
| 2027 | 26.053 | 0.92866 | 0.74065 | 0.41468 |
| 2026 | 25.445 | 0.90958 | 0.72474 | 0.40636 |
| 2025 | 24.837 | 0.8905 | 0.70883 | 0.39805 |
| 2024 | 24.229 | 0.87142 | 0.69291 | 0.38974 |
| 2023 | 23.621 | 0.85234 | 0.677 | 0.38143 |

**Fig. 8.27** Forecasting the upcoming five years for the USA, using linear regression analysis

REPUBLIC OF CHINA

| | GDP | MSB | RDB | FDI |
|---|---|---|---|---|
| 2027 | 20.757 | 0.34663 | 0.73386 | 0.33556 |
| 2026 | 19.953 | 0.33341 | 0.7052 | 0.32561 |
| 2025 | 19.148 | 0.32019 | 0.67655 | 0.31566 |
| 2024 | 18.344 | 0.30696 | 0.6479 | 0.30571 |
| 2023 | 17.54 | 0.29374 | 0.61924 | 0.29575 |

**Fig. 8.28** Forecasting the upcoming five years for the China, using linear regression analysis

RUSSIAN FEDERATION

| | GDP | MSB | RDB | FDI |
|---|---|---|---|---|
| 2027 | 2.5478 | 0.10245 | 0.048182 | 0.024289 |
| 2026 | 2.4732 | 0.099308 | 0.047274 | 0.024398 |
| 2025 | 2.3985 | 0.096166 | 0.046366 | 0.024508 |
| 2024 | 2.3238 | 0.093024 | 0.045458 | 0.024618 |
| 2023 | 2.2492 | 0.089881 | 0.044549 | 0.024727 |

**Fig. 8.29** Forecasting the upcoming five years for the Russia, using linear regression analysis

This final step illustrates the results acquired for Germany (refer to Fig. 8.31).

The relative distance approach is employed alongside the geometric mean formula, as specified in Eqs. 8.1 and 8.2, to ascertain the ranking order of the five countries. The examination of the results concluded that the United States of America achieved the most favourable outcome in comparison to the other candidate countries.

| JAPAN | GDP | MSB | RDB | FDI |
|---|---|---|---|---|
| 2027 | 5.1962 | 0.052194 | 0.1858 | 0.045125 |
| 2026 | 5.1826 | 0.051864 | 0.18412 | 0.043407 |
| 2025 | 5.1691 | 0.051534 | 0.18245 | 0.04169 |
| 2024 | 5.1556 | 0.051204 | 0.18077 | 0.039972 |
| 2023 | 5.142 | 0.050874 | 0.1791 | 0.038254 |

**Fig. 8.30** Forecasting the upcoming five years for the Japan, using linear regression analysis

| GERMANY | GDP | MSB | RDB | FDI |
|---|---|---|---|---|
| 2027 | 4.7652 | 0.058905 | 0.14787 | 0.089407 |
| 2026 | 4.6756 | 0.05777 | 0.14508 | 0.088689 |
| 2025 | 4.5861 | 0.056634 | 0.14228 | 0.08797 |
| 2024 | 4.4966 | 0.055499 | 0.13949 | 0.087252 |
| 2023 | 4.407 | 0.054364 | 0.1367 | 0.086534 |

**Fig. 8.31** Forecasting the upcoming five years for the Germany, using linear regression analysis

Simultaneously, considering the ranking order, the percentage disparity between the USA and China is 30%, between China and Russia is 88%, between Russia and Japan is 65%, and between Japan and Germany is 13%, thereby illustrating the significant differences among the candidate countries. The subsequent picture illustrates the implementation of the relative distance approach and the calculation of the geometric mean for each candidate for the year 2027 (refer to Fig. 8.32).

| YEAR 2027: | GDP | MSB | RDB | FDI | d1 | d2 | d3 | d4 | d_negat | rank |
|---|---|---|---|---|---|---|---|---|---|---|
| USA | 26.053 | 0.92866 | 0.74065 | 0.41468 | 100 | 100 | 100 | 100 | 100 | 1 |
| CHINA | 20.757 | 0.34663 | 0.73386 | 0.33556 | 79.672 | 37.326 | 99.083 | 80.92 | 69.878 | 2 |
| RUSSIA | 2.5478 | 0.10245 | 0.048182 | 0.024289 | 9.7793 | 11.032 | 6.5053 | 5.8572 | 8.0072 | 5 |
| JAPAN | 5.1962 | 0.052194 | 0.1858 | 0.045125 | 19.944 | 5.6203 | 25.086 | 10.882 | 13.226 | 4 |
| GERMANY | 4.7652 | 0.058905 | 0.14787 | 0.089407 | 18.29 | 6.3431 | 19.964 | 21.561 | 14.949 | 3 |

**Fig. 8.32** Calculation of the relative distance and the geometric mean for the year 2027

**YEAR 2026:**

| | GDP | MSB | RDB | FDI | d1 | d2 | d3 | d4 | d_negat | rank |
|---|---|---|---|---|---|---|---|---|---|---|
| **USA** | 25.445 | 0.90958 | 0.72474 | 0.40636 | 100 | 100 | 100 | 100 | 100 | 1 |
| **CHINA** | 19.953 | 0.33341 | 0.7052 | 0.32561 | 78.415 | 36.655 | 97.304 | 80.127 | 68.804 | 2 |
| **RUSSIA** | 2.4732 | 0.099308 | 0.047274 | 0.024398 | 9.7196 | 10.918 | 6.5229 | 6.004 | 8.0291 | 5 |
| **JAPAN** | 5.1826 | 0.051864 | 0.18412 | 0.043407 | 20.368 | 5.702 | 25.405 | 10.682 | 13.324 | 4 |
| **GERMANY** | 4.6756 | 0.05777 | 0.14508 | 0.088689 | 18.375 | 6.3513 | 20.018 | 21.825 | 15.027 | 3 |

**Fig. 8.33** Calculation of the relative distance and the geometric mean for the year 2026

**YEAR 2025:**

| | GDP | MSB | RDB | FDI | d1 | d2 | d3 | d4 | d_negat | rank |
|---|---|---|---|---|---|---|---|---|---|---|
| **USA** | 24.837 | 0.8905 | 0.70883 | 0.39805 | 100 | 100 | 100 | 100 | 100 | 1 |
| **CHINA** | 19.148 | 0.32019 | 0.67655 | 0.31566 | 77.096 | 35.956 | 95.446 | 79.3 | 67.68 | 2 |
| **RUSSIA** | 2.3985 | 0.096166 | 0.046366 | 0.024508 | 9.657 | 10.799 | 6.5412 | 6.1569 | 8.0503 | 5 |
| **JAPAN** | 5.1691 | 0.051534 | 0.18245 | 0.04169 | 20.812 | 5.7871 | 25.74 | 10.473 | 13.423 | 4 |
| **GERMANY** | 4.5861 | 0.056634 | 0.14228 | 0.08797 | 18.465 | 6.3599 | 20.073 | 22.1 | 15.108 | 3 |

**Fig. 8.34** Calculation of the relative distance and the geometric mean for the year 2025

The same calculation formulas are utilised for the relative distance and the geometric mean to ascertain the results for the year 2026 (refer to Fig. 8.33).

The same method is employed for the year 2025 (refer to Fig. 8.34).

The formulas are utilised for the subsequent year, 2024, to ascertain the values (see Fig. 8.35).

The relative distance method and the geometric mean are employed to determine the values for the final year, 2023, based on the analysis (refer to Fig. 8.36).

A systematic ranking was established to facilitate the visualisation of each country's status in this investigation. The results are sorted and presented according to their rankings (see Fig. 8.37).

The final phase will involve the testing component of the program, utilising a deep learning technique through the implementation of a RNN with LSTM layers. This system design is tailored to this program strategy, as it effectively manages temporal sequences relevant to this analysis. The verification process entails numerous tests to demonstrate the program's functionality; tests were conducted on the files accessible in the United States of America, and the results for each variable inside the system

**YEAR 2024:**

| | GDP | MSB | RDB | FDI | d1 | d2 | d3 | d4 | d_negat | rank |
|---|---|---|---|---|---|---|---|---|---|---|
| **USA** | 24.229 | 0.87142 | 0.69291 | 0.38974 | 100 | 100 | 100 | 100 | 100 | 1 |
| **CHINA** | 18.344 | 0.30696 | 0.6479 | 0.30571 | 75.711 | 35.226 | 93.503 | 78.438 | 66.503 | 2 |
| **RUSSIA** | 2.3238 | 0.093024 | 0.045458 | 0.024618 | 9.5912 | 10.675 | 6.5603 | 6.3164 | 8.0706 | 5 |
| **JAPAN** | 5.1556 | 0.051204 | 0.18077 | 0.039972 | 21.278 | 5.8759 | 26.089 | 10.256 | 13.524 | 4 |
| **GERMANY** | 4.4966 | 0.055499 | 0.13949 | 0.087252 | 18.559 | 6.3688 | 20.132 | 22.387 | 15.192 | 3 |

**Fig. 8.35** Calculation of the relative distance and the geometric mean for the year 2024

YEAR 2023:

| | GDP | MSB | RDB | FDI | d1 | d2 | d3 | d4 | d_negat | rank |
|---|---|---|---|---|---|---|---|---|---|---|
| USA | 23.621 | 0.85234 | 0.677 | 0.38143 | 100 | 100 | 100 | 100 | 100 | 1 |
| CHINA | 17.54 | 0.29374 | 0.61924 | 0.29575 | 74.255 | 34.463 | 91.468 | 77.538 | 65.27 | 2 |
| RUSSIA | 2.2492 | 0.089881 | 0.044549 | 0.024727 | 9.522 | 10.545 | 6.5804 | 6.4828 | 8.09 | 5 |
| JAPAN | 5.142 | 0.050874 | 0.1791 | 0.038254 | 21.769 | 5.9687 | 26.455 | 10.029 | 13.626 | 4 |
| GERMANY | 4.407 | 0.054364 | 0.1367 | 0.086534 | 18.657 | 6.3782 | 20.193 | 22.687 | 15.28 | 3 |

**Fig. 8.36** Calculation of the relative distance and the geometric mean for the year 2023

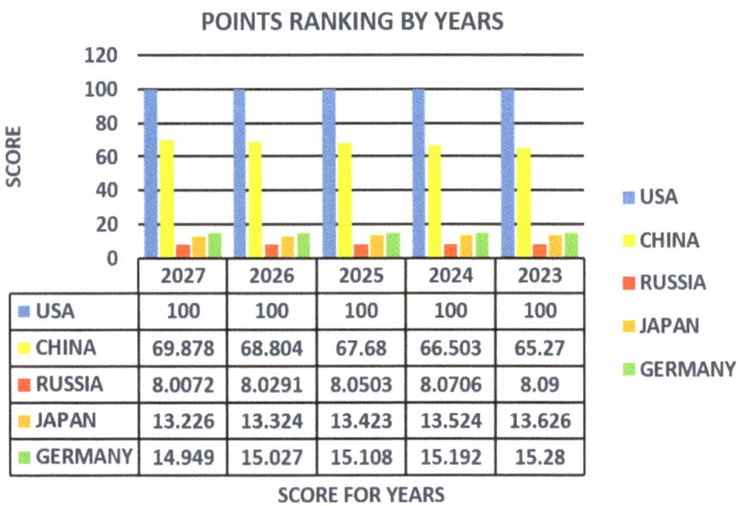


**Fig. 8.37** Ranking of results based on ranks

were examined. The system's variables include the countries' gross domestic product (GDP), military spending budget (MSB), research and development budget (RDB), and foreign direct investment budget (FDI).

It should be mentioned that Root Mean Square Error (RMSE) represents the root mean square error, being a measure of the discrepancy between the values predicted by the model and the actual values of the test data. For a regression problem (such as in economic forecasting), the RMSE is calculated as:

$$RMSE = \sqrt{\frac{1}{n}\sum_{i=1}^{n}(y_i - \hat{y}_i)^2} \tag{8.8}$$

where:

- n is the total number of test examples;
- $y_i$ represents the actual value of the test examples;
- $\hat{y}_i$ represents the value predicted by the model for the test examples.

A smaller RMSE value indicates higher model accuracy, as it shows that the model predictions are closer to the actual values.

In machine learning, particularly in neural networks, the term *Loss* denotes a mathematical function that measures the difference or error between the values predicted by a model and the actual or anticipated values in the dataset. This loss function is an essential indicator that directs the model's training process, offering an assessment of its performance.

The goal of training a machine learning model is to minimize the value of this loss function, that is, to reduce the error between the model's predictions and the observed reality. In the training process, the algorithm adjusts the model parameters (e.g., the weights in a neural network) to achieve this goal.

In the context of neural networks, the Mean Squared Error (MSE) *loss* function is often utilised for regression problems:

$$MSE = \frac{1}{n}\sum_{i=1}^{n}(y_i - \hat{y}_i)^2 \tag{8.9}$$

This method resembles the RMSE formula, except it omits the extraction of the square root. During the training process, the model adjusts its internal parameters to minimize this *loss* function, which is an essential component for improving the accuracy of its predictions. This adjustment process is iterative and relies on feedback from the prediction error assessment, allowing the model to learn and gradually adapt to the complexity of the data.

The following figure indicates the training evolution for US GDP data. This provides a visual representation of how the model optimizes over training cycles. For example, the value of the *loss* function fluctuates and the model parameters are changed to achieve greater accuracy. This visualization is a crucial tool for understanding the performance and behaviour of the model during training (refer to Fig. 8.38).

The responses of the system for the candidate country the USA and the GDP variable are presented in the following illustration (refer to Fig. 8.39).

Response Plot is a graph that illustrates the response of the model as a function of an independent variable. It is useful for visualising how the model reacts to data variability. The Predicted versus Actual Plot is a graph that compares the model's predicted values to the actual values. Each point represents an observation and a tight distribution around an ideal diagonal line indicates an accurate prediction. Distribution of Residuals, this presents the distribution of residuals, which are the differences between predicted and actual values. A distribution that is symmetrical and close to zero indicates an effective model. Residual Plot is a graph that displays the residuals on the vertical axis as a function of the independent variable on the horizontal axis. An effective model must have residuals that are randomly distributed about the zero axis, devoid of discernible patterns. Cumulative Absolute Error Plot shows the accumulated sum of absolute errors as a function of time or number of observations. It is beneficial for assessing the continuous performance of a model when further

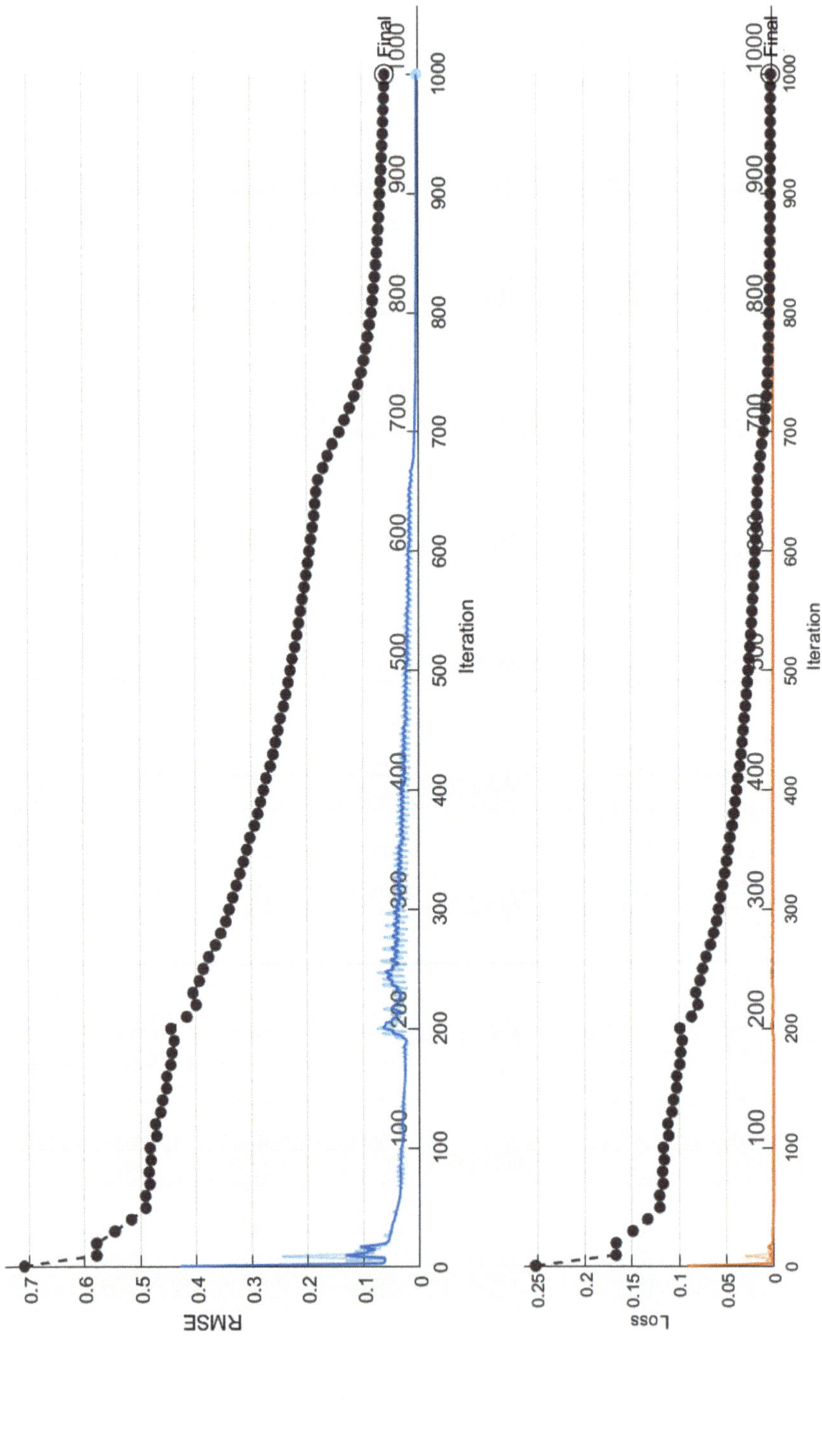


**Fig. 8.38** The graph illustrating the training for the data from the US—GDP

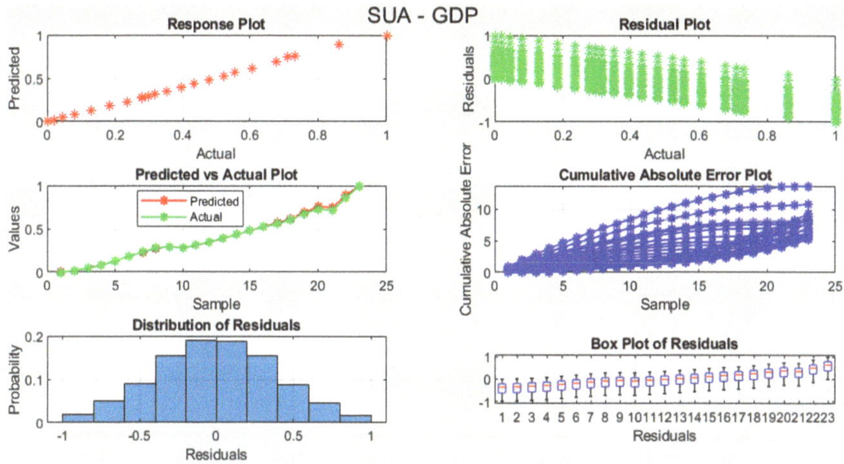


**Fig. 8.39** The system response for the USA with the GDP variable

data is incorporated. Box Plot of Residuals is a graph that displays the distribution of residuals as a box and whisker, highlighting the median, quartiles and possible extreme values. It assists in identifying asymmetries and potential outliers in the residuals.

The following figure depicts a monitoring table of the model training for the USA with the GDP variable (see Fig. 8.40). The table is condensed and generated each time the program executes, as it serves an informative role and aids in visualising the necessary information for the training process. Epoch represents the total number of passes through the training data set. The current number of iterations in an epoch is specified in the Iteration column in Fig. 8.40. Time Elapsed represents the total duration of training up to that moment. Mini-batch RMSE is the root mean square error value for the current training data. Validation is utilised for validation data. Mini-batch Loss is the value of the loss function associated with the current training data. Validation Loss represents the value of the loss function associated with the validation data. The Base Learning Rate denotes the learning rate employed at that specific time during training.

This is the graph that illustrates the training mode for the next variable in the system called MSB (refer to Fig. 8.41).

The meaning of each graph has been explained above. This graphic aims to assess the closeness of the prediction line to the original data line. The Predicted versus Actual Plot graph demonstrates that the prediction was successful and the neural network effectively learnt the data pattern. The subsequent graph depicts this figure (refer to Fig. 8.42).

This is the graph illustrating the training mode for the RDB variable (see Fig. 8.43).

This graph illustrates responses from different perspectives. This scenario examines if the system successfully learnt the data template for the RDB variable. From

Training on single GPU.

| Epoch | Iteration | Time Elapsed (hh:mm:ss) | Mini-batch RMSE | Validation RMSE | Mini-batch Loss | Validation Loss | Base Learning Rate |
|---|---|---|---|---|---|---|---|
| 1 | 1 | 00:00:01 | 0.50 | 0.55 | 0.1258 | 0.1524 | 0.0010 |
| 10 | 10 | 00:00:01 | 0.24 | 0.47 | 0.0293 | 0.1117 | 0.0010 |
| 20 | 20 | 00:00:02 | 0.05 | 0.46 | 0.0014 | 0.1065 | 0.0010 |
| 30 | 30 | 00:00:02 | 0.13 | 0.43 | 0.0091 | 0.0914 | 0.0010 |
| 40 | 40 | 00:00:03 | 0.05 | 0.42 | 0.0011 | 0.0903 | 0.0010 |
| 50 | 50 | 00:00:03 | 0.03 | 0.42 | 0.0005 | 0.0875 | 0.0010 |
| 900 | 900 | 00:00:42 | 0.02 | 0.44 | 0.0001 | 0.0967 | 0.0002 |
| 910 | 910 | 00:00:42 | 0.01 | 0.44 | 5.7732e-05 | 0.0974 | 0.0002 |
| 920 | 920 | 00:00:42 | 0.02 | 0.44 | 0.0001 | 0.0975 | 0.0002 |
| 930 | 930 | 00:00:43 | 0.01 | 0.44 | 6.9836e-05 | 0.0967 | 0.0002 |
| 940 | 940 | 00:00:43 | 0.02 | 0.44 | 0.0001 | 0.0966 | 0.0002 |
| 950 | 950 | 00:00:44 | 0.01 | 0.44 | 5.4807e-05 | 0.0973 | 0.0002 |
| 960 | 960 | 00:00:44 | 0.02 | 0.44 | 0.0001 | 0.0973 | 0.0001 |
| 970 | 970 | 00:00:45 | 0.01 | 0.44 | 5.3045e-05 | 0.0967 | 0.0001 |
| 980 | 980 | 00:00:45 | 0.01 | 0.44 | 5.6671e-05 | 0.0970 | 0.0001 |
| 990 | 990 | 00:00:46 | 0.01 | 0.44 | 5.1347e-05 | 0.0970 | 0.0001 |
| 1000 | 1000 | 00:00:46 | 0.01 | 0.44 | 5.1857e-05 | 0.0975 | 0.0001 |

**Fig. 8.40** Training illustration

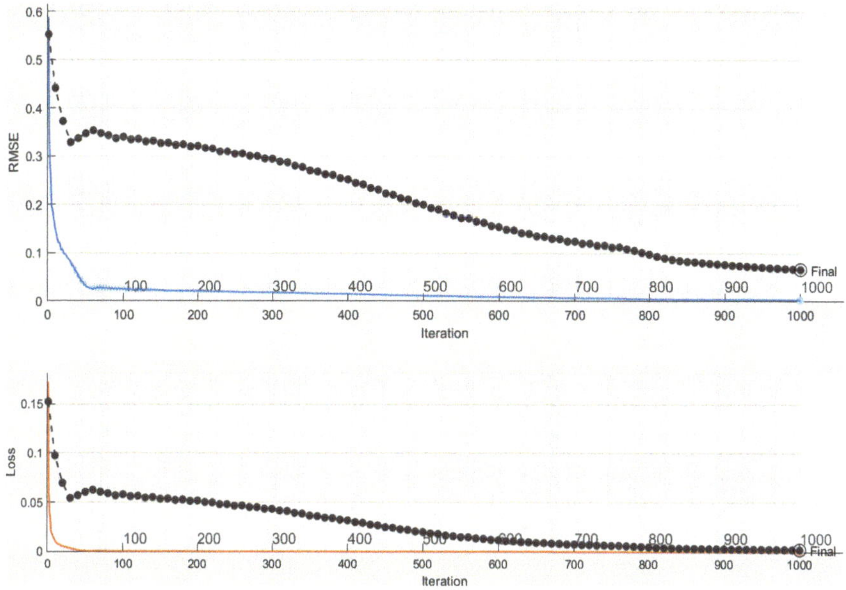


Fig. 8.41 The graph illustrating the training for the data from the US—MSB

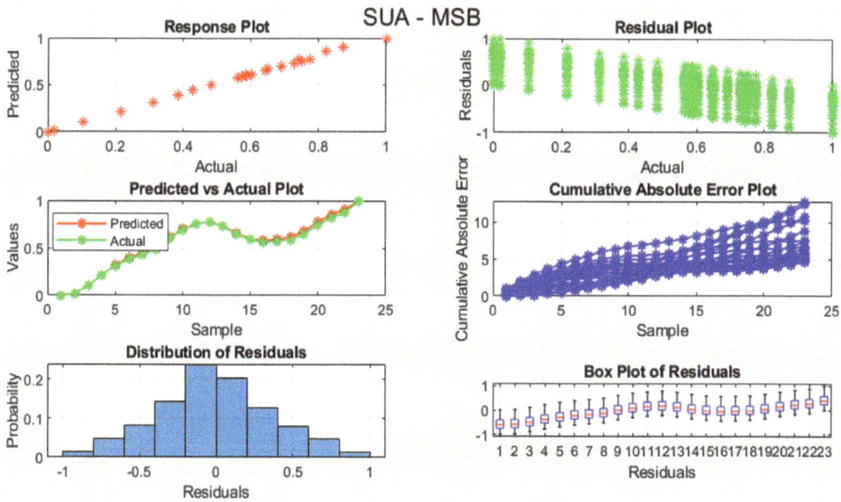


Fig. 8.42 The system response for the USA with the MSB variable

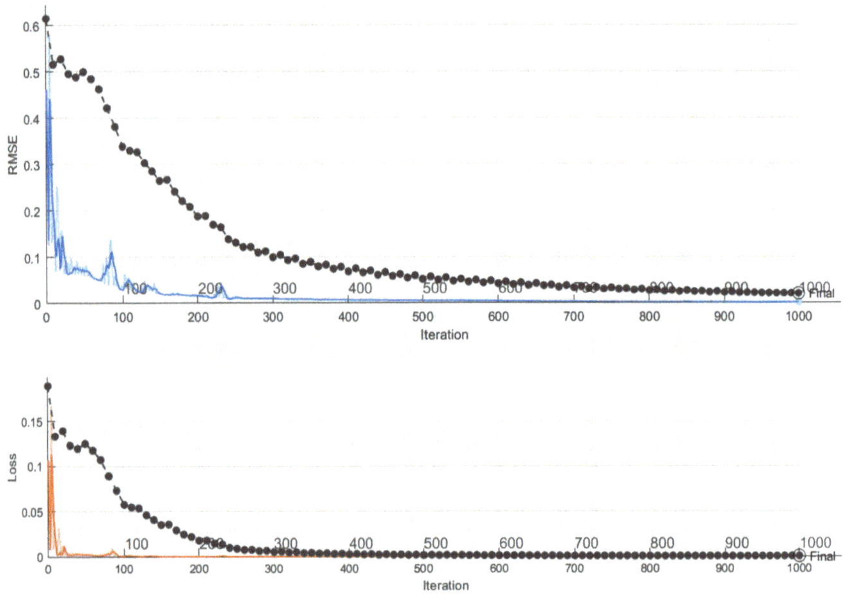


**Fig. 8.43** The graph illustrating the training for the data from the US—RDB

this illustration, it can be noticed that the data has been retained by the system (refer to Fig. 8.44).

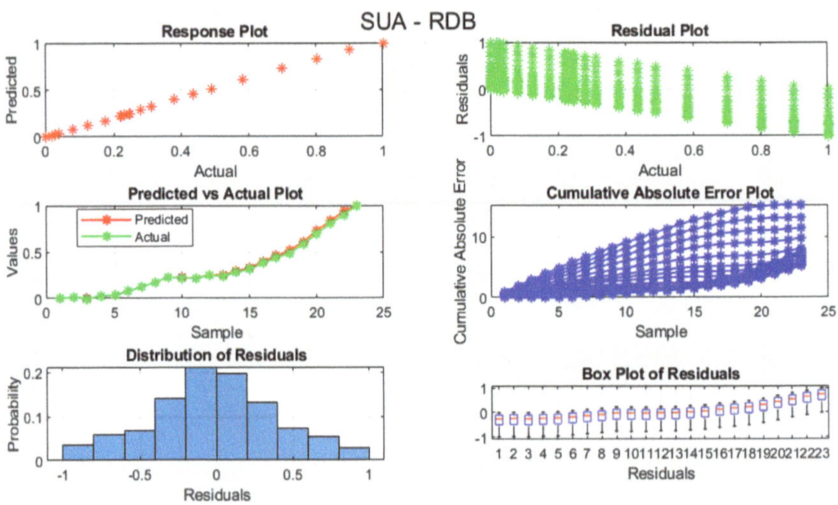


**Fig. 8.44** The system response for the USA with the RDB variable

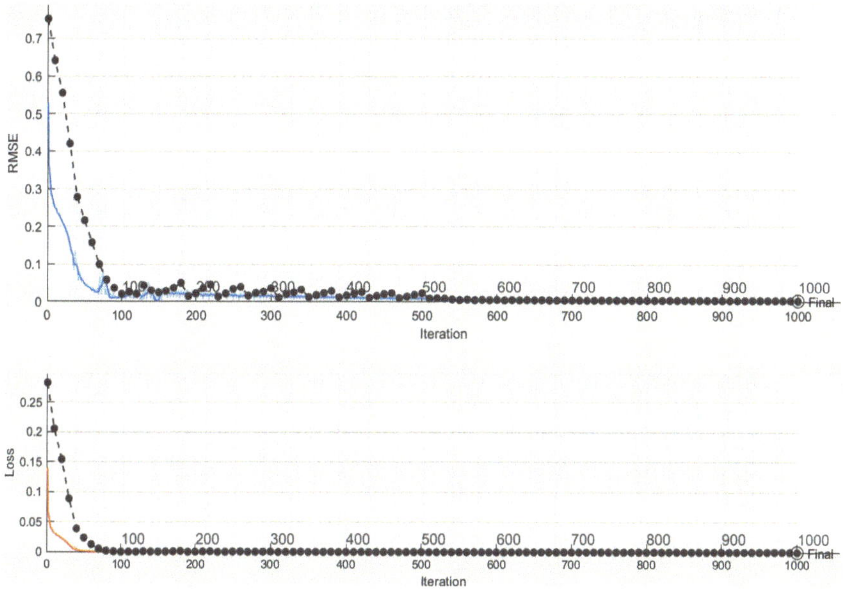


**Fig. 8.45** The graph illustrating the training for the data from the US—FDI

The following graph shows an illustration of the data training process for the FDI variable. The light blue line on the graph represents the system's attempt to learn the pattern. Moreover, it can be observed that the lines are close (see Fig. 8.45).

This graph illustrates the responses at different stages of the analysis. In this instance, the emphasis was on the system's capability to learn the data template for the FDI variable. In the figure positioned below, it can be noticed that the data has been retained by the system (see Fig. 8.46).

The tests continue to determine the values and the other variables within the other countries.

Ultimately, the mathematical inference method, linear regression method, and deep learning method are compared to the actual values recorded to ascertain which method is most closely aligned with the prediction methods employed in this research. Consequently, the GDP, MSB, RDB, and FDI data for each country, obtained from specialised resources, will be examined for the year 2023 to see if the recommended choices in this study align with the actual facts.

The chart below indicates the initial values stated by specialist sources for the United States of America in the year 2023. Due to the lack of published data, the focus is only on the US. This research validates the three discussed prediction methods. The mathematical inference method, notwithstanding its limited efficacy, approximates the initial values of the GDP variable by 4.5%, at MSB by 14.29%, at RDB by 2.11% and FDI by 10, 16%. The linear regression method has a much better performance and approaches the initial values of the GDP variable by 10.87, with MSB by 4.41%, with RDB by 16.01% and with FDI by 28.08%. Furthermore, the last method that

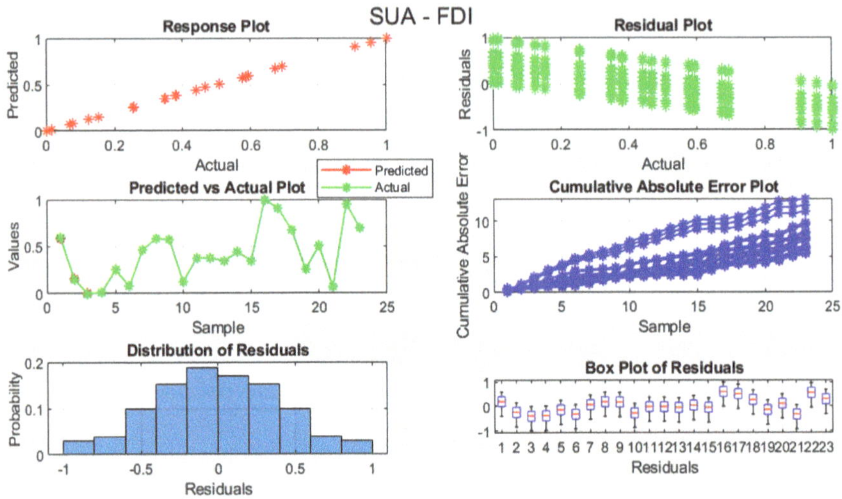


**Fig. 8.46** The system response for the USA with the FDI variable

implements a deep learning neural network approaches the original values of the Gross Domestic Product variable by 1.13%, at MSB by 1.35%, at RDB by 0.74% and FDI by 9.27%. The deep learning method is significantly closer to the true values; nonetheless, it is necessary to note that additional hardware resources are required to attain a higher accuracy in data prediction.

The graph above illustrates the outcomes derived from the implemented methods (refer to Fig. 8.47).

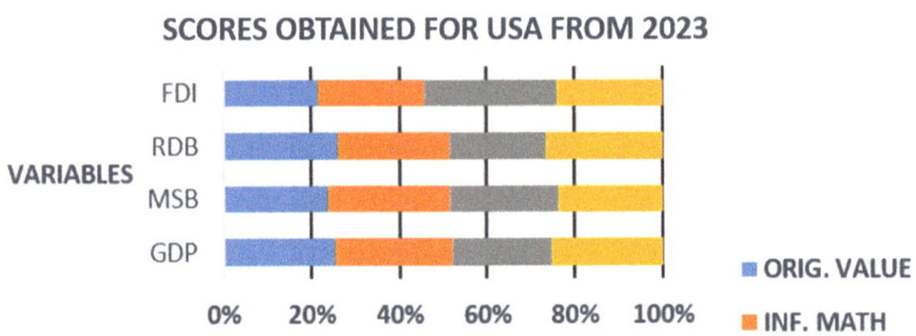


| | GDP | MSB | RDB | FDI |
|---|---|---|---|---|
| ORIG. VALUE | 26.5 | 0.816 | 0.806 | 0.274 |
| INF. MATH | 27.75 | 0.952 | 0.789 | 0.305 |
| LINE. REGRES | 23.621 | 0.852 | 0.677 | 0.381 |
| DEEP LEARN. | 26.2 | 0.805 | 0.812 | 0.302 |

SCORE

**Fig. 8.47** The score obtained by the USA for the year 2023

## 8.6 Conclusions and Future Paths Advanced by Further Developments

The research incorporated FLS, alongside mathematical models such as mathematical inference and linear regression, to examine and forecast the developmental paths for the United States of America, the People's Republic of China, the Russian Federation, Japan, and Germany. Deep learning techniques were employed to enhance the predictive accuracy and robustness of the analysis. Moreover, the study integrated aspects of XAI to ensure that the decision-making process of these complex models remains transparent and understandable, thereby facilitating a deeper insight into the underlying mechanisms driving the predictions. FLS may serve as a tool that can be incorporated into intelligent systems to emulate decision-making processes grounded in human perception. The aim of the research was to forecast the developmental progress of the candidate countries, particularly in terms of embracing new technologies within their military and research sectors, which could have consequences for the potential establishment of their dominance. The system combined several capabilities, being able to realize the method of finding future development indices by applying calculation methods based on data prediction. Additionally, the system had the ability to determine future values through mathematical inference, linear regression and by training a deep learning type neural network, thus being able to more easily observe the state of development of each country.

Furthermore, there has been a substantial emphasis on economic expansion and progress. Elements such as GDP per capita, the accumulation of past investments, and techno-logical innovation have received significant interest. Contemporary states consistently prioritise the enhancement of various factors, concentrating on goals such as elevating the savings rate, broadening access to knowledge, fostering creativity in scientific domains, and advancing innovation, all aimed at facilitating sustainable development. Within a globalized framework, the rivalry among countries centres on their capacity to achieve swift progress in critical domains and sustain a consistent rate of development. The competition for influence and power internationally has brought attention to world leaders that effectively leverage their ample resources, ultimately consolidating their status on a global scale. These leaders recognise the importance of rational use of resources to ensure long-term sustainability and stability.

Countries endeavour to optimise their developmental potential through several strategies that promote scientific research, foster an innovative environment, and enhance access to advanced information. Consequently, the foundations for a continual and sustainable evolution, in accordance with global progress objectives, are established.

In this view, global leadership is not restricted to the effective management of resources; it also encompasses the creation of an environment that is conducive to the advancement of intellectual and technological capabilities. The ability to convert these resources into lasting competitive advantages is therefore a vital measure of any country's place within the intricate geopolitical landscape of the 21st century.

This program serves as an outstanding instrument, offering several advantages in assessing a country's development coefficient, particularly for military defence, economic advancement, and technical innovation. The system is meant to retrospectively assess existing data and to employ advanced logic for predicting future values of key variables based on their historical trends. This approach reveals an exploration into artificial intelligence, offering an original framework for evaluating and predicting state dynamics within the context of global competitiveness. Furthermore, its development occurs within a context of evolving international order, a multifaceted transition primarily driven by the significant emergence of modern countries distinguished by extensive assets and unique development strategies.

This application transcends mere examination of developmental aspects; it serves as a platform for intricate research and forecasting, significantly influencing the contest for regional hegemony and the evolution of the global geopolitical environment. This technique provides a fresh perspective on the history and interconnections of countries representing an unprecedented method for researching and implementing artificial intelligence within a geopolitical framework.

The proposed prediction system, despite its significant performance, encounters inherent limitations, and this study emphasises critical risk variables in macroeconomic forecasts. These variables include a country's energy reliance or self-sufficiency, specific economic trends, strategic partnerships, and collaborative assistance within the international community, among others.

A vital element is the risk rating of a country, as determined by rating organisations such as Standard and Poor's, Fitch, or Moody's. These indicators offer insight into the stability and investor confidence in the economy of a country, hence complicating the evaluation of macroeconomic projections.

The acquisition and utilisation of conclusive data is crucial to guarantee the accuracy and relevance of predictions. This system is fundamentally reliant on the quality and precision of accessible data, and meticulous analysis and implementation within the system yield dependable outcomes.

It is worthwhile to emphasise that, despite these limitations, the system may be enhanced by integrating additional datasets. This extension may provide a more holistic picture and definitive outcomes, enabling the system to adjust to dynamic shifts in the macroeconomic and geopolitical landscape.

## Appendix 8.1

Algorithm RegressionAnalysis

```
ALGORITHM RegressionAnalysis
    // Initialize option for country selection
    option <- 1
    // Read data from Excel based on the selected
country
    SELECT option
        CASE 1: T <- ReadTable('1USA_data_2000-2022.xlsx')
        CASE 2: T <- ReadTable('1CHN_data_2000-2022.xlsx')
        CASE 3: T <- ReadTable('1RUS_data_2000-2022.xlsx')
        CASE 4: T <- ReadTable('1JAP_data_2000-2022.xlsx')
        CASE 5: T <- ReadTable('1GER_data_2000-2022.xlsx')
    END SELECT
    // Calculate mean and standard deviation for
historical data variables
    xm <- Mean(T['YEARS'])
    xs <- StdDev(T['YEARS'])
    ym_GDP <- Mean(T['GDP'])
    ys_GDP <- StdDev(T['GDP'])
    ym_MSB <- Mean(T['MSB'])
    ys_MSB <- StdDev(T['MSB'])
    ym_RDB <- Mean(T['RDB'])
    ys_RDB <- StdDev(T['RDB'])
    ym_FDI <- Mean(T['FDI'])
```

```
ys_FDI <- StdDev(T['FDI'])
// Calculate regression coefficients for all
variables
// Calculate regression coefficients for GDP
variable
xd_GDP <- T['YEARS']
yd_GDP <- T['GDP']
xm_GDP <- Mean(xd_GDP)
ym_GDP <- Mean(yd_GDP)
n_GDP <- Sum((xd_GDP-xm_GDP)*(yd_GDP-ym_GDP))
d_GDP <- Sum((xd_GDP-xm_GDP)^2)
b1_GDP <- n_GDP/d_GDP
b0_GDP <- ym_GDP-b1_GDP*xm_GDP
// Calculate regression coefficients for MSB
variable
xd_MSB <- T['YEARS']
yd_MSB <- T['MSB']
xm_MSB <- Mean(xd_MSB)
ym_MSB <- Mean(yd_MSB)
n_MSB <- Sum((xd_MSB-xm_MSB)*(yd_MSB-ym_MSB))
d_MSB <- Sum((xd_MSB-xm_MSB)^2)
b1_MSB <- n_MSB/d_MSB
b0_MSB <- ym_MSB-b1_MSB*xm_MSB
// Calculate regression coefficients for RDB
variable
xd_RDB <- T['YEARS']
yd_RDB <- T['RDB']
xm_RDB <- Mean(xd_RDB)
ym_RDB <- Mean(yd_RDB)
n_RDB <- Sum((xd_RDB-xm_RDB)*(yd_RDB-ym_RDB))
d_RDB <- Sum((xd_RDB-xm_RDB)^2)
b1_RDB <- n_RDB/d_RDB
b0_RDB <- ym_RDB-b1_RDB*xm_RDB
// Calculate regression coefficients for FDI
variable
xd_FDI <- T['YEARS']
yd_FDI <- T['FDI']
xm_FDI <- Mean(xd_FDI)
ym_FDI <- Mean(yd_FDI)
n_FDI <- Sum((xd_FDI-xm_FDI)*(yd_FDI-ym_FDI))
d_FDI <- Sum((xd_FDI - xm_FDI)^2)
b1_FDI <- n_FDI/d_FDI
b0_FDI <- ym_FDI-b1_FDI*xm_FDI
// Estimate variable values for the desired year
using linear regression
y <- 2023
y_GDP <- b0_GDP+b1_GDP*year
```

```
    y_MSB  <- b0_MSB+b1_MSB*year
    y_RDB  <- b0_RDB+b1_RDB*year
    y_FDI  <- b0_FDI+b1_FDI*year
    // Display results
    OUTPUT "Example of application for the year 2023
(UNITED STATES OF AMERICA):"
    OUTPUT "Variable GDP:"
    OUTPUT "The regression coefficient is:"+b1_GDP
    OUTPUT "The intercept is:"+b0_GDP
    OUTPUT "The estimated value of the prediction for
year "+y+" is:"+y_GDP
    OUTPUT "Variable MSB:"
    OUTPUT "The regression coefficient is:"+b1_MSB
    OUTPUT "The intercept is:"+b0_MSB
    OUTPUT "The estimated value of the prediction for
year "+y+" is:"+y_MSB
    OUTPUT "Variable RDB:"
    OUTPUT "The regression coefficient is:"+b1_RDB
    OUTPUT "The intercept is:"+b0_RDB
    OUTPUT "The estimated value of the prediction for
year "+y+" is:"+yRDB
    OUTPUT "Variable FDI:"
    OUTPUT "The regression coefficient is:"+b1_FDI
    OUTPUT "The intercept is:"+b0_FDI
    OUTPUT "The estimated value of the prediction for
year "+y+" is:"+yFDI
    // Display values for all variables in a table
    OUTPUT "Calculation applied to the list of years:"
    y <- [2023,2024,2025,2026,2027]
    y_GDP  <- b0_GDP+b1_GDP*y
    y_MSB  <- b0_MSB+b1_MSB*y
    y_RDB  <- b0_RDB+b1_RDB*y
    y_FDI  <- b0_FDI+b1_FDI*y
    // Display variable values in a table with column
names and row names
    T <- CreateTable(y_GDP,y_MSB,y_RDB,y_FDI)
    T.SetVariableNames({'GDP','MSB','RDB','FDI'})
    T.SetRowNames(ConvertToString(y))
    T.FlipRows()
    // Display country-specific header
    SELECT option
       CASE 1: OUTPUT "UNITED STATES OF AMERICA"
       CASE 2: OUTPUT "REPUBLIC OF CHINA"
       CASE 3: OUTPUT "RUSSIAN FEDERATION"
       CASE 4: OUTPUT "JAPAN"
       CASE 5: OUTPUT "GERMANY"
```

```
        DEFAULT:  OUTPUT  "The  selected  value  does  not
exist"
    END SELECT
    OUTPUT T
    // Export table to Excel
    SELECT option
        CASE 1: c_name <- 'USA'; f <- 'File_USA.xlsx'
        CASE 2: c_name <- 'CHN'; f <- 'File_CHN.xlsx'
        CASE 3: c_name <- 'RUS'; f <- 'File_RUS.xlsx'
        CASE 4: c_name <- 'JAP'; f <- 'File_JAP.xlsx'
        CASE 5: c_name <- 'GER'; f <- 'File_GER.xlsx'
        DEFAULT:  OUTPUT  "The  selected  value  does  not
exist"
    END SELECT
    // Create export table
    et <- CreateTable(T.GDP,T.MSB,T.RDB,T.FDI)
    et.SetRowNames(T.GetRowNames())
    et.SetVariableNames({'GDP','MSB','RDB','FDI'})
    et.AddColumn('C',              ReplicateValue(c_name,
et.GetRowCount()))
    et            <-            ReorderColumns(et,
['C','GDP','MSB','RDB','FDI'])
    // Export table to Excel file
    WriteTableToExcel(et,  f, true)
    OUTPUT "Table has been saved to Excel file: " + f
END ALGORITHM
```

## Appendix 8.2

Algorithm CalculateRelativeDistancesAndGeometricMean

```
ALGORITHM CalculateRelativeDistancesAndGeometricMean
    // Clear console
    ClearConsole()
    // Set the option for the year
    option <- 1
    // Read data from Excel file based on the option
    SWITCH option
        CASE 1: year <- 2027
        CASE 2: year <- 2026
        CASE 3: year <- 2025
        CASE 4: year <- 2024
        CASE 5: year <- 2023
    END SWITCH
    //USA Reading data from the file
```

```
f_USA <- 'File_USA.xlsx';
data_USA <- readtable(f_USA);
// Converting the 'Row' column to numeric
data_USA.Row <- str2double(data_USA.Row);
// Filtering data for the year 2027
USA_ydata <- data_USA(data_USA.Row == year, :);
// Data for USA
USA_GDP <- USA_ydata.GDP;
USA_MSB <- USA_ydata.MSB;
USA_RDB <- USA_ydata.RDB;
USA_FDI <- USA_ydata.FDI;
//CHN Reading data from the file
f_CHN <- 'File_CHN.xlsx';
data_CHN <- readtable(f_CHN);
// Converting the 'Row' column to numeric
data_CHN.Row <- str2double(data_CHN.Row);
// Filtering data for the year 2027
CHN_ydata <- data_CHN(data_CHN.Row == year, :);
// Data for CHN
CHN_GDP <- CHN_ydata.GDP;
CHN_MSB <- CHN_ydata.MSB;
CHN_RDB <- CHN_ydata.RDB;
CHN_FDI <- CHN_ydata.FDI;
//RUS Reading data from the file
f_RUS <- 'File_RUS.xlsx';
data_RUS <- readtable(f_RUS);
// Converting the 'Row' column to numeric
data_RUS.Row <- str2double(data_RUS.Row);
// Filtering data for the year 2027
RUS_ydata <- data_RUS(data_RUS.Row == year, :);
// Data for RUS
RUS_GDP <- RUS_ydata.GDP;
RUS_MSB <- RUS_ydata.MSB;
RUS_RDB <- RUS_ydata.RDB;
RUS_FDI <- RUS_ydata.FDI;
//JAP Reading data from the file
f_JAP <- 'File_JAP.xlsx';
data_JAP <- readtable(f_JAP);
// Converting the 'Row' column to numeric
data_JAP.Row <- str2double(data_JAP.Row);
// Filtering data for the year 2027
JAP_ydata <- data_JAP(data_JAP.Row == year, :);
// Data for JAP
JAP_GDP <- JAP_ydata.GDP;
JAP_MSB <- JAP_ydata.MSB;
JAP_RDB <- JAP_ydata.RDB;
JAP_FDI <- JAP_ydata.FDI;
```

```
//GER Reading data from the file
f_GER <- 'File_GER.xlsx';
data_GER <- readtable(f_GER);
// Converting the 'Row' column to numeric
data_GER.Row <- str2double(data_GER.Row);
// Filtering data for the year 2027
GER_ydata <- data_GER(data_GER.Row == year, :);
// Data for GER
GER_GDP <- GER_ydata.GDP;
GER_MSB <- GER_ydata.MSB;
GER_RDB <- GER_ydata.RDB;
GER_FDI <- GER_ydata.FDI;
// Calculating relative distances and geometric mean
v_USA <- [USA_GDP,USA_MSB,USA_RDB,USA_FDI];
v_CHN <- [CHN_GDP,CHN_MSB,CHN_RDB,CHN_FDI];
v_RUS <- [RUS_GDP,RUS_MSB, RUS_RDB,RUS_FDI];
v_JAP <- [JAP_GDP,JAP_MSB, JAP_RDB,JAP_FDI];
v_GER <- [GER_GDP,GER_MSB, GER_RDB,GER_FDI];
// Calculate the maximum values for each variable
across all countries
Imax_v <- max([v_USA;v_CHN;v_RUS;v_JAP;v_GER]);
rd_USA<-(USA_GDP,USA_MSB,USA_RDB,USA_FDI)/Imax_v*100
rd_CHN<-(CHN_GDP,CHN_MSB,CHN_RDB,CHN_FDI)/Imax_v*100
rd_RUS<-(RUS_GDP,RUS_MSB,RUS_RDB,RUS_FDI)/Imax_v*100
rd_JAP<-(JAP_GDP,JAP_MSB,JAP_RDB,JAP_FDI)/Imax_v*100
rd_GER<-(GER_GDP,GER_MSB,GER_RDB,GER_FDI)/Imax_v*100
// Function to calculate geometric mean
function GM(data)
    product <- 1
    n <- length(data)

    // Calculate the product of values
    for i <- 1 to n
        product <- product*(data[i]/100)^(1/4)

    // Calculate the geometric mean
    gm <- product*100

    return gm
end function
gm_USA <- GM(rd_USA)
gm_CHN <- GM(rd_CHN)
gm_RUS <- GM(rd_RUS)
gm_JAP <- GM(rd_JAP)
gm_GER <- GM(rd_GER)
// Display results in console
r <- ['USA','CHINA','RUSSIA','JAPAN','GERMANY']
```

```
    GDP_v  <-  [USA_GDP,CHN_GDP,RUS_GDP,JAP_GDP,GER_GDP]
    MSB_v  <-  [USA_MSB,CHN_MSB,RUS_MSB,JAP_MSB,GER_MSB]
    RDB_v  <-  [USA_RDB,CHN_RDB,RUS_RDB,JAP_RDB,GER_RDB]
    FDI_v  <-  [USA_FDI,CHN_FDI,RUS_FDI,JAP_FDI,GER_FDI]
    d1<-
[GDP_v[1],rd_USA[1];GDP_v[2],rd_CHN[1];GDP_v[3],rd_RUS[
1];GDP_v[4],rd_JAP[1];GDP_v[5],rd_GER[1]]
    d2<-
[MSB_v[1],rd_USA[2];MSB_v[2],rd_CHN[2];MSB_v[3],rd_RUS[
2];MSB_v[4],rd_JAP[2];MSB_v[5],rd_GER[2]]
    d3<-
[RDB_v[1],rd_USA[3];RDB_v[2],rd_CHN[3];RDB_v[3],rd_RUS[
3];RDB_v[4],rd_JAP[3];RDB_v[5],rd_GER[3]]
    d4<-
[FDI_v[1],rd_USA[4];FDI_v[2],rd_CHN[4];FDI_v[3],rd_RUS[
4];FDI_v[4],rd_JAP[4];FDI_v[5],rd_GER[4]]
    d_not  <-  [gm_USA,gm_CHN,gm_RUS,gm_JAP,gm_GER]
    //  Create  table  T
    T<-
CreateTable([GDP_v,MSB_v,RDB_v,FDI_v,d1[:,2],d2[:,2],d3
[:,2],d4[:,2],d_not],r,['GDP','MSB','RDB','FDI','d1','d
2','d3','d4','d_not'])
    //  Add  'rank'  column
    T.rank  <-  NaN(Size(T,  1),  1)
    //  Find  the  index  of  the  'd_negat'  column
    idx_d_n  <-  FindIndex('d_not',    T.GetColumnNames())
    //  Assign  values  to  the  'rank'  column
    s_idx  <-  Sort(T[:,  idx_d_negat],   'descend')
    T.rank[s_idx]   <-  1:numel(s_idx)
    //  Display  results  in  console
    SWITCH  option
       CASE  1:  OUTPUT  "YEAR  2027:"
       CASE  2:  OUTPUT  "YEAR  2026:"
       CASE  3:  OUTPUT  "YEAR  2025:"
       CASE  4:  OUTPUT  "YEAR  2024:"
       CASE  5:  OUTPUT  "YEAR  2023:"
    END  SWITCH
    OUTPUT  T
END  ALGORITHM
```

## Appendix 8.3

Algorithm DeepLearningBasedPrediction

```
ALGORITHM  DeepLearningBasedPrediction
```

```
   // Clear console, clear all variables, and close all
figures
   ClearConsole()
   ClearAllVariables()
   CloseAllFigures()
   // Select files:USA=1,CHN=2,RUS=3,JAP=4,GER=5
   f <- 1
   // Select variable:GDP=1,MSB=2,RDB=3,FDI=4
   c <- 1
   // Configure file variables
   SWITCH f
      CASE 1: fTrain <- '1USA_data_2000-2022.xlsx'
              fTest <- 'TEST1USA_data_2000-2022.xlsx'
              ft <- 'SUA'
      CASE 2: fTrain <- '1CHN_data_2000-2022.xlsx'
              fTest <- 'TEST1CHN_data_2000-2022.xlsx'
              ft <- 'CHN'
      CASE 3: fTrain <- '1RUS_data_2000-2022.xlsx'
              fTest <- 'TEST1RUS_data_2000-2022.xlsx'
              ft <- 'RUS'
      CASE 4: fTrain <- '1JAP_data_2000-2022.xlsx'
              fTest <- 'TEST1JAP_data_2000-2022.xlsx'
              ft <- 'JAP'
      CASE 5: fTrain <- '1GER_data_2000-2022.xlsx'
              fTest <- 'TEST1GER_data_2000-2022.xlsx'
              ft <- 'GER'
      OTHERWISE: ERROR 'Invalid value'
   END SWITCH
   // Load training and test data based on the column
and file variables
   SWITCH c
   // First column - GDP
      CASE 1: trainD <- xlsread(fTrain, 'B2:B24')
              testD <- xlsread(fTest, 'B2:B24')
              ct <- ' GDP'
   // Second column - MSB
      CASE 2: trainD <- xlsread(fTrain, 'C2:C24')
              testD <- xlsread(fTest, 'C2:C24')
              ct <- ' MSB'
   // Third column - RDB
      CASE 3: trainD <- xlsread(fTrain, 'D2:D24')
              testD <- xlsread(fTest, 'D2:D24')
              ct <- ' RDB'
   // Fourth column - FDI
      CASE 4: trainD <- xlsread(fTrain, 'E2:E24')
              testD <- xlsread(fTest, 'E2:E24')
              ct <- ' FDI'
```

```
      OTHERWISE: ERROR 'Invalid value'
   END SWITCH
   // Reverse the order of lines
   trainD <- FlipUp(trainD)
   testD <- FlipUp(testD)
   // Split data for training, validation, and testing
   [trainInd,valInd,testInd] <- DivideData(size(trainD,
1), 0.7, 0.15, 0.15)
   // Training data
   XTrain <- trainD(trainInd, 1)
   YTrain <- trainD(trainInd, 1)
   // Validation data
   XValid <- trainD(valInd, 1)
   YValid <- trainD(valInd, 1)
   // Test data
   XTest <- testD
   YTest <- testD
   // Data normalization
   XTrain <- NormalizeData(XTrain, 'range')
   YTrain <- NormalizeData(YTrain, 'range')
   XValid <- NormalizeData(XValid, 'range')
   YValid <- NormalizeData(YValid, 'range')
   XTest <- NormalizeData(XTest, 'range')
   YTest <- NormalizeData(YTest, 'range')
   // LSTM architecture definition
   inputSize <- 1
   numResponses <- 1
   numHiddenUnits <- 500
   layers <- [
      sequenceInputLayer(inputSize)
      lstmLayer(numHiddenUnits,'OutputMode','sequence')
      lstmLayer(numHiddenUnits,'OutputMode','sequence')
      fullyConnectedLayer(numResponses)
      regressionLayer
   ]
   opts <- trainingOptions('adam', ...
      'MaxEpochs', 1000, ...
      'GradientThreshold', 0.01, ...
      'InitialLearnRate', 0.001, ...
      'ExecutionEnvironment', 'auto', ...
      'Plots', 'training-progress', ...
      'ValidationData',{Transpose(XValid),
Transpose(YValid)}, ...
      'ValidationFrequency', 10, ...
      'MiniBatchSize', 16, ...
      'LearnRateSchedule', 'piecewise', ...
      'LearnRateDropFactor', 0.9, ...
```

```
        'LearnRateDropPeriod',    50, ...
        'L2Regularization',    0.0001
    )
    // Model training
    net<-TrainNetwork(Transpose(XTrain),
Transpose(YTrain),   layers,  opts)
    // Prediction
    YPred  <- Predict(net,   Transpose(XTest))
    // De-normalization   (return  to the original  scale)
    YPred   <-  RescaleData(Transpose(YPred),      'InputMin',
min(YTest),   'InputMax',   max(YTest))
    YTest   <-  RescaleData(Transpose(YTest),      'InputMin',
min(YTest),   'InputMax',   max(YTest))
    // Plotting  results
    Figure()
    tf <- ConcatenateStrings(ft,     ' -',  ct)
    // Add the title  for the entire  figure
    Suptitle(tf)
    // Response  Plot
    Subplot(3,   2,  1)
    Plot(YTest,   YPred,  'r*')
    Xlabel('Actual')
    Ylabel('Predicted')
    GridOn()
    Title('Response   Plot')
    // Residual  Plot
    Subplot(3,   2,  2)
    Residuals   <- YTest  - YPred
    Plot(YTest,   Residuals,   'g*')
    Xlabel('Actual')
    Ylabel('Residuals')
    GridOn()
    Title('Residual   Plot')
    // Predicted  vs Actual  Plot
    Subplot(3,   2,  3)
    Plot(YPred,   'r-*',  'LineWidth',   1)
    HoldOn()
    Plot(YTest,   'g-*',  'LineWidth',   1)
    Xlabel('Sample')
    Ylabel('Values')
    GridOn()
    Legend('Predicted',    'Actual')
    Title('Predicted   vs Actual  Plot')
    // Cumulative  Absolute  Error  Plot
    Subplot(3,   2,  4)
    CumulativeError   <-  Cumsum(Abs(Residuals))
    Plot(CumulativeError,    'b-*',  'LineWidth',   1)
```

```
    Xlabel('Sample')
    Ylabel('Cumulative Absolute Error')
    GridOn()
    Title('Cumulative Absolute Error Plot')
    // Distribution of Residuals Plot
    Subplot(3, 2, 5)
    Histogram(Residuals, 'Normalization', 'probability')
    Xlabel('Residuals')
    Ylabel('Probability')
    GridOn()
    Title('Distribution of Residuals')
    // Box Plot of Residuals
    Subplot(3, 2, 6)
    Boxplot(Residuals)
    Xlabel('Residuals')
    GridOn()
    Title('Box Plot of Residuals')
END ALGORITHM
```

# References


1. Gentile G et al (2021) A history of the third offset, 2014–2018, R. Corporation, Ed., RAND Corporation
2. P. a. i. p. g. h. b. i. d. b. 2018, "J. Cassano," August 2018
3. Liu P, Nie G, Mohamed MMA (2022) Causality between technological innovation and economic growth: evidence from the economies of developing countries
4. U. N. & W. Report, "These Countries Have the Most Technological Expertise".
5. Statista (2023) Statista. U.S. defense outlays and forecast as a percentage of GDP 2000–2033, 18 05 2023. https://www.statista.com/statistics/217581/outlays-for-defense-and-forecast-in-the-us-as-a-percentage-of-the-gdp/
6. Li A (2023) An application of extenics, spatial factors, and natural resource market in China: the role of artificial intelligence and geopolitical risk. Resour Policy 81(2023):103289. https://doi.org/10.1016/j.resourpol.2022.103289
7. Estrada MAR, Park D, Staniewski M (2023) Artificial Intelligence (AI) can change the way of doing policy modelling. J Policy Model 45:1099–1112. https://doi.org/10.1016/j.jpolmod.2023.11.005
8. Dwivedi YR et al (2021) Artificial intelligence (AI): Multidisciplinary perspectives on emerging challenges, opportunities, and agenda for research, practice and policy. Int J Inf Manag 57:101994. https://doi.org/10.1016/j.ijinfomgt.2019.08.002
9. Briscoe E, Fairbanks J (2020) Artificial scientific intelligence and its impact on national security and foreign policy. Published for the Foreign Policy Research Institute by Elsevier Ltd. https://doi.org/10.1016/j.orbis.2020.08.004
10. Bengt-Åke Lundvall B, Cecilia Rikap C (2022) China's catching-up in artificial intelligence seen as a co-evolution of corporate and national innovation systems. Res Policy 51:104395. https://doi.org/10.1016/j.respol.2021.104395
11. Yan J, Leidner D, Uchenna Peters U (2024) Global techno-politics: a review of the current status and opportunities for future research. Int J Inf Manag 75:102729. https://doi.org/10.1016/j.ijinfomgt.2023.102729

12. Wirtz WB, Weyerer CJ, Ines Kehl I (2022) Governance of artificial intelligence: a risk and guideline-based integrative framework. Gov Inf Q 39:101685. https://doi.org/10.1016/j.giq.2022.101685
13. Feijoo C, Kwon Y, Bauer MJ, Bohlin E, Howell B, Jain R, Potgieter P, Vu K, Whalley J, Xia J (2020) Harnessing artificial intelligence (AI) to increase wellbeing for all: the case for a new technology diplomacy. Telecommun Policy 44:101988. https://doi.org/10.1016/j.telpol.2020.101988
14. Zhao P, Ga Y, Sun X (2022) How does artificial intelligence affect green economic growth?—Evidence from China. Sci Total Environ 834:155306. https://doi.org/10.1016/j.scitotenv.2022.155306
15. Kim D, Song Y, Kim S, Lee S, Wu Y, Shin J, Lee D (2023) How should the results of artificial intelligence be explained to users?—Research on consumer preferences in user-centered explainable artificial intelligence. Technol Forecast Soc Chang 188:122343. https://doi.org/10.1016/j.techfore.2023.122343
16. Fatima S, Desouza CK, Dawson SG, Denford SJ (2022) Interpreting national artificial intelligence plans: a screening approach for aspirations and reality. Econ Anal Policy 75:378–388. https://doi.org/10.1016/j.eap.2022.04.012
17. Petrella BS, Miller C, Cooper B (2021) Russia's artificial intelligence strategy: the role of state-owned firms. Published for the Foreign Policy Research Institute by Elsevier Ltd. https://doi.org/10.1016/j.orbis.2020.11.005
18. Horowitz CBM, Kahn L, Mahoney C (2020) The Future of military applications of artificial intelligence: a role for confidence-building measures? Published for the Foreign Policy Research Institute by Elsevier Ltd. https://doi.org/10.1016/j.orbis.2020.08.003
19. Chen P, Chu Z, Zhao M (2024) The road to corporate sustainability: the importance of artificial intelligence. Technol Soc 76:102440. https://doi.org/10.1016/j.techsoc.2023.102440
20. Korneeva E, Salge TO, Teubner T, Antons D (2023) Tracing the legitimacy of artificial intelligence: a longitudinal analysis of media discourse. Technol Forecast Soc Chang 192:122467. https://doi.org/10.1016/j.techfore.2023.122467
21. Kanzola AM, Papaioannou K, Petrakis P (2024) Unlocking society's standings in artificial intelligence. Technol Forecast Soc Chang 200:123106. https://doi.org/10.1016/j.techfore.2023.123106
22. Mansfield E (1988) Innovation, technology and the economy. Manag Sci 34
23. Zhu X (2012) Understanding China's growth: past, present, and future. J Econ Perspect 26(4):103–124
24. Bogoviz AV (2017) Modernization of the approach to usage of region's budget resources in the conditions of information economy development. Eur Res Stud J XX(3B):570–577
25. Dale KM, Jorgenson W (2005) Information technology and the Japanese economy 19(4):460–481
26. Siebert MSH (2021) Technology and economic performance in the German economy
27. William JPA, Perry J (2014) Ensuring a strong U.s. defense for the future: the national defense panel review of the 2014 quadrennial defense review. United States Institute of Peace
28. Manoj J (2022) China's military-civil fusion strategy, the US response, and implications for India
29. Tay K (2020) China's military looks to civilians to boost innovation
30. Meernik J (2000) Modeling international crises and the political use of military force by the USA. J Peace Res 37(5):547–562
31. Sasaki H (2016) Special contribution leading the way: IoT innovation in Japan and the world. Fujitsu Sci Tech J (FSTJ) 52(4)
32. Javaid M et al (2022) Understanding the adoption of industry 4.0 technologies in improving environmental sustainability. Sustain Oper Comput 3:203–217
33. Vijayakumar A (2023) Potential impact of artificial intelligence on the emerging world order. F1000Research 1186
34. Zadeh LA (2008) Is there a need for fuzzy logic? 178(13):2751–2779

35. Trillas LEE (2015) Fuzzy logic: an introductory course for engineering students, vol 320. Springer
36. Wang JTR (2021) Exploring the coupling and forecasting of financial development, technological innovation, and economic growth, vol 163
37. OECD (2023) FDI flows, Available on line at: https://data.oecd.org/fdi/fdi-flows.htm. Available: https://data.oecd.org/fdi/fdi-flows.htm. Accessed 18 Oct 2023
38. W. B. R. (2023) Data for United States, China, Russia, Japan, Germany Available on line at: https://data.worldbank.org/?locations=CN-US-RU-JP-DE. https://data.worldbank.org/?locations=CN-US-RU-JP-DE. Accessed 18 Oct 2023
39. OECD (2023) OECD Data, Available on line at: https://stats.oecd.org/Index.aspx?DataSetCode=GERD_SOF. https://stats.oecd.org/Index.aspx?DataSetCode=GERD_SOF. Accessed 25 Oct 2023
40. Dattner AHI (2018) Modern statistical tools for inference and prediction of infectious diseases using mathematical models 27(7)
41. Enescu MC, Nicolăescu CG, Stirbu C (2023) Intelligent system of analysis of the level of gain of development hegemony for great powers. In: 15th international conference on electronics, computers and artificial intelligence (ECAI)

# Chapter 9
# Explainable Artificial Intelligence for Decision-Making Processes in Active Distribution Grids

**Gheorghe Grigoras and Bogdan-Constantin Neagu**

**Abstract** The Distribution Network Operators (DNOs) operate and plan the active distribution grids (ADGs), which are composed of tens of thousands of electric distribution substations and millions of end-users (consumers and prosumers). With the help of the smart metering infrastructure, the data collected from these points must be analysed by experts (Decision Makers—DMs) who know the structure and operating regimes of the ADGs very well. They rely on processing techniques and their experience and knowledge to make the best decisions. However, due to the chosen algorithms in the data processing, which use mathematical models built on some assumptions, DMs should justify their decisions based on the identified solutions and the technical and economic impact on the ADGs. The DMs began using Artificial Intelligence (AI) techniques as they looked to integrate them into their workflows. These can perform remarkably well in various processing tasks, which are in direct contrast to the capabilities of the DMs in these areas. Over the past few years, many AI algorithms have been developed to improve machine learning models' explainability, which is a challenge that has prompted many researchers and developers to propose new approaches to improve this field. The Explainable Artificial Intelligence (XAI) concept has been implemented to help DMs understand how these AI models operate. The incorporation of AI techniques represents an additional task to human-computer interaction as it raises the possibility that XAI will significantly impact the cognitive and behavioral aspects of assisted decision-making. In the chapter, an XAI-based system proposed as a framework that aims to provide explanations of how AI models work is developed to identify the feasible solutions for the planning and operation issues due to the integration of the active end-users (prosumers) from the ADGs, such as the estimation of the energy efficiency through loss level. The identified solution through the novel proposed framework can help DNOs make

G. Grigoras (✉) · B.-C. Neagu
Faculty of Electrical Engineering, Power Engineering Department, Gheorghe Asachi Technical University of Iasi, Iasi, Romania
e-mail: gheorghe.grigoras@academic.tuiasi.ro

B.-C. Neagu
e-mail: bogdan.neagu@tuiasi.ro



N. Bizon and B. Appasani (eds.), *Explainable Artificial Intelligence for Trustworthy Decisions in Smart Applications*, https://doi.org/10.1007/978-3-031-97007-8_9

profitable investments to save as much energy as possible by implementing effective strategies in areas with technical issues.

**Keywords** Explainable artificial intelligence · Decision-making processes · Active distribution grids · Machine learning · Clustering · Decision trees

## Abbreviations

| | |
|---|---|
| ADG | Active Distribution Grid |
| AI | Artificial Intelligence |
| ANN | Artificial Neural Network |
| DGO | Distribution Grid Operator |
| DL | Deep Learning |
| DM | Decision Maker |
| DT | Decision Tree |
| EDG | Electric Distribution Grid |
| EV | Electric Vehicle |
| ML | Machine Learning |
| NTL | Non-Technical Losses |
| XAI | Explainable artificial intelligence |

## 9.1 Introduction

### *9.1.1 The Motivation*

The "active distribution grid" (ADG) concept has been complex from the beginning, defined as "a distribution network with operation and control capabilities and with distributed or decentralized energy within it" [1]. As the energy transition takes place, the complexity of these grids will continue to rise. The large-scale integration of the prosumers at the low voltage level and distributed generation sources at the medium voltage level, the emergence of storage systems, and electric mobility are the factors that have contributed the most to the complexity of the ADGs. Because of them and others referring to the age of the equipment/devices, distribution operators are having a hard time developing new assets and managing their networks. Thus, the Distribution Grid Operators (DGOs) must operate their networks at a tighter limit while dealing with the increasing uncertainty and cyber-physical complexity. Although the DGOs can accelerate the transition toward the ADGs, addressing this challenge is complicated due to high investments in the technical and communication infrastructures. Therefore, the decision-makers (DM) are the key players in developing and implementing strategies to transition of the current EDGs toward the ADGs. They need to be able to perform their duties efficiently and effectively. Due to thc increasing complexity of the decision-making process, human error has been identified as a risk factor when implementing the strategies. In addition, various studies focused on the limits of decision-making, which could help improve the capabilities of DMs [2]. According to behavioral economist and Nobel laureate in economics Daniel Kahneman, wherever human judgment exists, there will also be disturbing factors [3]. In other words, the experts can be susceptible to disturbances/ noises and diverse biases in the decision-making process. They need to be engaged in their actions to understand the reasoning behind a decision and to help them identify the factors that influence their decision and distinguish between disturbances and essential variables [4].

Machine learning (ML) and deep learning (DL) are becoming more prevalent in the quick transition toward the ADGs. These techniques have advantages over traditional approaches regarding accuracy, efficiency, and noise immunity [5]. Unfortunately, these models often act as opaque black boxes, raising most of the time ethical and moral questions. For instance, if the DMs do not understand the reasons behind the decisions made by the ML and DL-based systems, they might be unable to recognize possible biases or errors [6]. However, some challenges can prevent ML and DL from being successful despite their large-scale use. One is the lack of clear explanations for how these models interpret and make decisions based on real-world information [5]. For instance, if a decision is made using an ML model, then the effects of disturbances can be felt. Developing a learning algorithm beyond memorizing the training data can help produce a more complex pattern and some correct outputs. Figure 9.1 shows the relationship between variance and bias about the generalization error [4].

The DMs need to understand explainability and interpretability to explain for their decisions to DGOs and other stakeholders. These algorithms are usually inaccurate

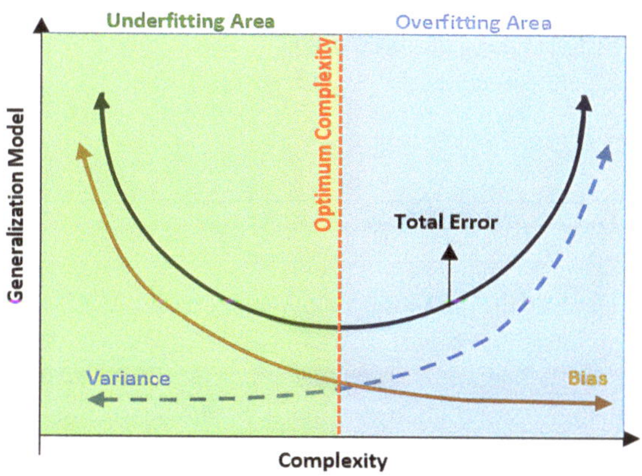


**Fig. 9.1** Dependence between model complexity and generalization error considering contribution of bias and variance

or imperfect and can inherit prejudice and bias from their human creators. Unregulated developments within AI have primarily focused on maximizing efficiency and profitability instead of human priorities and values [7].

The explainability field is important in fulfilling transparency requirements and promoting interaction between AI and DMs, helping to develop and maintain learning-based systems [4]. Explainable artificial intelligence (XAI) is a concept that seeks to provide developers, researchers, and users with a better understanding of how machine learning models work. The National Institute of Standards and Technology defines XAI as systems that offer explanations of their outputs and processes. They should follow four principles [7].

1. Explanations should contain supporting evidence or explanations;
2. Meaningful explanations should be able to be easily understood by the target audience;
3. Explanation accuracy should reflect the reason for the output;
4. The limits associated with the knowledge must be set for an optimal operation.

The characteristics that can help AI algorithms be trusted are interpretability and accuracy [6], see Fig. 9.2. These factors can also offer safety, reliability, robustness, and responsibility. The methods presented in the figure can detect and make accurate predictions when working with complex data. However, in many cases, their ability to interpret these relationships is reduced by the process that they learn and discover. The accuracy and interpretability of ML methods lead to a trade-off from the DM frequently. The complexity of the used approach should depend on the application. Not least, the objective associated with the decision-making strategy should determine whether the interpretability and accuracy are worth using one method or another.

XAI can help solve the issues associated with the transition toward the ADGs by developing strategies that let DMs understand and trust the systems' decision-making capabilities. XAI works toward bridging the gap between DMs' comprehension of AI algorithms and complexity. Thus, the DMs can assess the dependability and fairness of AI systems thanks to XAI's disclosure of the inner workings of AI models, promoting accountability and trust [6].

### *9.1.2 The State-of-The Art*

Regarding the operation and planning of the ADGs, the decisions are made by experts who know the tiniest details of the distribution process. Due to the complexity of the decisions, often based on ML, DMs may have a hard time trusting the recommendations and decisions made by the algorithms. This situation is especially true in applications that require high reliability levels. In the last few years, many new principles and techniques have been proposed to improve the understanding of ML models. Although the development of learning algorithms has significantly enhanced

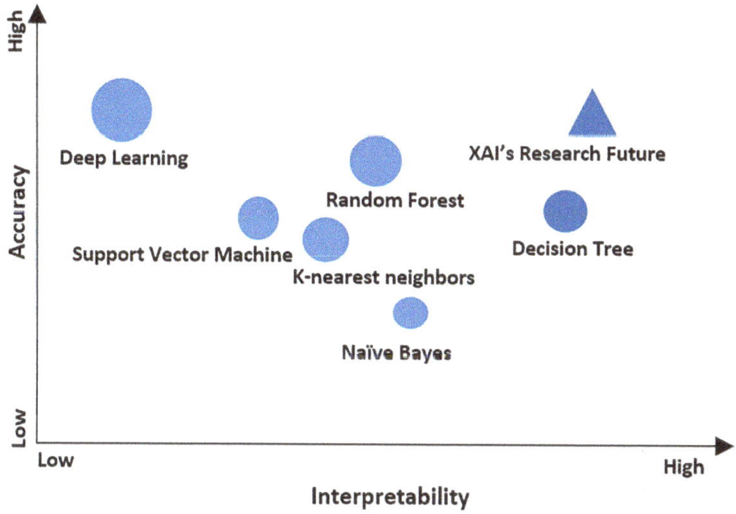


**Fig. 9.2** Comparing different machine learning methods based on the dependence between interpretability and accuracy

the knowledge of how models influence decisions, explainability remains a relatively new area of research, see Fig. 9.3 [5].

In the power system field, AI helps processing the very large databases to identify operation patterns used in developing optimal control strategies.

One of the most common techniques used in developing applications of DL systems is the Shapley additive explanation method presented in [7]. This method analyzes the ML model's results based on the most significant features and can be used in the ADGs to identify and control faults. Although there has been much research on using XAI in power systems, most of it focuses on transmission networks. The complexity of ADGs may prevent the extension of this technology. The quality of training populations for XAI techniques can affect their performance because the

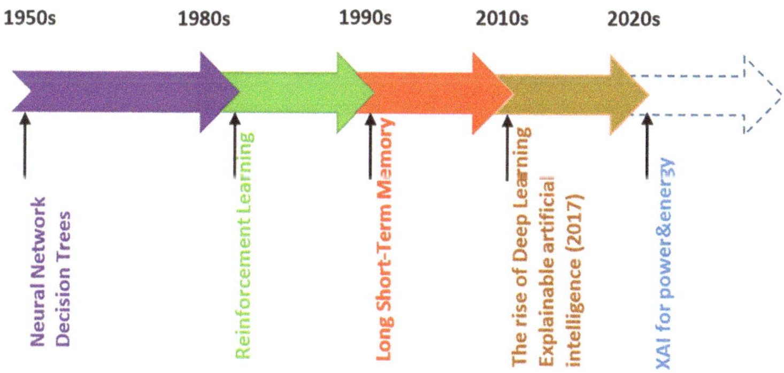


**Fig. 9.3** The evolution of XAI until use in the power system

limited data collected during the training process could introduce biases in the model. A potential solution could be based on Physics-guided neural networks and graph learning. Developers of these systems should use realistic assumptions regarding the data quality. Another issue can be associated with the distinction between certification and interpretation. There is no consensus on the appropriate evaluation and explanation of the distinction between interpretation and certification [7].

The reference [6] provided an overview of the various XAI techniques designed to improve the decision-making transparency. One of these is the model-agnostic XAI. This method focused on post hoc interpretability, considering the advantages of rule extraction and feature significance analysis. Thus, the DMs can understand how AI systems make decisions. For instance, if the training or design procedures are changed, the DMs can add interpretability attributes. Other examples of XAI techniques that can improve the interpretability of decisions include rule-based models and attention mechanisms.

Lee et al. proposed an XAI-based analysis and evaluation method that can help predict the power quality of the EDGs [8]. The increasing number of linked distribution facilities and the worsening of the power quality affect the continuity of the power supply. Thus, the EDGs should monitor and supervise continuously to minimize the impact of these factors. Predicted load fluctuations could reduce the impact of distributed power supply on the EDG's variability.

The study performed by Nwafor et al. [9] developed an approach to reducing non-technical losses (NTL) by analyzing the data collected from both the customer's consumption and operational processes. The system called NTLCONVNET consists of a set of three one-dimensional convolutional neural networks with varying depths. There has been limited research on the decision rationale behind the selection of features in the prediction of NTL. To improve the model's explainability, the researchers used the Shapley additive explanations kernel and the tree-based explainers. The researchers used a ranking framework to rank the variables across different models. They found that the staff-related factors were significantly associated with NTL.

When addressing the issues that may arise due to the adverse effects of electric vehicle charging on the power distribution system, it is important to predict the network's variability based on some factors that affect the system, such as the building and maintenance of facilities. The impact of electric vehicle (EV) charging on the AEGs has been analysed in [10] using a load prediction model. Lee et al. studied the effects of this issue on various factors, such as power loss and voltage stability. They used the feature engineering approach to transform the data collected from the univariate time series for EV charging into a time series with a multivariate dataset. XAI techniques such as SHAP and LIME have been chosen to explain the results. A machine learning framework that can explain the various factors that affect the reactive power of EDGs has been proposed in [11]. An efficient model has been used to estimate the Shapley values to minimize the computational burden. The framework is model-agnostic, which means it can be used to explain other types of ML models.

As a vital energy source, solar power has become a significant component of the solution to reducing carbon emissions. However, its integration into the ADGs is still in its early stages due to the unpredictable nature of solar power. The development of an AI framework used to improve the decision-making processes in solar power distribution has been proposed in [12]. This method allowed the system to predict the optimal energy yield and the solar power distribution. The study analysed the influence of various parameters, such as temperature and solar irradiance, using four different regression models. The Random Forest Regressor has been identified as the best model. The XAI techniques have been proposed to address challenges in solar power generation, such as the degradation of the storage battery and the error estimation in the performance of the solar panels. The findings of this study demonstrated how XAI can make it easy to develop a future of sustainable energy that's more resilient and efficient. Ebrahimi and Rastegar proposed in [13] an ML-based model easily implemented to improve the efficiency of switch placement in power distribution networks. Previous studies have been based on reliability-focused optimization techniques to find the optimal placement of switches. However, the number of tests conducted on these models increased exponentially with the complexity of the problem. Thus, the first step in implementing the proposed model is to identify the features associated with the candidate points. Then, a learning model explored the link between the candidate points and the optimal locations for the switch installation. After training, the learning model can identify the optimal strategy for the switch placement. According to the results obtained from the simulation process, the proposed model has been 29 times faster than the traditional mathematical model.

The increasing number of renewable power sources, such as solar and wind power, has resulted in the rapid emergence of distributed generators over the past decade. Although photovoltaic (PV) panels are the fastest-growing technology, they can also cause various operational issues when integrated into the ADGs. One of these is associated with the voltage fluctuations caused by the high PV penetration. Thus, Utama et al. [14] used an artificial neural network (ANN) to analyze the inputs and outputs of AC optimal power flow solutions, addressing the issue of reactive power dispatch in distributed PV systems. In the case of a decentralized system, the SHAP method to identify the non-local state measurements that can influence the system's operation has been used. In this case study, the researchers compared the performance of two ANN controllers against that of a baseline control strategy. The two controllers exhibited better performance in terms of their efficiency and line congestion. The proposed controllers achieved a reactive power control performance that is equivalent to that of an ACOPF-level system. They also reduced the computational burden and data privacy.

Figure 9.4 presents the synthetic classification of the various XAI research directions based on the state-the-art associated with the smart grid field [4]. The research is accelerated by crossing the directions conducted in more and more studies, mainly in recent years. However, the analysis of the studies highlighted that the XAI is primarily focused on developing methods that rely on labelled data. Unfortunately, this approach can be very challenging to implement in unsupervised learning due to the lack of explicit labels. The DM can address the following two questions:

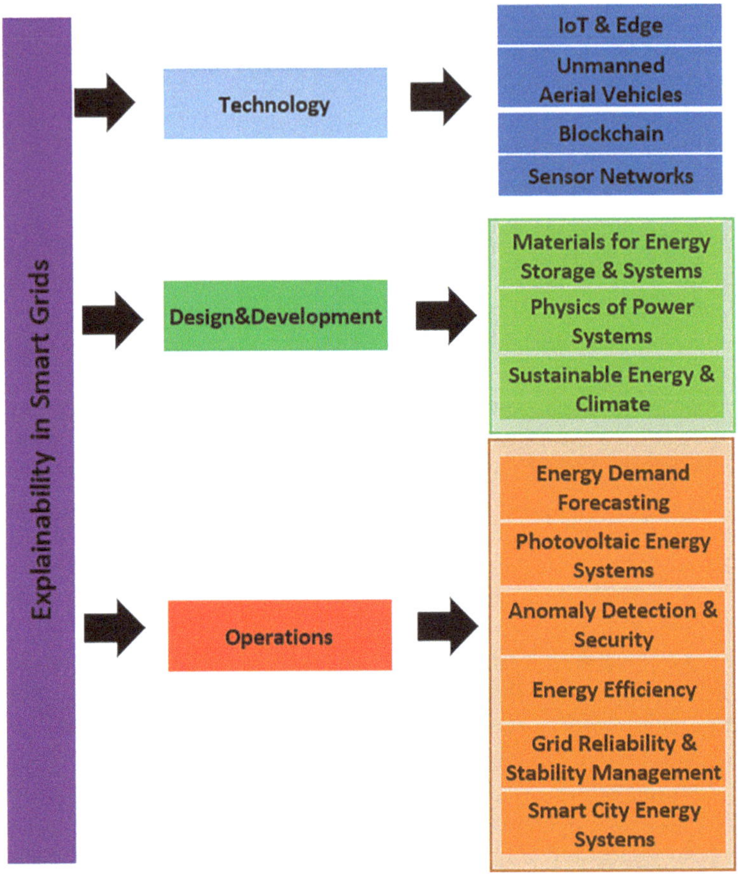


**Fig. 9.4** Applications of XAI in the smart grids

1. How can the unsupervised learning and XAI methods be employed to create meaningful explanations for automatically detected patterns associated with the technical issues identified in the planning and operation of the ADGs?
2. Can the explanations help us understand the detected patterns?

The study aims to find responses to these two questions based on a case analyze related to an issue from the ADGs associated with energy loss estimation. A methodology built for XAI applied in the case of unsupervised learning, including a decision tree-based method developed for K-means clustering, has been proposed.

The chapter has three other sections in its structure. Section 9.2 details the stages included in the theoretical framework for developing XAI methodology clustering-based unsupervised learning. Section 9.3 presents a case study where the performance of the proposed framework has been demonstrated to solve a technical issue faced by the DGOs inside the ADGs regarding energy loss estimation. Section 9.4 includes discussions and conclusions.

## 9.2 A Framework Integrating Decision Tree-Based Explainable Clustering Used in Decision Making Process

Unlike supervised learning, which is represented by the labelled data, unsupervised learning is characterized by identifying unknown patterns in the collected information [15, 16]. Thus, this makes the decision-making strategies more complex and challenging to explain. In addition, the semantic meaning of the patterns revealed should be interpreted, which poses significant XAI challenges.

Due to the complexity of unsupervised learning, few effective methods exist to develop XAI. To address this issue, an efficient framework provable for clustering is proposed. The framework contains a decision tree-building procedure to establish a clustering strategy for explainable partition of a dataset.

The proposed approach has three stages:

1. Obtain the patterns with the K-Means clustering algorithm and label the elements according to the cluster's assigned data.
2. Building the decision tree.
3. Explaining the partition of a dataset.

The details regarding each stage are presented in the following sections.

### 9.2.1 *K-Means Clustering Algorithm*

K-means clustering has been proposed to obtain an optimal partition of the data. This method that involves partitioning data points into subsets using the minimize sum of squares criterion. The algorithm tries to find similarities between clusters and data points [17, 18]. The stages associated with the algorithm' framework are the following, see Fig. 9.5.

**Stage 0**. Calculate the maximum number of partitions characterized through the number of clusters $K$ [19]:

$$K_{\max} = \sqrt{N} \tag{9.1}$$

where
N the number of the elements from the database

**Stage 1**. Introducing the number of clusters for each partition $K$, $K = 1, \ldots K_{max}$;

**Stage 2**. Set the centroids associated with each cluster k from the analysed partition $K$, $k = 1, \ldots, K$;

**Stage 3**. Allocating each element from the database to the nearest centroids based on the used distance. K-means clustering can support various distances:

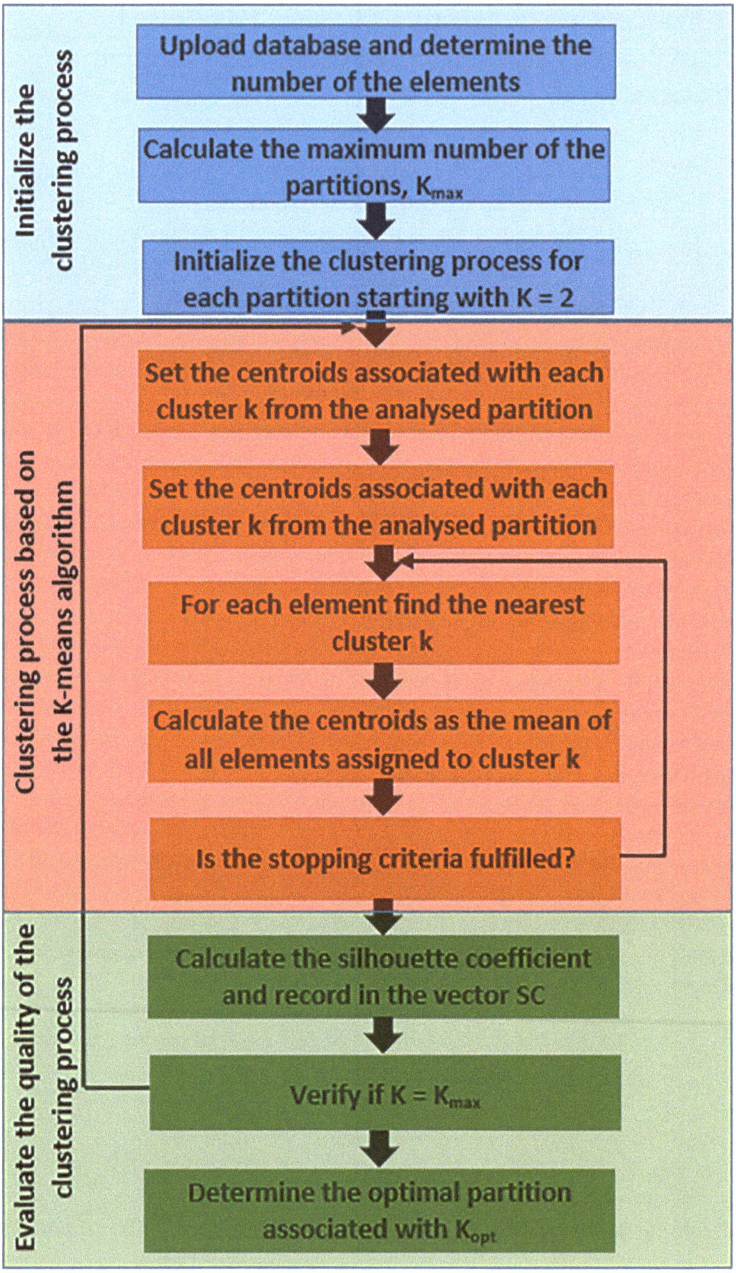


**Fig. 9.5** The steps of the K-means clustering algorithm

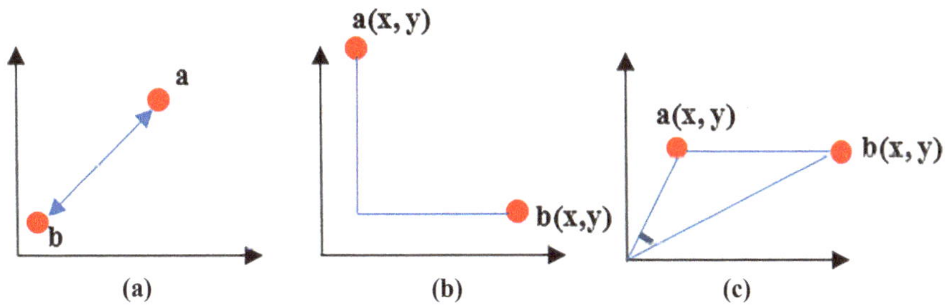


**Fig. 9.6** The distances supported by the K-Means Clustering algorithm: **a** Euclidean distance, **b** Manhattan distance, and **c** Cosine distance

Euclidean distance, Manhattan distance, squared Euclidean distance, or Cosine distance, see Fig. 9.6.

The distance between two data points is usually determined by calculating the *Euclidean distance*. For instance, if two elements, $a$ and $b$, are characterized by two or more features, their distance is equivalent to the straight-line distance in the Euclidean space.

$$d = \sqrt{\sum\nolimits_{i=1}^{n} (a_i - b_i)^2} \tag{9.2}$$

*Squared Euclidean Distance* is a copy of the Euclidean distance formula without the square root.

$$d = \sum_{i=1}^{n} (a_i - b_i)^2 \tag{9.3}$$

The distance between two points on the right angles is the *Manhattan distance*. It is an absolute value, meaning that the negative ones won't affect it.

$$d = \sum_{i=1}^{n} \left|b_{i,x} - a_{i,x}\right| + \left|b_{i,y} - a_{i,y}\right| \tag{9.4}$$

In this case, the Cosine distance between two vectors that joined the origin point is measured by taking the angle between them.

$$d = \frac{\sum_{i=1}^{n} (b_i - a_i)}{\sum_{i=1}^{n} (b_i)^2 \times \sum_{i=1}^{n} (a_i)^2} \tag{9.5}$$

**Stage 4**. Calculate again the centroids as the mean of all the elements assigned to each cluster $k$, $k = 1, \ldots, K$;

$$c_k = \frac{1}{|M_k|} \sum_{x_k \in M_k} x_k \, , \quad k = 1, \ldots, K \tag{9.6}$$

Where$M_k$ is the set of all elements assigned to the kth cluster, and $c_k$ represents the centroid of kth cluster.

**Stage 5**. Reiterate Stages 3 and 4 until the stop criterion is fulfilled.

**Stage 6**. The quality of the partitions is verified through an internal test. The most frequently used test involves the calculation of the silhouette coefficient (SC) [19]:

$$SC_{P_k} = \frac{l_i - o_i}{\max(l_i \, , \, o_i)} \tag{9.7}$$

$$l_i = \min_{k \neq i} \frac{1}{|k|} \sum_{j \in k} d(i, j) \tag{9.8}$$

$$o_i = \frac{1}{|k| - 1} \sum_{\substack{j \in k \\ i \neq k}} d(i, j) \tag{9.9}$$

where $l_i$—the separation between clusters (it indicates the degree to which they don't overlap), $o_i$—cohesion of the clusters (it measures the similarity of the elements within the cluster), $|k|$—the number of elements from the cluster $k$, $d(i,j)$—the distance between the elements $i$ and $j$.

**Stage 7**. The silhouette coefficient, SC, determines the optimal partition. It corresponds to the SC's maximum value.

## *9.2.2 Decision Trees*

DMs use the Decision trees (DTs) to solve complex regression and classification problems. This tree-structured approach considers the various features of a dataset and its decision rules. The two nodes' categories in a decision tree are the leaf and the decision nodes. The former is the output of the decisions, while the latter has multiple branches [20, 21].

The tree is constructed iteratively, with each step resulting from selecting of the best features and a threshold for minimizing the errors. The number of leaf nodes in a tree is the maximum that it can be before the value exceeds the optimal partition value. It can also generate a tree with a large size for better clustering assignments and more flexibility in the model's decisions. DTs provide solutions based on the given dataset's features, typically combining with other methods. Like a tree, the

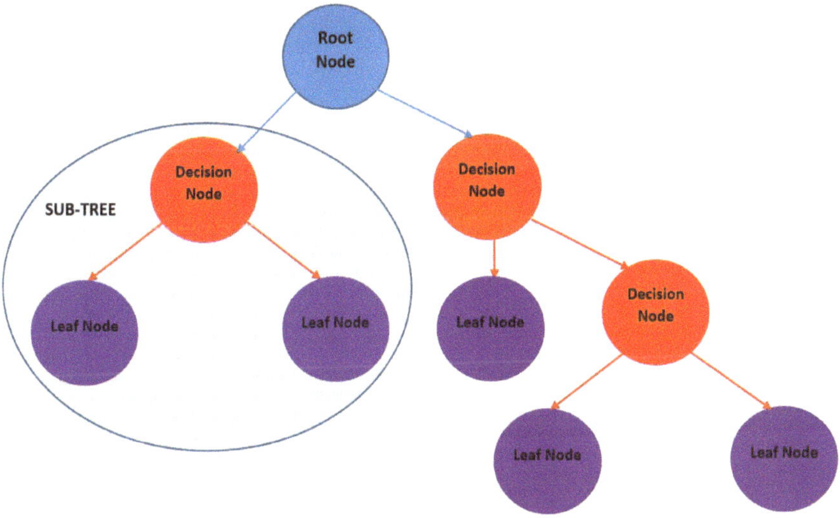


**Fig. 9.7** The structure of a decision tree

root node grows towards other branches, called sub-trees (containing the decision nodes and leaf nodes), and creates a structure resembling a tree, see Fig. 9.7.

The classification and regression algorithm (CART) is the most often used to build it. A decision tree merely asks a question, and its answer determines the classification sub-trees within it.

There are many algorithms in machine learning, and choosing the best one for a given problem is the most important thing to remember when creating a model. There are two main reasons to use a decision tree [20].

- It is a type of cognitive representation that mimics the thinking process of the DMs.
- Its structure resembles a tree, making it easy to understand its logic.

For instance, if the task of a decision tree is to classify a given dataset, its algorithm starts at the root. It then compares the values of the root attribute and the record attribute, and after that, it jumps to the next branch. The algorithm then moves further up the tree. It compares the value of the record attribute with that of the next sub-node and continues until it reaches the leaf.

The stages of the classification process using the DT are presented in the following, see Fig. 9.8.

**Stage 1**. DT should start at the root node, which integrates all elements from the database.
**Stage 2**. Use the attribute selection measure to find the best features in the database.
**Stage 3**. Split the DT into sub-trees, including potential values for the best features.
**Stage 4**. After building the DT, extract the most significant features.

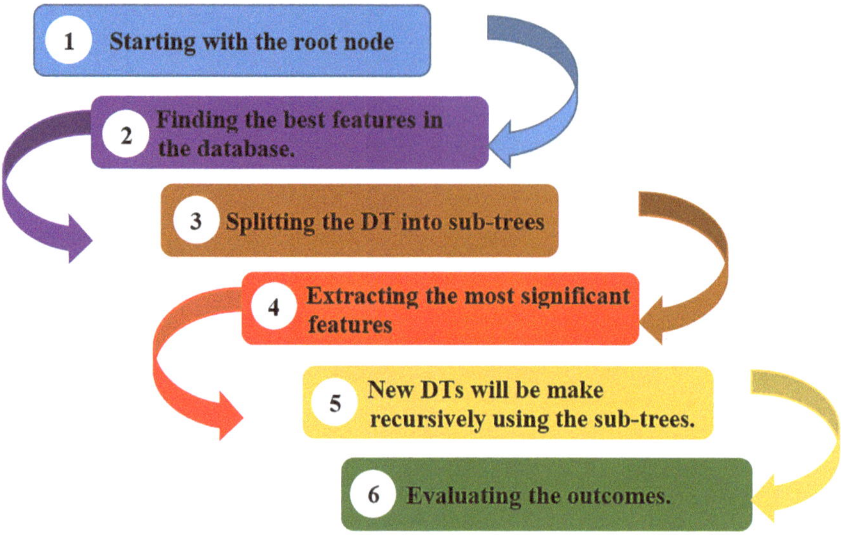


**Fig. 9.8** The stages of the classification process using the DT

**Stage 5.** New DTs will be made recursively using the sub-trees. The DM can continue until the DT can no longer classify the nodes.

**Stage 6.** Once the DM has the expected outcomes, it is determined which decision is the best. The highest expected value may not always be as wanted. The preferred values in DT analysis come from a probability algorithm. It is up to the DM to determine how the outcomes of the DT are evaluated.

One of the most common issues that DMs encounter when implementing a DT is selecting the best features for the decision nodes and the root node. This process is based on a feature selection measure. Two techniques are used to perform this process [22].

1. **Information Gain (IG)**: This metric indicates the changes in a dataset's entropy after the DT has been segmented according to a feature. IG can determine how a feature gives us information about a class. The value of IG dictates the nodes' split and the building of the tree.
2. **Gini index (GI)**: This measure specifies the purity or impurity in building a decision tree in the CART algorithm. A low value of the GI is preferred over a high one in the feature selection process. The CART algorithm only produces binary splits. GI is used to create these types of splits.

Regarding the optimization of the DT, pruning is one of the most critical techniques used in ML. This technique removes the parts of a decision tree that do not provide enough predictive power to improve its ability to perform better. It can also make a decision tree smaller by removing unwanted nodes.

In some cases, the DM uses pre-pruning to stop the development of a DT before it gets too complex. This technique can prevent the training data from getting overfitted, which can lead to poor performance. In other cases, after the DT is fully grown, post-pruning involves removing branches or nodes to improve the model's ability to generalize. Post-pruning can simplify the DT, maintaining its accuracy and improving its interpretability and performance by minimizing its complexity and mitigating overfit. This process can lead to more robust models that generalize to better data.

### *9.2.3 Explaining the Partition Trees*

In AI, classification involves assigning points to predefined classes or categories. AI techniques can help us, we can gain a deeper understanding of a model's decision-making process. The following are commonly used XAI methods for this task.

**Feature Importance**: Various methods determine the importance of features in a classification process. These methods assign weights and scores to different features to help us understand their relative importance.

**Rule Extraction**: Methods that extract human-readable rules from a classification framework attempt to provide a more complete representation of how the system makes decisions.

**Local Explanations**: A local explanation method is one of the most common techniques for providing a more accurate representation of a model's decision-making process. This method focuses on explaining the model's various features.

## 9.3 Case Study. Energy Losses Estimation

One of the most challenging problems in optimally designing and operating the AEDNs is calculating energy losses. This process involves considering the various factors that affect the operating conditions.

Besides the rated voltage, other factors such as the length of the feeders, the number of electric distribution substations (EDSs), the distributed generation sources, and the load requested by the end-users play a significant role in a comprehensive understanding of the operating regimes. Even if a feeder has more power losses than others, it does not necessarily mean it is operating out of its steady state [23].

All the above-highlighted aspects emphasize the importance of energy loss estimation for DNOs in increasing the performance of ADGs. If all technical characteristics are known, a deeper analysis of the various components of the ADGs can be performed.

### 9.3.1 Implementation of the Framework Integrating Decision Tree-Based Explainable Clustering Used in Energy Losses Estimation

This paragraph presents a framework based on the methodology proposed in Sect. 9.2. It is an original approach since other methods did not use the integrating decision tree-based explainable clustering used in the decision-making.

The first step considers the identification process of the technical factors that influence the operation of the ADGs. These refer to the number of EDSs, the rated power of the distribution transformers, and the length and type of line sections of the ADGs. The next step integrates the classification of the feeders from the ADG based on the K-Means clustering algorithm. At the end of the clustering process, an optimal number of the clusters associated with the feeders with similar characteristics is obtained. The third step involves building a DT, which will analyze the characteristics of each cluster of the feeders from the ADG. Finally, a deeper understanding of the obtained clusters is performed using the feature importance-based XAI method. Based on the obtained clusters, the annual energy losses in an ADG with more feeders with the formula:

$$\Delta W_{AEDN} = \sum_{k-1}^{K} f_k av_{W_k} + \sqrt{\sum_{k=1}^{K} f_k \left(sd_{W_k}\right)^2}, \; in\,[\text{MWh}] \tag{9.10}$$

where

$f_k$—the number of the feeders from the cluster $k$;

$av_{Wk}$—the average energy losses corresponding to the feeders from the cluster $k$, $k = 1, \ldots, K$.

$sd_k$—the standard deviation of annual energy losses corresponding to the feeders from the cluster $k$;

$K$—the number of feeders' clusters from the AEDNs.

The proposed framework has been tested using data collected from a Romanian ADG with 140 urban MV feeders. The input variables used in this study included the number of power transformers, the length of each section, and the total number of EDSs.

From the total number of feeders, 25 were removed in the preprocessing step due to missing data related to their primary characteristics. These characteristics included the length and sections of their branches and the rated power of the power transformers. The input data from the remaining 125 feeders have been recorded in a database.

The DM calculates each feeder's annual energy losses using the loss time method, and they are then grouped into representative clusters based on the K-mean algorithm. The optimal partition resulted from the clustering process performed in the range $K = 2, \ldots, K_{max}$, where $K_{max} = 11$ (calculated with the relation (9.1)), has been $K_{opt} = 5$, for which the global silhouette coefficient had the highest value (SC = 0.749), as can be seen in Fig. 9.9.

Table 9.1 presents the statistical parameters of the features associated with each cluster obtained with the K-means clustering algorithm. Figure 9.10 presents the energy losses associated with the feeders from each cluster depending on the features included in the clustering process.

The next stage includes building the decision tree. The classification and regression tree method, CART is used to generate decision tree. This method combines the historical data with other information to form a decision tree. Before it can be used, a number of patterns must be identified. In this case, the clustering method was utilized to determine the number of patterns. The model was tested with a 30% sample of the database and learned about 70% of the database, composed of 85 feeders.

The graph in Fig. 9.11 shows the decision tree generated for the learning database. The leaf and internal nodes represent the outcomes and tests, respectively.

The importance of the feature is calculated to understand the obtained model. The value of the features is computed based on the reduction in the number of impurities at the node. The sample count also calculates the probability of reaching the same node. The most significant features are usually found in the upper branches. A single feature can contribute to reducing impurities in the decision tree.

Next, DM can check training set accuracy as a benchmark and feature importance. Figure 9.12 shows that the importance ranking starts with Loading Level and then Length and Installed Power.

The confusion matrix can also provide an overview of the model's performance using the test database containing 40 feeders. The obtained results are integrated

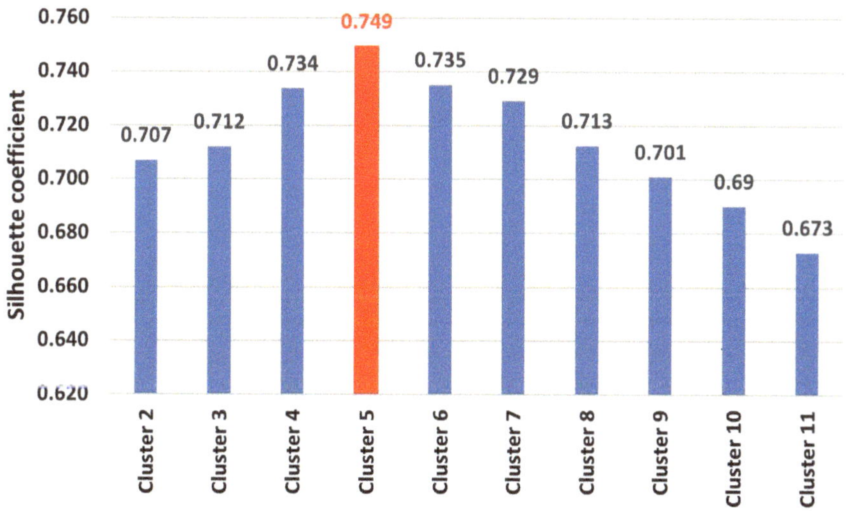


**Fig. 9.9** The values of the global silhouette coefficient in the range $K = 2, \ldots, K_{max}$

**Table 9.1** The statistical parameters of the clusters

| Features | Statistical parameters | Clusters | | | | |
|---|---|---|---|---|---|---|
| | | C1 | C2 | C3 | C4 | C5 |
| Number | – | 8 | 42 | 13 | 9 | 13 |
| Length | av [km] | 7.8 | 3.1 | 6.4 | 4.2 | 2.2 |
| | sd [km] | 2.2 | 1.0 | 1.4 | 1.1 | 0.9 |
| Installed power | av [kVA] | 6620 | 3574 | 3430 | 3276 | 1490 |
| | sd [kVA] | 931 | 949 | 990 | 654 | 737 |
| Loading level | av [%] | 26 | 24 | 45 | 77 | 73 |
| | sd [%] | 7 | 10 | 17 | 7 | 20 |
| Energy losses | av [MVA] | 240 | 116 | 127 | 168 | 51 |
| | sd [MVA] | 36 | 33 | 35 | 52 | 18 |

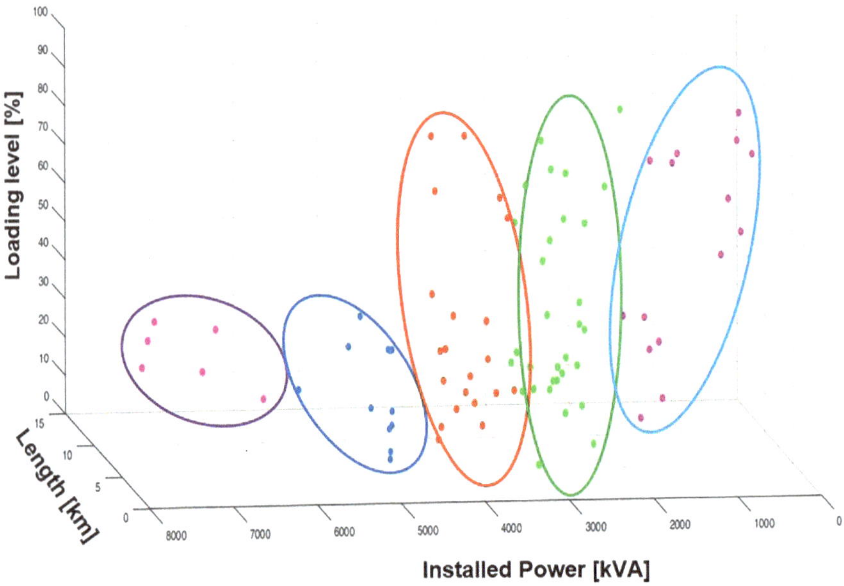


**Fig. 9.10** The energy losses associated with the feeders from each cluster depending on the features included in the clustering process (loading level, length, and installed power)

into a confusion matrix (see Table 9.2) containing actual classifications and predictions. Thus, the DM can evaluate the model's performance. Figure 9.13 presents the accuracy rate of the classification for each cluster.

The energy losses in the ADNs were estimated using the relation (9.10) and the data from the classification process. Only a 3.6% difference can be observed between the energy losses obtained with the proposed framework and the classical method used by the Romanian DNO (as seen in Fig. 9.14).

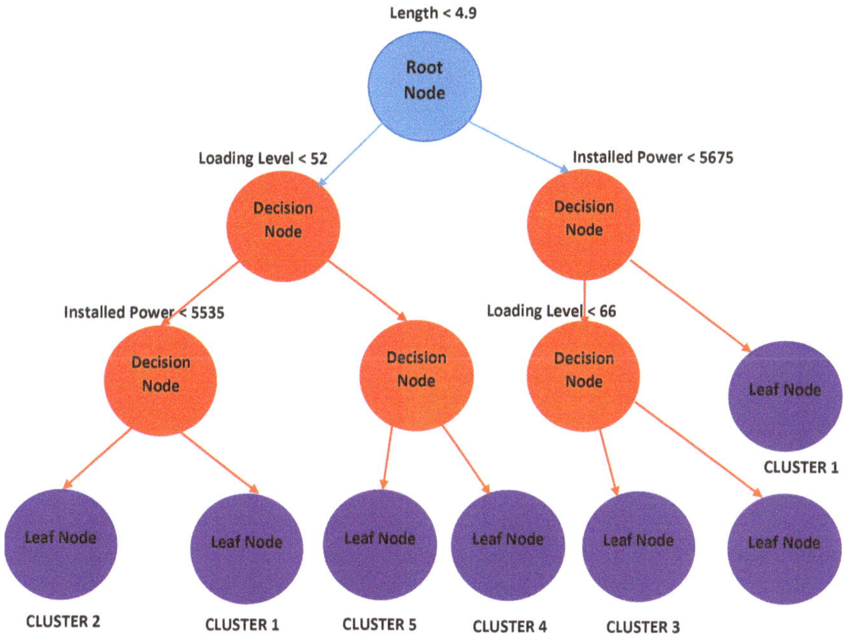


**Fig. 9.11** The decision tree associated with the energy loss estimation

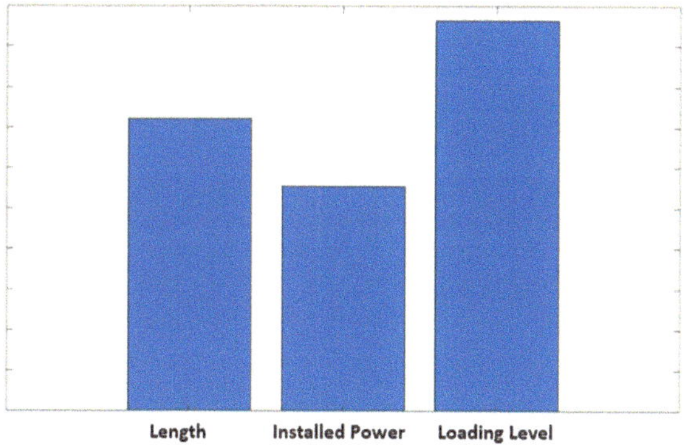


**Fig. 9.12** The feature importance in the classification process

**Table 9.2** The confusion matrix obtained for the test database

| C1 | C2 | C3 | C4 | C5 |
|---|---|---|---|---|
| 4/4 | 0/0 | 0/0 | 0/0 | 0/0 |
| 0/0 | 14/15 | 1/0 | 0/0 | 0/0 |
| 0/0 | 2/0 | 12/14 | 0/0 | 0/0 |
| 0/0 | 0/0 | 0/0 | 0/0 | 0/0 |
| 0/0 | 1/0 | 0/0 | 0/0 | 5/6 |

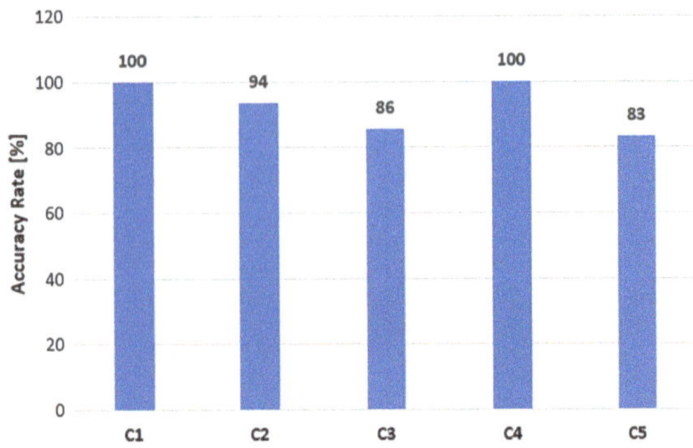


**Fig. 9.13** The accuracy rate of the classification for each cluster

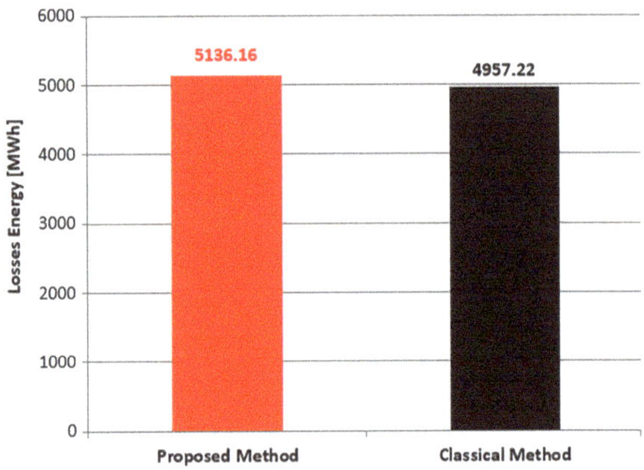


**Fig. 9.14** Comparison between the energy losses obtained with the proposed and the classical method

This value is very good in the context of many uncertainties regarding the required load in the nodes, the powers injected by the distributed generation sources, and the primary characteristics of the feeders.

## *9.3.2 Discussions*

The impact of input variables (predictors) on estimating energy loss can be analysed using SHAP. The Decision Maker can use the Shapley values of predictors or query points to interpret which predictors have the largest (or smallest) impact on the magnitude of model output [24, 25]. Predictions are represented as class scores for the proposed classification model corresponding with the feeders used in the energy loss estimation. The Shapley value of a predictor for a query point explains the deviation of the prediction for that query point from the average prediction attributable to the predictor. Therefore, the mean absolute Shapley value of a given predictor's query points measures its significance in model predictions. Figure 9.15 illustrates the

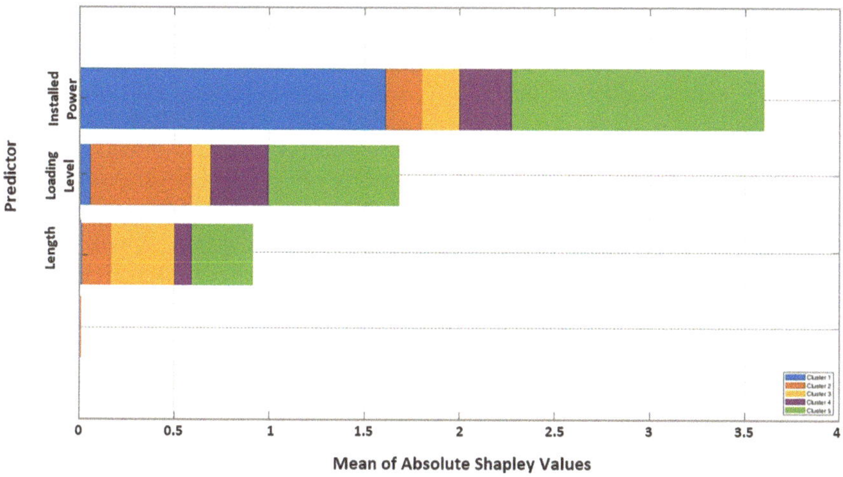


**Fig. 9.15** The mean of absolute shapley values across the five clusters

**Table 9.3** The mean absolute shapley values of the input variables obtained for each cluster

| Input variable | Cluster 1 | Cluster 2 | Cluster 3 | Cluster 4 | Cluster 5 |
|---|---|---|---|---|---|
| Length | 0.002 | 0.156 | 0.336 | 0.091 | 0.328 |
| Installed power | 1.612 | 0.192 | 0.195 | 0.279 | 1.321 |
| Loading level | 0.058 | 0.535 | 0.100 | 0.305 | 0.688 |

global importance factors of the three input variables (length, installed power, and loading level), providing a deep insight into estimating energy losses.

The mean of absolute Shapley values for each input variable in the global importance estimation data are displayed and ranked by importance. Table 9.3 presents the mean absolute Shapley values indicated.

Additionally, Figs. 9.16, 9.17, 9.18, 9.19 and 9.20 illustrate the significance of each input variable across the five clusters.

Thus, the study considered not only global outputs but also individual clusters. Based on the obtained results, the length of the feeders is the least significant, while the installed power of the transformers is the most critical.

## 9.4 Conclusions

The rapid emergence and evolution of deep learning and machine learning techniques have led to the development of automated systems for various applications from EDGs. These techniques have multiple advantages over traditional methods, such as their efficiency and accuracy. One of the biggest challenges that can prevent

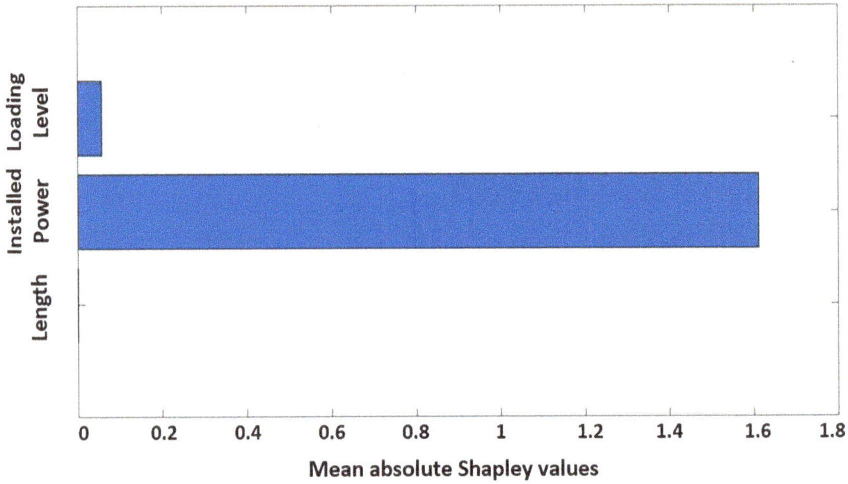


**Fig. 9.16** The mean of absolute shapley values of input variables assigned with cluster C1

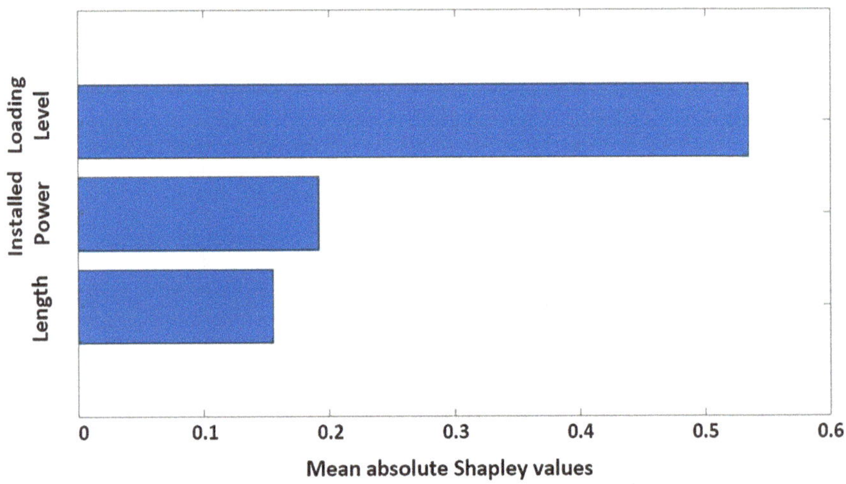


**Fig. 9.17** The mean of absolute shapley values of input variables assigned with cluster C2

the success of deep learning and machine learning-based systems is the lack of transparency regarding the reasons behind their decisions. One of the most common issues that can affect the success of a machine learning system is the disturbance caused by its decision-making. Developing an efficient learning algorithm that doesn't rely on memorizing training data can help improve the model's accuracy and produce more complex patterns.

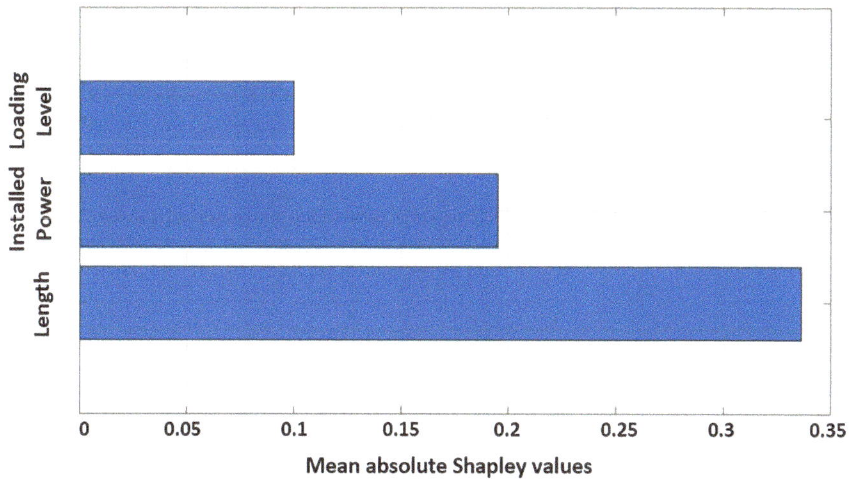


**Fig. 9.18** The mean of absolute shapley values of input variables assigned with cluster C3

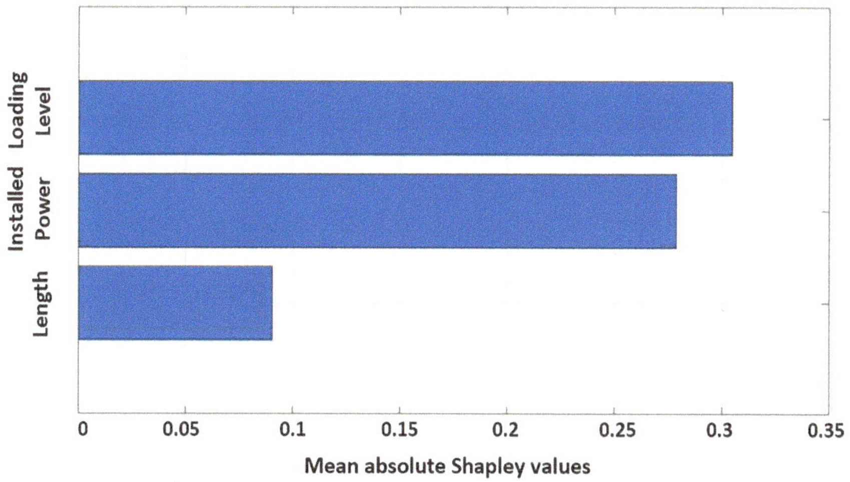


**Fig. 9.19** The mean of absolute shapley values of input variables assigned with cluster C4

Developers and designers of AI systems should provide the DMs with clear explanations of their decisions. Unfortunately, many of the algorithms used in this technology are not ideal and can lead to biases or prejudice. In this context, explainability has become essential for promoting transparency and facilitating interaction between DMs and AI. It aims to provide a better understanding of how ML models work. In most cases, the DMs cannot interpret relationships properly due to the process involved in learning and discovering. This aspect is often a disadvantage when implementing machine learning methods. Thus, the DMs should evaluate the accuracy

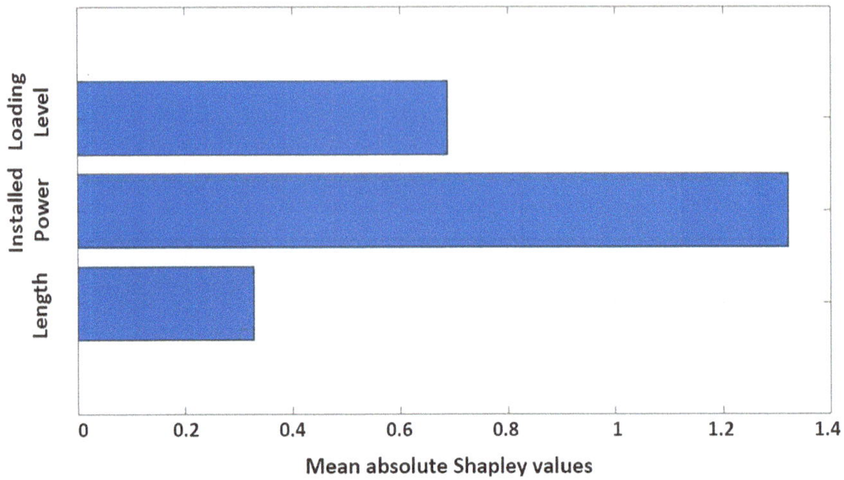


**Fig. 9.20** The mean of absolute shapley values of input variables assigned with cluster C5

and completeness of the depending on the application. Finally, the decision-making process should also be considered when using the XAI's models.

Regarding the ADNs, DMs make decisions regarding the planning and operation stages, having the necessary knowledge to understand the system's details. The XAI applications from the ADGs, these are focused on developing methods with labelled data. However, XAI can be hard to implement in an unsupervised setting due to the absence of explicit labels.

Thus, the use of XAI and unsupervised learning methods for detecting patterns in the ADNs has been addressed in this chapter as a meaningful tool for the decisions made regarding the operation and planning activities. An efficient three-stage-based framework has been proposed to address the estimation of energy losses. The framework contains a building procedure of a decision tree to establish a clustering strategy for the explainable partition of a dataset. The cognitive capabilities of XAI's model have been better than those of the DMs. These helped to improve the ML model's performance and find problems in training data.

However, the development of XAI models applied in the ADGs is still in its early stages. Many other issues need to be resolved in the ADGs with the help of XAI, such as understanding the solutions to be easy for the DMs. The solutions can help DGOs save money on energy by implementing sustainable practices in areas with technical issues.

## References

1. D'Adamo C, Jupe J, Abbey C (2009) Global survey on planning and operation of active distribution networks—update of CIGRE C6.11 working group activities. In: CIRED 2009 20th international conference and exhibition on electricity distribution Part 1, Prague, Czech Republic
2. Marot A, Rozier A, Dussartre M, Crochepierre L, Donnot B (2022) Towards an AI assistant for power grid operators. In: 1st international conference on hybrid human-artificial intelligence, Amsterdam, The Netherlands
3. Kahneman D, Frueh S (2022) Try to design an approach to making a judgment; don't just go into it trusting your intuition. Issues Sci Technol 38(3):23–26
4. Ortigossa ES, Gonçalves T, Nonato LG (2024) Explainable artificial intelligence (XAI)—from theory to methods and applications. IEEE Access 12:80799–80846
5. Machlev R, Heistrene L, Perl M, Levy KY, Belikov J, Mannor S, Levron Y (2022) Explainable artificial intelligence (XAI) techniques for energy and power systems: review. Chall Oppor Energy AI 9:100169
6. Praveenraj DDW, Victor M, Vennila C, Alawadi AH, Diyora P, Avudaiappan T (2023) Exploring explainable artificial intelligence for transparent decision making. In: International conference on newer engineering concepts and technology (ICONNECT-2023), New Delhi, India
7. Alsaigh R, Mehmood R, Katib I (2003) AI explainability and governance in smart energy systems: a review. Front Energy Res 11
8. Lee H, Lim H, Lee B (2023) Explainable AI-Based approaches for power quality prediction in distribution networks considering the uncertainty of renewable energy. In: 27th international conference on electricity distribution (CIRED 2023), Rome, Italy
9. Nwafor O, Okafor E, Aboushady A, Nwafor C, Zhou C (2023) Explainable artificial intelligence for prediction of non-technical losses in electricity distribution networks. IEEE Access 11:73104–73115
10. Lee H, Lim H, Lee B (2022) Analysis of EV Charging Load Impact on Distribution Network Using XAI Technique, in CIRED Porto Workshop 2022: E-mobility and power distribution systems, June 2–3. Porto, Portugal
11. Liao W, Schäfer B, Qin D, Zhang G, Wang Z, Yang Z (2024) An explainable framework for machine learning-based reactive power optimization of distribution network. In: 23rd power systems computation conference (PSCC 2024), Paris, France
12. Nallakaruppan MK, Shankar N, Bhuvanagiri PB, Padmanaban S, Khan SB (2024) Advancing solar energy integration: unveiling XAI insights for enhanced power system management and sustainable future. Ain Shams Eng J 15(6):102740
13. Ebrahimi M, Mohammad R (2024) Towards an interpretable data-driven switch placement model in electric power distribution systems: an explainable artificial intelligence-based approach. Eng Appl Artif Intell 129:107637
14. Utama C, Meske C, Schneider J, Ulbrich C (2022) Reactive power control in photovoltaic systems through (Explainable) artificial intelligence. Appl Energy 328:120004
15. Sarker IH (2021) Machine learning: algorithms, real-world applications and research directions. Sn Comput. Sci. 2:160
16. Grigoras G, Cartina G, Gavrilas M (2009) Using of clustering techniques in optimal placement of phasor measurements units. In: 9th WSEAS/IASME international conference on electric power systems, high voltage, electric machines, Genova, Italy
17. Miraftabzadeh SM, Colombo CG, Longo M, Foiadelli F (2023) K-Means and alternative clustering methods in modern power systems. IEEE Access 11:119596–119633
18. Grigoras G, Cartina G, Rotaru F (2010) Using K-means clustering method in determination of the energy losses levels from electric distribution systems. In: 12th WSEAS international conference on mathematical methods and computational techniques in electrical engineering, Timisoara, Romania
19. Batool F, Hennig C (2021) Clustering with the average silhouette width. Comput Stat Data Anal 158:107190

20. Mienye ID, Jere N (2024) A survey of decision trees: concepts, algorithms, and applications. IEEE Access 12
21. Grigoras G, Scarlatache F (2013) Energy losses estimation in electrical distribution networks with a decision trees-based algorithm. In: 8th international symposium on advanced topics in electrical engineering (ATEE), Bucharest, Romania
22. Lina CL, Fan CL (2019) Evaluation of CART, CHAID, and QUEST algorithms: a case study of construction defects in Taiwan. J Asian Arch Build Eng 18(6):539–553
23. Grigoras G, Cartina G, Bobric EC (2010) Strategies for power/energy saving in distribution networks. Adv Electr Comput Eng 10(2):61–64
24. Alvarez-Garcia M, Ibar-Alonso R, Arenas-Parra M (2024) A comprehensive framework for explainable cluster analysis. Inf Sci 663:120282
25. Ning Y, Ong ME, Chakraborty B, Goldstein BA, Ting DS, Vaughan R, Liu N (2022) Shapley variable importance cloud for interpretable machine learning. Patterns 3,100452
26. Clement T, Nguyen HTT, Kemmerzell N, Abdelaal M, Stjelja D (2023) Coping with data distribution shifts: XAI-Based adaptive learning with SHAP clustering for energy consumption prediction. In: Liu T, Webb G, Yue L, Wang D (eds) AI 2023: Advances in artificial intelligence. AI 2023. Lecture Notes in Computer Science, 14472, Springer, Singapore
27. Grigoras G, Scarlatache F (2015) Processing of smart meters data for peak load estimation of consumers. In: 9th international symposium on advanced topics in electrical engineering (ATEE), Bucharest, Romania

# Chapter 10
# Explainable AI for Building Trustworthy Transformer-Based Abstractive Summarization Models

**Hiranya Panawenna, Kasun Gayashan Hettihewa, Sandareka Wickramanayake, and Dulani Meedeniya**

**Abstract** Abstractive summarization (AS) is a natural language processing (NLP) task that generates a concise and coherent summary of a given document by rephrasing or paraphrasing the content, capturing the essential information rather than directly extracting sentences or phrases from the source text. AS is utilized in multiple mission-critical downstream tasks across various domains, including healthcare and law. Nevertheless, the existing state-of-the-art AS models are based on black-box deep learning models, such as Transformers, and they cannot explain why specific facts were included in the summary while others were omitted. This chapter proposes a novel framework for explaining which facts have been excluded from a summary by a given AS model and the rationale behind the selections. First, the effectiveness of existing feature attribution methods in explaining the decisions of AS models is analyzed. Then, a new framework, Fact Omission Explanation (FOE), is introduced, which utilizes a feature attribution method to analyze the fact-selection process of a given AS model and generate a linguistic explanation of the facts excluded and their respective reasons. The proposed framework was assessed using the Pub-Med dataset, PEGASUS (a state-of-the-art AS model), and quantitative metrics. The results demonstrate that FOE produces relevant and comprehensive explanations that elucidate the omission of key facts by AS models, thereby enhancing the trustworthiness of summarization systems.

**Keywords** Abstractive text summarization · Natural language processing · Smart applications · Explainability · Artificial intelligence

---

H. Panawenna · K. G. Hettihewa · S. Wickramanayake (✉) · D. Meedeniya
Department of Computer Science and Engineering, University of Moratuwa, Moratuwa, Sri Lanka
e-mail: sandarekaw@cse.mrt.ac.lk

H. Panawenna
e-mail: pasadie.23@cse.mrt.ac.lk

K. G. Hettihewa
e-mail: kasungh@uom.lk

D. Meedeniya
e-mail: dulanim@cse.mrt.ac.lk

 
N. Bizon and B. Appasani (eds.), *Explainable Artificial Intelligence for Trustworthy Decisions in Smart Applications*, https://doi.org/10.1007/978-3-031-97007-8_10

## 10.1 Introduction

Text summarization, a key task in Natural Language Processing (NLP), involves generating concise summaries from input documents. It is classified into Extractive Summarization (ES) and Abstractive Summarization (AS) [1]. ES directly extracts segments from the source document to summarize. In contrast, AS may rephrase the main ideas of the text using new words and sentence structures, resembling how humans summarize the text. AS is widely applied in mission-critical fields like healthcare [2, 3], law [4, 5], and finance [6, 7].

However, current state-of-the-art AS models rely on complex architectures, such as Transformers [8], that function as "black boxes," making it challenging for users to understand why certain facts are included or excluded in the summary. It is imperative to understand the fact selection process of an AS model to trust the generated summary in mission-critical domains, such as healthcare, where doctors analyze treatment options, law, where attorneys examine relevant documents or case reports, and finance, where analysts review reports and news to make strategic decisions. Unlike ES, AS paraphrases the source content rather than directly copying it, making it more challenging to identify which facts are selected or omitted and to gain insight into the reasoning behind these decisions. While several existing works have applied eXplainable Artificial Intelligence (XAI) techniques to derive insights into AS models [4, 9], none of them have analyzed the excluded facts and the reasons behind those exclusions.

This chapter proposes the Fact Omission Explanation (FOE) framework to explain why the AS model omits specific key facts in the input document during summarization. First, the effectiveness of the existing XAI techniques is analyzed, in particular the feature attribution methods such as attention visualizations [10, 11], Gradient × Input (GI) [12], and Gradient-weighted Class Activation Mapping (Grad-CAM) [13], in deriving insights into the AS models' fact selection process. Next, a new framework is developed that identifies important facts the AS model has excluded from the summary, along with the rationale for these selections. The proposed framework first identifies the sentences to which the AS model has given the least priority in generating the summary using a feature attribution method. Next, the input sentences with low relevance to the summary but containing key phrases from the input document are identified. KeyBERT [14] is utilized to identify the key phrases. Finally, the input document, the generated summary, and the sentences identified above are fed into a Large Language Model (LLM) to generate the explanations illustrating which facts have been omitted from the summary and the reasons for those omissions.

Since understanding the omission of key facts is more important in long document summarization, the experiments to assess the proposed framework were conducted using the PubMed dataset [15], which contains research articles in the medical domain, and the BigBirdPegasus model [16]. The initial analysis revealed that both Attention Analysis and the GI method were effective in identifying the tokens of the input document that influenced the summary. However, as the GI method requires much less computational power than attention analysis, it was utilized as the feature attribution method in the proposed framework. The GPT-4 [17] model was used to generate the explanations. The experimental results indicate that the proposed approach can generate explanations that reflect the AS model's reasoning behind the omission of certain key factors.

The main contributions of this work are in:

1. Identifying the effective feature attribution methods to obtain insights into AS models, and
2. Proposing a framework to generate user-friendly explanations revealing the reasons why some key facts are omitted from the summary in AS.

The rest of the chapter is structured as follows. Section 10.2 describes related work. Section 10.3 presents a comparative analysis of several feature attribution methods that aid in understanding AS models, particularly Transformer-based AS models. Section 10.4 outlines the primary contribution of this chapter: a framework for explaining which facts have been excluded from the summary by a given AS model, along with the rationale behind the selections. The methodology, experiments, and results are then presented. This is followed by a discussion in Sect. 10.5 and a conclusion on Sect. 10.6.

## 10.2 Related Work

### *10.2.1 Abstractive Summarization*

AS aims to generate concise and coherent summaries capturing key information from the source text. Sequence-to-sequence (Seq2Seq) models, based on Recurrent Neural Networks [18], which were initially introduced for machine translation, have been adapted for AS. Zhou et al. [19] introduced Selective Encoding for Abstractive Sentence Summarization, extending the Seq2Seq framework. Later, attention mechanisms were integrated into Seq2Seq models [20–23] to focus on relevant parts of the source text while generating the summary, improving the model's ability to capture long-range dependencies and produce coherent summaries. Since plain Seq2Seq methods tend to be factually incorrect and repetitive, See et al. have introduced a hybrid pointer-generator network [24] that copies words from the source text while producing novel words through the generator. Several other researchers have also approached AS as a two-step process of sentence selection and content generation [25–27]. Another perspective had been a hierarchical approach for document structural compression and coverage [28].

Following the introduction of Transformers by Vaswani et al. [8], several advanced transformer-based architectures were adapted for AS. BERT [29] is a pre-trained transformer model designed for natural language understanding, leveraging bidirectional context to create rich, contextual embeddings. Zhang et al. [30] have adapted BERT for AS by combining it with transformer-based decoders. BART [31] has been built on transformer architecture with a dual objective of denoising autoencoding and causal language modeling, allowing it to excel in AS tasks by combining bidirectional understanding with flexible Seq2Seq capabilities. PEGASUS [32], explicitly designed for AS, introduces a unique gap-sentence generation approach during pre-training. By removing sentences from input documents and creating gap-filled passages, PEGASUS learns robust contextual representations optimized for summarization. Raffel et al. [33] have addressed AS as a Text-to-Text Transfer learning task within a unified framework for any text-processing problem using the Text-to-Text Transfer Transformer (T5).

Researchers have also explored diverse training techniques to improve AS, such as SimCLS [34], a method for contrastive learning of AS that performs metric-oriented training via a generate-then-evaluate two-stage framework. Several other works have explored frameworks for generating multiple candidate summaries for AS and determining their qualities [35, 36].

### *10.2.2 Explainable Artificial Intelligence*

Recently, XAI has attracted a lot of attention as explaining the decisions of AI models to their users is crucial, especially when such models are deployed in mission-critical applications such as healthcare. Feature attribution is one of the primary XAI approaches that explain a model's decision by indicating the influence of each input feature (e.g., a word) on the model's decision. Some of the widely used feature attribution methods for NLP include Attention Visualizations, Grad-CAM [13], GI [12], and Rationalization [37].

**Attention Visualization**: Attention visualization is an XAI technique for interpreting attention-based models such as transformers. Attention mechanisms in Deep Neural Networks (DNNs) assign weights to input elements (e.g., words in a sentence) based on their relative importance in influencing the model's decision. Visualizing these weights facilitates the understanding of what the model attends to when making its decision [38–40]. Transformers employ two attention mechanisms: self-attention and cross-attention. In Seq2Seq tasks such as summarization, cross-attention is used in the decoder to align the input sequence with the output sequence being generated. The influence of each token in the input sequence towards the output sequence can be understood by visualizing the cross-attention.

**Grad-CAM**: Grad-CAM was introduced for Convolutional Neural Network (CNN)-based computer vision tasks. Grad-CAM generates heatmaps that indicate which parts of the input image contributed most to a CNN's prediction by utilizing gradients and activation output of the selected layer. The heatmap is computed as shown in Eq. 10.1:

$$L^c_{Grad-CAM} = ReLU\left(\sum_k \alpha^c_k A^k\right) \tag{10.1}$$

where $c$ is the class of interest, $A^k$ represents the activations of the $k$-th feature map, and $\alpha_k^c = \frac{1}{Z}\sum_i \sum_j \frac{\partial y^c}{\partial A_{ij}^k}$ is the gradient-based weight for the feature map $k$ of size $Z$ in a given convolutional layer with respect to the decision $y^c$. $A_{ij}^k$ is the activation at spatial location i, j of the feature map. *ReLU* ensures only positive contributions are visualized.

**Gradient x Input**: GI utilizes the gradients of the model's output with respect to the inputs to calculate the influence of each input feature towards the model prediction. Consider an input $x = \{x_1, x_2, ..., x_n\}$. To calculate the influence of the $i$ th input feature, GI first calculates the gradients of model output, $f(x)$, with respect to the input feature $x_i where\ i \in \{1, 2, ..., n\}$ and then multiply it by the input feature itself as shown in Eq. 10.2:

$$Gradient \times Input(GI) = x_i.\frac{\partial f(x)}{\partial x_i} \tag{10.2}$$

The gradient tells how sensitive $f(x)$ is to changes in $x_i$. Multiplying by $x_i$ scales the importance based on the actual feature value, filtering out noisy gradients for near-zero inputs.

### 10.2.3 *Explainable Artificial Intelligence for Abstractive Summarization*

Several works have explored the use of attention to understand AS models. Norkute et al. [4] have derived "attention highlights" on the input text, highlighting the key areas from which the abstractive summary is derived. A user study they conducted has demonstrated that user trust in the DNNs increased after the attention highlights feature was implemented. This study used a Pointer-generator network. Baan et al. [9] analyzed the transformer-based AS models. They used self-attention and cross-attention to identify the different interpretable parts of the document that each attention head focuses on. This study has focused on interpreting the attention heads rather than explaining to the user which key information was extracted from the source text in generating the summary and why. Saha et al. [41] proposed Summarization Programs (SP), using Neural Modular Trees to enhance interpretability in AS. Each tree maps summary sentences (root nodes) to source sentences (leaf nodes) through modular operations such as fusion, compression, and paraphrasing. They introduced SP-SEARCH to identify SPs for human summaries and used them to train seq2seq models that generate and execute SPs to produce summaries. However, the approach sacrifices some factual consistency and coherence compared to state-of-the-art models, such as BART and PEGASUS.

While the above studies attempted to enhance the explainability of AS models, none of them have focused on explaining what facts have been omitted and why, in natural language, to support end users in their decision-making processes.

### 10.2.4 Comparative Analysis of Feature Attribution Methods for Abstractive Summarization Models

In this section, a comparative analysis is conducted to evaluate the effectiveness of widely used feature attribution methods in providing explanations for AS models. Feature attribution methods provide insights to recognize which parts of the input sequence are most influential in shaping the summary. This helps verify that the model attends to relevant and important information in the input text for summary generation. The feature attribution methods used in this analysis are attention visualizations for transformers, Grad-CAM [13], and GI [12].

Suppose $X = \{x_1, x_2, ..., x_n\}$ denotes the input embeddings of the text with sequence length $n$, and $H = \{h_1, h_2, ..., h_n\}$ represents the set of encoder hidden states where $h_i \in R^d$ is the hidden state of $x_i$ from the final layer of the encoder and $d$ is the embedding dimension. Also, suppose $Y = \{y_1, y_2, ..., y_m\}$ denotes the set of class probabilities output by the decoder for the tokens in the summary with sequence length $m$.

Grad-CAM, originally developed for CNNs, is adapted for transformer-based AS models as described below. First, the influence of $x_i$ towards $y_j$, denoted as $G_{ij}$ is calculated as follows:

$$\alpha_{ij} = \left\| \left( \frac{\partial y_j}{\partial h_{i1}}, \frac{\partial y_j}{\partial h_{i2}}, ..., \frac{\partial y_j}{\partial h_{id}} \right) \right\| \tag{10.3}$$

$$G_{ij} = \left\| \mathrm{Re}\,LU\left(\alpha_{ij} \times h_i\right) \right\| \tag{10.4}$$

Then, the Grad-CAM output $G$ is the set of all $G_{ij}$ values between each pair of input and output tokens, $G = \{G_{11}, G_{12}, ..., G_{nm}\}$.

GI is adapted for transformer-based AS models as described below. Frist, the influence of $x_i$ towards $y_j$, denoted as $GI_{ij}$ is calculated as follows

$$\eta_{ij} = \frac{\partial y_j}{\partial x_i} \tag{10.5}$$

$$GI_{ij} = \eta_{ij} \otimes x_i, \tag{10.6}$$

where $\otimes$ denotes the element-wise multiplication.

The output $GI$ is the set of $GI_{ij}$ values between each pair of input and output tokens given by $GI = \{GI_{11}, GI_{12}, ..., GI_{nm}\}$.

The visualizations generated using cross-attention, Grad-CAM, and GI highlight the importance of each input token in contributing to the generation of each output token.

### 10.2.5 *Implementation Details*

The experiments in this study are conducted using the PEGASUS [32] model, which is a state-of-the-art AS model trained on the XSum [27] benchmark dataset. The XSum dataset consists of over 226,000 online news articles. The experiments were conducted on Google Colab, utilizing a system configuration of 12.7 GB RAM and 107.7 GB disk space. Due to resource limitations, only the first 100 samples from the XSum dataset, which contain fewer than 2200 characters, were selected for analysis.

### 10.2.6 *Results*

This section presents the saliency maps generated to demonstrate the influence of input tokens towards the decision-making of the PEGASUS [32] model using attention values, Grad-CAM, and GI methods. Visualizations for three randomly selected samples from the XSum [27] dataset are shown below.

**Sample 1** is a news article extracted from BBC sports [42].

**Summary**: Novak Djokovic and his doubles partner Viktor Troicki have been knocked out of the Olympic Games in the first round (Figs. 10.1, 10.2 and 10.3).

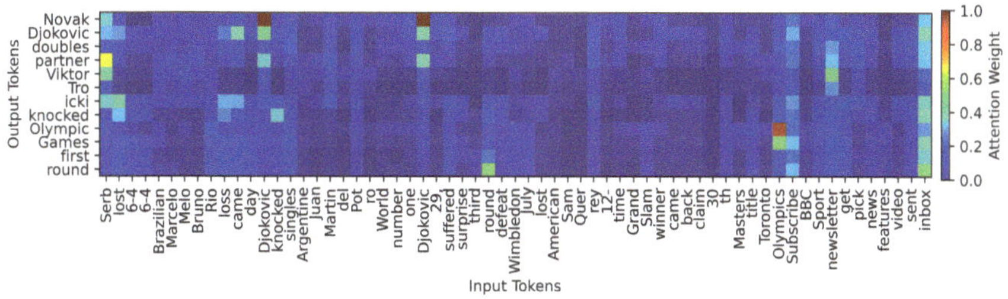


**Fig. 10.1** Attention visualization for sample 1

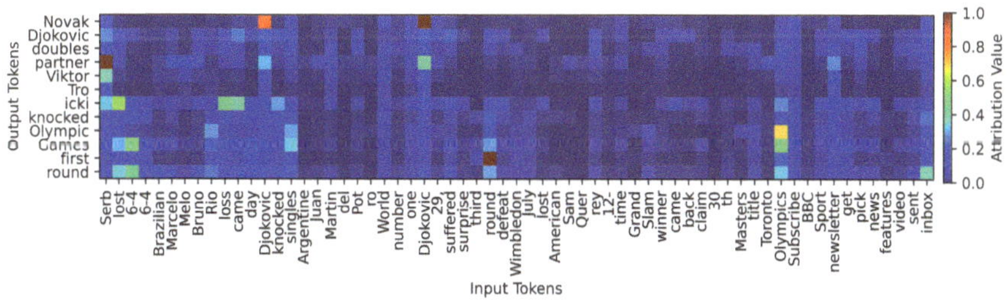


**Fig. 10.2** Grad-CAM visualization for sample 1

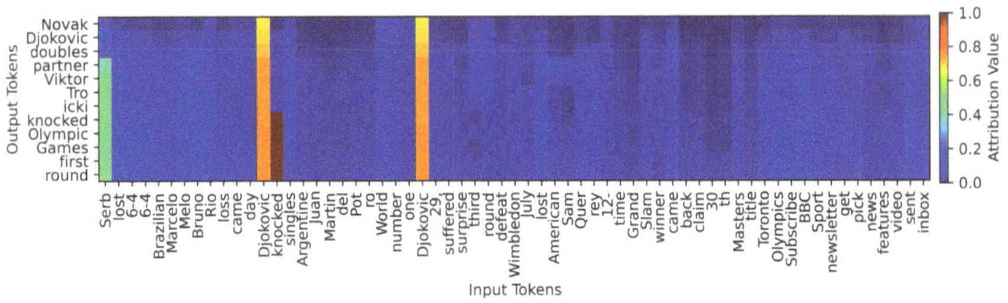


Fig. 10.3 GI visualization for sample 1

According to the visualization, there are a few common words that have achieved higher importance in both Grad-CAM and GI visualizations, such as "Djokovic", "round", and "Olympic". Furthermore, most of these common high-score tokens also appear in the generated summary. In the heatmap of attention visualization, most common words that appear in Grad-CAM and GI heatmaps have obtained higher attention values, which are shown in dark colour. It is notable that in cross-attention visualization, the words "subscribe" and "inbox" have gained slightly higher attention for all the tokens in the output sequence. However, those words do not make a significant contribution to the summary of the input text. It is clear that those particular words have been almost neglected by giving a very low value in GI visualization, which is based on the initial embedding layer.

**Sample 2** is a news article extracted from BBC news [43].

**Summary**: Footage has emerged of the moment a van carrying boxer Floyd Mayweather was set on fire outside a hotel (Figs. 10.4, 10.5 and 10.6).

In the illustrations, the common salient tokens from Grad-CAM and GI, which have appeared in the generated summary can be observed. Those are "Mayweather", "outside", "van", "set", "fire", and "hotel". Furthermore, it can be recognized that some of the words in that particular list have gained higher attention in attention visualizations. It is also noted that the common words captured from all the illustrations have a comparatively higher impact on the overall meaning of the generated summary.

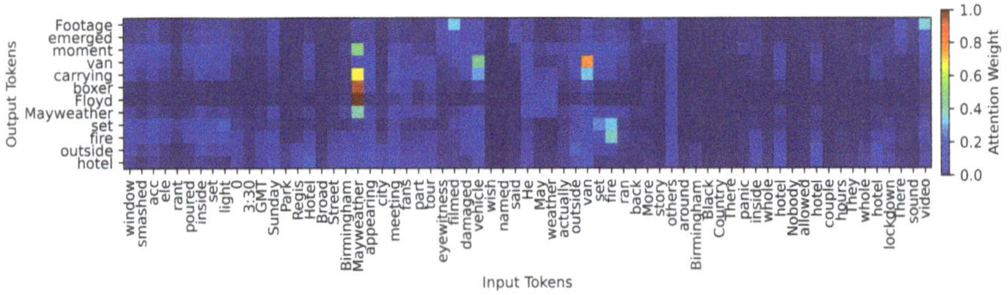


Fig. 10.4 Attention visualization for sample 2

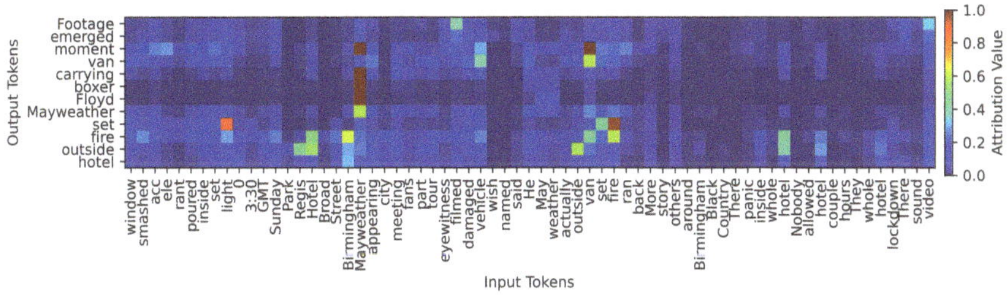


**Fig. 10.5** Grad-CAM visualization for sample 2

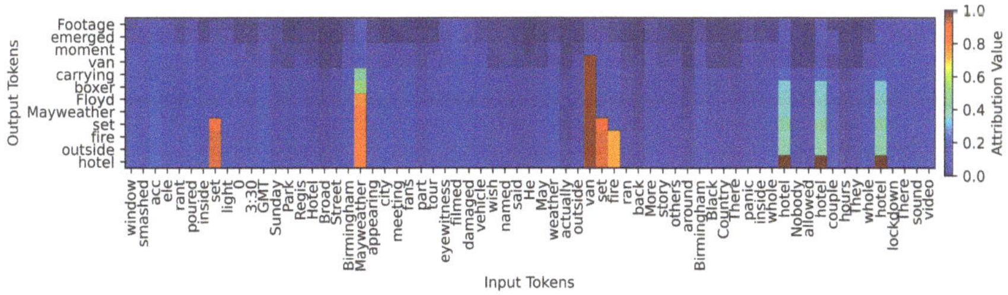


**Fig. 10.6** GI visualization for sample 2

**Sample 3** is a news article extracted from BBC news [44].

**Summary**: A man has admitted attempting to rob a shop after CCTV footage of the attack was shown in court (Figs. 10.7, 10.8 and 10.9).

Notably, a significant amount of common words with higher values appear in all the illustrations. A few of them are included in the summary which are having higher importance to generate the overall meaning of the summary.

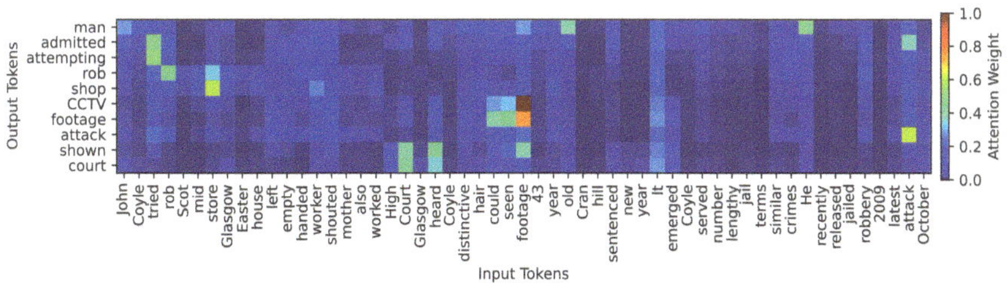


**Fig. 10.7** Attention visualization for sample 3

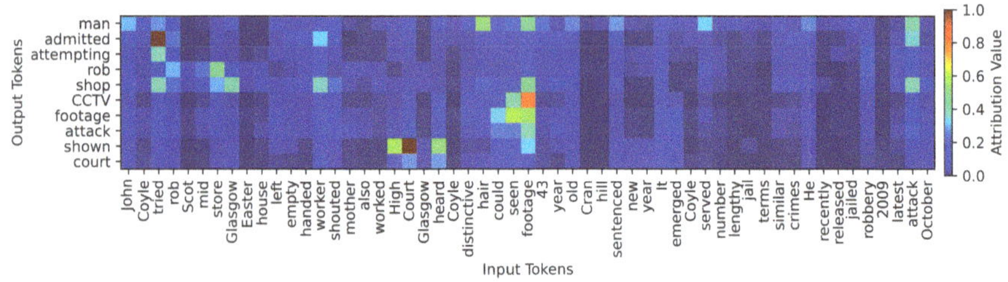


**Fig. 10.8** Grad-CAM visualization for sample 3

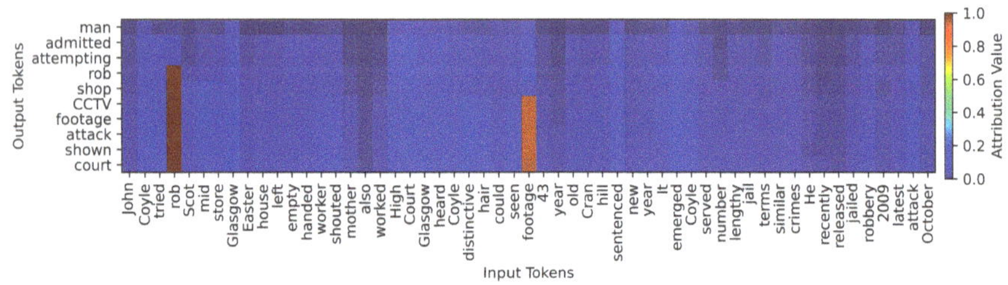


**Fig. 10.9** GI visualization for sample 3

### *10.2.7 Evaluation of Feature Attribution Methods*

To conduct the evaluations, the importance scores of each input token were aggregated across all output tokens. *Max* function was used to aggregate the importance scores. Then, the top k input tokens with the highest importance scores were selected, and the overlap between the top k tokens and the generated summary was calculated using BERT Score [45] to evaluate each feature attribution method. The generated summary was used as the reference. This experiment was conducted using 100 samples from the XSum [27] dataset. Two experiments were conducted to calculate the overlap for the top 5 tokens and top 10 tokens with the summary, and the results are presented in Tables 10.1 and 10.2. Stop words and punctuations were removed before calculating the overlap.

The similarity of the top k words generated by XAI methods and the generated summary demonstrated a significantly lower score. The experiments are limited to evaluating the similarity for the top 5 words and 10 words. It is noticeable that the similarity with the top 10 words shows a higher score compared to the similarity

**Table 10.1** BERT scores comparing top 5 tokens of Grad-CAM, GI, and attention with the generated summary

| | Precision | Recall | F1 score |
|---|---|---|---|
| Grad-CAM | 0.5685 | 0.4857 | 0.5225 |
| GI | 0.6274 | 0.5235 | 0.5693 |
| Attention | 0.6103 | 0.5061 | 0.5521 |

**Table 10.2** BERT scores comparing top 10 tokens of Grad-CAM, GI, and attention with the generated summary

| | Precision | Recall | F1 score |
|---|---|---|---|
| Grad-CAM | 0.5718 | 0.5521 | 0.5608 |
| GI | 0.5972 | 0.5882 | 0.5916 |
| Attention | 0.6026 | 0.5814 | 0.5909 |

with the top 5 words. It's challenging to capture the overall meaning of the generated summary only with the top 5 or 10 words. Furthermore, the Grad-CAM and GI methods have used only the encoder for the explanations. The summary that emerged from decoder cross-attention patterns is not illustrated by Grad-CAM and GI methods. This can be the reason for those two methods to have slightly lower similarity scores compared to the attention method.

### *10.2.8 Discussion*

In this study, experiments were conducted with XAI methods that can be adapted to explain AS models. The PEGASUS model was used as the AS model, which is a state-of-the-art model for AS. Grad-CAM and GI, which were originally applied to computer vision and neural networks to explain the black box nature of deep learning models, were adapted for the AS model.

During the study, better insights were obtained after removing stop words from the visualizations, as it was observed that some stop words with higher importance and attention scores had no impact on the explainability of the internal function of the summarization model. Furthermore, punctuations were removed to further refine the illustration to deliver a better explanation. Most of the text generation and summarization models introduce tokens like "</s>", "<pad>", "[CLS]", "<|endoftext|>" in the generated output. In some situations, those tokens get higher attention or importance without delivering meaningful insights for the explanations [46]. Those tokens were excluded from the visualizations due to the insignificant impact on the explainability of the summarization model. XAI methods evaluation experiments were conducted to prove the effectiveness of feature attribution methods in explaining the summarization model's decision-making.

Overall, the study has demonstrated the significance of XAI techniques used in the experiments in providing insight into the inner workings of an AS model. In particular, the illustrations of Grad-CAM, GI, and Attention depict a significant correlation, and their high-value tokens appear in the generated summary most of the time. Moreover, the Attention visualizations give insight into word combinations and ordering to generate meaningful sentences to deliver the summary.

## 10.3 Fact Omission Explanation (FOE) Framework

This section introduces the Fact Omission Explanation (FOE) framework, which builds upon the findings presented in Sect. 10.3 to explain the omission of key facts by AS models. Section 10.4.1 provides the technical details of the proposed framework. The framework is designed to generate explanations that are factually consistent, complete, relevant, and coherent. This framework is an extension of the work in [47].

Section 10.4.2 outlines the experimental setup used to evaluate the framework's effectiveness, focusing on the relevance and completeness of the generated explanations. Section 10.4.3 presents the results and conclusions derived from these experiments.

### *10.3.1 Overview of the Proposed FOE Framework*

The FOE framework consists of three primary stages. First, a feature attribution method is employed to identify sentences in the input document that the AS model assigns minimal relevance during the summary generation. Next, sentences with low relevance to the generated summary but containing key phrases from the input document are extracted, employing KeyBERT [14] for key phrase identification. Finally, the input document, the generated summary, and the identified sentences are provided as input to a Large Language Model (LLM) to generate explanations, highlighting omitted facts and providing potential reasons for their exclusion. An overview of the FOE framework is shown in Fig. 10.10.

The overall algorithm of the proposed FOE framework is shown in Algorithm 10.1. It accepts a Transformer-based AS model *M*, an input document *D*, an *LLM*, and the *Key-phrase Extraction Model (KPM)* as inputs. First, *D* is fed to *M* to generate

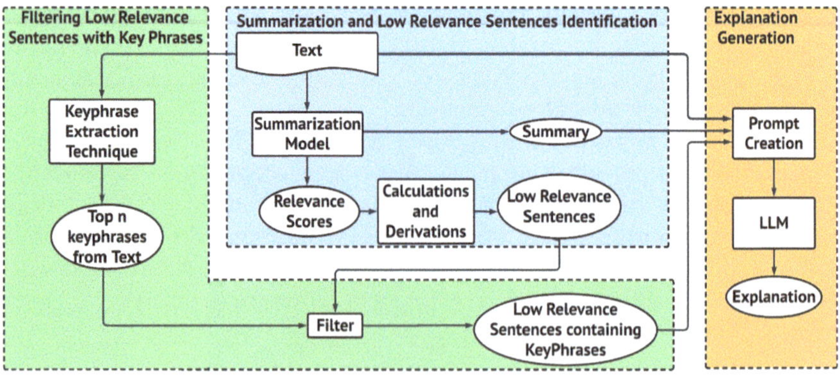


**Fig. 10.10** The overview of the proposed fact omission explanation framework

the abstractive summary, *S*. Next, Algorithm 10.2 is called to obtain the Low Relevance Sentences *(LRS)* in the input document that received low relevance from *M* in generating the summary. Afterwards, Algorithm 10.3 is called to extract the Key Phrases in *D (KPD)* and filter out the set of low-attention sentences containing any of the key phrases in the input document, denoted as *LRS-KPD*. Next, given *D, S* and *LRS-KPD*, prompt Θ is created. Finally, the LLM generates an explanation *E* as the output guided by the prompt.

**Algorithm 10.1: Fact Omission Explanation (FOE)**

**Input:** Transformer-based AS Model (*M*), Input Document (*D*), Key-phrase Extraction Model (*KPM*), Large Language Model (*LLM*)
**Output:** Explanation for omitted facts (*E*)

1: *E* ← {} {Initialization}
2: *S* ← *M*(*D*) {Generating summary of *D* using *M*}
3: *LRS* ← deriveLowRelevanceSentences(*M,D,S*) {Algorithm 10.2 }
4: *LRS-KPD* ← deriveLRS-KPD(*LRS, D, KPM*) {Algorithm 10.3}
5: *Θ* ← createPrompt(*D, S, LRS-KPD*)
6: *E* ← *LLM(Θ)* {Generate explanation for omitted facts}
7: Return *E*

The algorithm to derive LRS is shown in Algorithm 10.2. It accepts *M, D,* and *S* as the inputs. Suppose the set of tokens in *D* is denoted as $T_D$, and the set of sentences in *D* is denoted as $S_D$. Then using a feature attribution method *F* the relevance of each token $t_D \in T_D$ to the output summary *S* is calculated, denoted by $\alpha_{t_D}$. Based on those relevance scores, the threshold relevance $\eta$ is calculated as the mean of relevance scores of all the tokens in $T_D$. Next, a sentence-level relevance score for each sentence $s_D \in S_D$, is calculated. Finally, the set of sentences with relevance less than $\eta$ are selected as the *LRS*.

**Algorithm 10.2: Derive Low Relevance Sentences**

**Input:** Transformer-based AS Model (*M*), Input Document (*D*), Summary (*S*), Feature Attribution Method (*F*)
**Output:** Set of Low Relevance Sentences (*LRS*) - The input sentences with overall low relevance to the output

- $LRS \leftarrow \{\}$ Initialization
- $S_D \leftarrow$ the set of sentences in $D$
- $T_D \leftarrow$ the set of tokens in $D$
- $R_D \leftarrow$ the set of relevance scores corresponding to the tokens in $T_D$
- $n_D \leftarrow$ Number of tokens in $D$
- for $t_D$ in $T_D$ do
  - $\alpha_{t_D} \leftarrow F(t_D, S, M)$ {Calculating the relevance of $t_D$ towards $S$ using $F$}
  - $R_D \leftarrow R_D + \alpha_{t_D}$
- $\eta \leftarrow \mathrm{sum}(R_D)/n_D$ {Calculating the threshold of relevance scores as the mean}

{Selecting sentences with overall low relevance scores}

- for $s_D \in S_D$ do
  - $\beta = 0 \leftarrow$ Total Relevance Score of $s_D$ w.r.t. $S$
  - $n_{s_D}$ = getNumberOfTokens($s_D$)
  - for $t_D$ in $s_D$ do
    - $\alpha_{t_D} \leftarrow$ getRelevance($t_D$, $R_D$)
    - $\beta = \beta + \alpha_{t_D}$
  - end for
  - if $\beta / n_{s_D} < \eta$ then
    - $LRS = LRS + s_D$
- end for
- return *LRS*

Finally, the algorithm for deriving a set of low-attention sentences containing any of the key phrases in the input document, denoted as LRS-KPD is described in Algorithm 10.3. The algorithm accepts the Key-phrase Extraction Model (KPM), D, and LRS as inputs. Then, it passes *D* through KPM to obtain the key phrases in *D*, denoted by *KPD*. Next, for each sentence in LRS, the algorithm checks whether at least one key phrase is present in the sentence, and if so, that sentences are added to the set of LRS-KPD.

**Algorithm 10.3: Derive LRS-KPD**

**Input:** Key-phrase Extraction Model (*KPM*), Input Document (*D*), Low Relevance Sentences (*LRS*)
**Output:** Set of Low Relevance Sentences containing Key Phrases (*LRS-KPD*)

```
LRS-KPD ← {} Initialization
S_LRS ← the set of sentences in LRS
KPD ← KPM(D) {Extracting Key phrases in D using KPM}

{Selecting low relevance sentences containing key-phrases in the document:}
for s_LRS ∈ S_LRS do
    has_keyphrases ← FALSE
    for kpd ∈ KPD do
        if kpd in s_LRS then
            has_keyphrases ← TRUE
        end if
    end for
    if has_keyphrases then
        LRS-KPD = LRS-KPD + s_LRS
    end if
end for
return LRS-KPD
```

## 10.3.2 Experimental Setup

Given the utility of AS in summarizing lengthy documents in mission-critical domains, the PubMed dataset [48] (containing medical research articles) is used in these experiments. To be precise, a HuggingFace adaptation of the PubMed dataset is used [15]. Due to input length limitations of the original PEGASUS summarization model, it was not suitable for this dataset. Therefore, the BigBirdPegasus model available on HuggingFace [16] was used, which is an adaptation of BigBird Transformers introduced in [49] coupled with the Pegasus tokanizer method [32]. BigBird utilizes a sparse attention mechanism, combined with global attention and random attention, to efficiently handle input sequences. To make the experimentation process fast and efficient, the BigBirdPegasus model pretrained on PubMed was used, which was publicly available on HuggingFace [50].

The Pytorch library and HuggingFace Transformer models were used in this implementation. The experiments were conducted in a Google Colab environment with system RAM of 83.5 GB, GPU RAM of 40 GB, and Disk space of 235.7 GB. For experiments with Attention, a larger System RAM of 334.6 GB was used. Such a specification with high RAM was needed to conduct experiments on long documents.

Due to resource constraints, 50 randomly selected instances of the PubMed test dataset with less than 2000 words were used for the experiments.

Experiments were conducted with several prompt structures for ChatGPT as an ablation study, with the prompt containing:

1. $D + S + LRS\text{-}KPD$ (Algorithm 10.1)
2. $D + S + KPDiff$ (where, *KPDiff* are the keyphrases in $D$ but not in $S$)
3. $D + S + LRS\text{-}KPD + KPDiff$
4. $D + S + LRS$

The prompt for the method detailed in Algorithm 10.1 is depicted in Fig. 10.11.

Based on the findings in Sect. 10.3, GI method was selected as the feature attribution method to derive *LRS* in Algorithm 10.2, considering its good tradeoff between computational efficiency and evaluation metrics outlined in Tables 10.1 and 10.2. The KeyBert model [14], which performs keyword extraction based on BERT [29] was used as the keyphrase extraction technique in Algorithm 10.3. It extracts sub-phrases from the document and compares the cosine similarity between the sub-phrase embeddings and the full document embedding. The most similar sub-phrases are selected as key phrases. In the experiments, the candidate keyphrase length limit was set to three as that limit allowed for diversity while also retaining coherence. The top 30 key phrases from $D$ were extracted. Then the top 10 key phrases from $S$ were extracted to compare and obtain *KPDiff*.

The LLM used in the experiments was "ChatGPT-4o", currently the most advanced model released for researchers by OpenAI [17]. A key consideration in the

```
Prompt = [

        { system_prompt =
You are a Language Model explaining the decisions of an
Abstractive Summarization (AS) model.
The following information is given to you in a user prompt.
1. Text - This is the input passage
2. Summary - This is the summary generated by the AS model.
3. Low Relevance Sentences with Key Phrases (KP) - The AS model
has considered these input sentences as the least important
sentences. However, they contain KPs of the Text.

In your response mention what facts have been excluded in the
Summary compared to the Text and explain why those facts have
been excluded considering the main ideas of the Text and the
Summary.
Keep the output precise and concise. Do not exceed 100 words.
        }

        {user_prompt =
Text, Summary, Low Relevance Sentences containing KeyPhrases
        }
]
```

**Fig. 10.11** LLM prompt

```
Summary:

Enteric Duplications ( EDs ) are uncommon anomalies that can
occur at any point of the gastrointestinal tract . The small
intestine is the most common location ; retroperitoneum is an
extremely rare site . In general , diagnosed in the neonatal
period or during infancy , they are increasingly diagnosed
prenatally ; early prenatal detection is possible. There have
been seven reported cases of retroperitoneal ED cyst in the
english literature.
```

**Fig. 10.12** Sample summary

```
Low Relevance Sentences containing KeyPhrases:
        The small intestine is the most common location ;
     retroperitoneum is an extremely rare site.
        Prenatal ultrasound at 22 weeks of gestation
     objectified an abdominal cystic mass located in the left
     upper abdominal quadrant associated with fetal
     pyelectasis.
        Postnatal ultrasound found a retro peritoneal para -
     aortic liquid - filled mass measuring 60 mm 33 mm 22 mm.
        Magnetic resonance imaging ( MRI ) confirmed the
     presence of a retro peritoneal cyst occupying the upper
     left retro peritoneal space ; with mass effect
     displacing the left kidney down [ figure 1 ].
        Per operative finding was a retro peritoneal cyst
     above the left adrenal rarely an echo genic mass is
     found due to complication ( hemorrhage ).
```

**Fig. 10.13** Low relevance sentences containing key phrases

experiments was explanation length. Excessively short explanations may lack coherence, while overly long ones contradict the purpose of summarization. A 100-token limit was determined to be optimal, ensuring clarity while effectively covering the omitted facts.

Here are the user prompt components for method (1) above and the explanation derived from the LLM for a test example taken from a PubMed research article [51] (Figs. 10.12, 10.13 and 10.14).

### *10.3.3 Evaluation and Results*

To evaluate the proposed method, BertScore method [45] and Recall-Oriented Understudy for Gisting Evaluation (ROUGE) method [52] were utilized.

```
Explanation:
The summary excludes details about specific diagnostic findings,
such as the prenatal ultrasound showing a cystic mass in the
upper left abdominal quadrant, its postnatal size and location,
MRI results, and preoperative findings. These facts were excluded
as they focus on a detailed case presentation and measurements,
which, while useful for clinical insights, are not central to the
broader context of ED anomalies. The summary focuses on the
rarity, primary locations, and general diagnostic trends,
maintaining a high-level overview of EDs rather than delving into
specific case data.
```

**Fig. 10.14** Explanation

ROUGE is a widely used evaluation metric for assessing the quality of automatically generated summaries by comparing them to reference summaries. It quantifies the overlap of n-grams between the generated and reference summaries, with higher ROUGE scores indicating greater similarity. In this study, ROUGE-1 scores were utilized to measure unigram overlap between the reference and candidate texts. BertScore too, is an automatic evaluation metric for text generation. It computes token-level similarity between candidate and reference sentences using contextual embeddings. This is a more nuanced assessment of semantic similarity, capturing meaning beyond basic token overlap.

In the proposed adaptation to evaluate the quality of generated explanations, the explanation is considered as the candidate text while considering the set of sentences in $D$ containing *KPDiff* (defined in Sect. 10.4.3), as the reference text. Let us call this set of sentences $S_{KPDiff}$. Hence, a higher ROUGE score reflects that the explanation has a higher degree of overlap with the parts of the input document containing key information that are not there in the summary. Similarly, a higher BertScore score indicates that the explanation has a high semantic similarity with the input document's components containing key information omitted in the summary.

To evaluate whether the proposed method improves the explanations of omitted facts, the results were compared with a benchmark explanation. The benchmark explanation was generated using the same LLM but with only $D$ and $S$ as inputs, without additional insights into the model's inner workings. The set of ROUGE scores and BertScores thus obtained averaged for the 50 test cases selected for the experiments is summarized in Tables 10.3 and 10.4 respectively. Results are presented for two sets of experiments, one considering the GI method as the feature attribution method and the other considering Attention.

Table 10.3 shows that the highest ROUGE-1 precision, recall, and F1 were obtained by method (1). In Table 10.4, it can be observed that the highest BertScores precision, recall, and F1 were also obtained by method (1). Method (4) has consistently performed worse than the benchmark in almost all metrics (except Bertscore precision using the GI method).

**Table 10.3** ROUGE-1 scores comparing variants of the proposed method with the benchmark

| Feature attribution method | ROUGE-1 scores | Prompt design | | | | |
|---|---|---|---|---|---|---|
| | | (1) $D + S +$ *LRS-KPD* | (2) $D + S$ + *KPDiff* | (3) $D + S +$ *LRS-KPD* + *KPDiff* | (4) $D + S$ + *LRS* | Benchmark $(D + S)$ |
| GI | Precision | **0.5392** | 0.5182 | 0.5236 | 0.5111 | 0.5206 |
| | Recall | **0.1187** | 0.1118 | 0.1149 | 0.1112 | 0.1149 |
| | F1 | **0.1919** | 0.1813 | 0.1854 | 0.1801 | 0.1852 |
| Attention | Precision | **0.5661** | 0.5182 | 0.5274 | 0.5126 | 0.5206 |
| | Recall | **0.1199** | 0.1118 | 0.1175 | 0.1147 | 0.1149 |
| | F1 | **0.1951** | 0.1813 | 0.1887 | 0.184 | 0.1852 |

**Table 10.4** BertScores comparing variants of the proposed method with the benchmark

| Feature attribution method | BertScores | Prompt design | | | | |
|---|---|---|---|---|---|---|
| | | (1) $D + S +$ *LRS-KPD* | (2) $D + S$ + *KPDiff* | (3) $D + S +$ *LRS-KPD* + *KPDiff* | (4) $D + S$ + *LRS* | Benchmark $(D + S)$ |
| GI | Precision | **0.774** | 0.7723 | 0.7738 | 0.7725 | 0.7722 |
| | Recall | **0.8336** | 0.8291 | 0.8296 | 0.8292 | 0.83 |
| | F1 | **0.8027** | 0.7996 | 0.8007 | 0.7998 | 0.8 |
| Attention | Precision | **0.7741** | 0.7723 | 0.7738 | 0.7717 | 0.7722 |
| | Recall | **0.8353** | 0.8291 | 0.8303 | 0.8284 | 0.83 |
| | F1 | **0.8035** | 0.7996 | 0.801 | 0.799 | 0.8 |

Method (1) performing best can be attributed to the prompt providing only the necessary information to generate an explanation. Method (2) does not contain any insights into the summarization model workings, hence performing worse than (1). This reflects the importance of using an insight into the summarization process (using a feature attribution method), to obtain a better explanation from the LLM. Method (3) seems to have redundant information compared to (1), *KPDiff*, which worsens the performance. Method (4) overwhelms the LLM and results in the worst performance by providing the LLM with excessive unnecessary information, that is all the low relevance sentences in $D$ unfiltered. Given the substantial length of the input documents, it was essential to design a concise yet informative prompt. Method (1) achieves a good tradeoff between prompt length and informativeness, ensuring that the LLM can effectively process the input while maintaining explanation quality.

Moreover, it is notable that the results obtained using Attention as the feature attribution method were slightly higher than the results achieved with GI. However, the method of choice in these experiments was the GI method, considering its significantly greater computational efficiency compared to Attention.

Additionally, it can be observed at a glance that the scores are somewhat low in general, especially the ROUGE-1 scores. The reason for this is that both the ROUGE-1 and BERTScore evaluation metrics are inherently designed to evaluate summarization itself by comparing a candidate text with the reference text. However, when adapting those metrics to evaluate explanations in the proposed methodology, the explanation was compared with the set of sentences in the input document that contain key-phrases. Hence it is expected that the scores would be low in general, especially ROUGE-1 scores which compare the unigram overlap between the candidate and the reference.

Results show that the relevance (precision) and completeness (recall) of the explanations of omitted facts in long documents can be improved using the proposed method. Using ROUGE-1 scores and BertScores, they have been evaluated with two mechanisms: unigram overlap and cosine similarity of contextual embeddings. This shows that providing the LLM with insights into the text selection process of the summarization model aids the LLM in generating an explanation that closely represents the AS model's reasoning while successfully capturing which key details were omitted from the summary and why.

## 10.4 Discussion

### *10.4.1 Study Contributions*

In this chapter, an algorithm, FOE, is proposed that leverages feature attribution methods to gain insights into the inner workings of Transformer-based AS models. The FOE provides natural language explanations on what key facts were omitted and why. Given that omissions of key facts are particularly critical in the context of long document summarization, the experiments were focused on this setting. The BigBirdPegasus model [16] was used as the underlying summarization model, and the GPT-4 [17] model was utilized to generate the explanations. Experiments were conducted on the PubMed dataset [15], which comprises long medical research articles. The results show that the generated explanations are effective in terms of relevance, and completeness, as reflected in Tables 10.3 and 10.4.

The two main research contributions highlighted in Sect. 10.1 have been achieved via this study as described below:

1. Identifying the effective feature attribution methods to obtain insights into AS models—Experiments were conducted with several explainability techniques for transformer-based AS as discussed in Sect. 10.3. Based on this analysis, the GI method was selected as the foundation for the FOE framework.
2. Proposing a framework to generate user-friendly explanations revealing the reasons why some key facts are omitted from the summary in AS—Sect. 10.4.3 shows that the proposed explainable AS algorithm based on a feature attribution

method is capable of providing relevant, complete, and user-friendly explanations that clarifies why specific content was excluded from the AS output.

### 10.4.2 Comparison with the Existing Studies

As described in the related studies in Sect. 10.2.2, none of the existing work has explicitly focussed on explaining fact selection/omission of a transformer-based AS model in natural language, which is critical for applications requiring high transparency, such as legal or medical decision-making.

For instance, Norkute et al. [4] derived "attention highlights" on the input text, rather than providing explanations in natural language. However, this approach requires the user to manually review the highlights and infer which key facts were omitted and the reasons for their exclusion. Baan et al. [9] attempted to uncover the behavior of individual attention heads in Transformer-based models by mapping them to interpretable features, enhancing transparency at the model level. However, this lacked direct actionable insights for downstream users. The SP framework introduced by Saha et al. [41], went a step further to model the generative process of summaries, as a mode of enhancing explainability. However, despite its transparency, the summaries produced in this method were less effective than the state-of-the-art Transformer-based models such as PEGASUS. Additionally, this method too does not output explanations in natural language.

The proposed method addresses these limitations by offering feature attribution-based insights into which key facts were omitted from the summary and the reasons for their exclusion, all within a concise natural language explanation. Furthermore, this approach does not compromise the performance of the Transformer-based AS model, as it merely observes the inner workings of the model, extracts insights from them, and uses those insights to guide an LLM in generating useful and easy-to-understand explanations.

### 10.4.3 Open Challenges and Future Directions

A key open challenge in explainable AS is the absence of a standardized evaluation metric for explanations. Although ROUGE scores and BERTScore were employed to assess the effectiveness of the proposed method, these metrics are not explicitly designed to evaluate the quality of natural language explanations. Developing a dedicated metric for evaluating explanations is an essential avenue for future research.

In addition to the need for a dedicated evaluation metric, there is also a need to design datasets that include explanations for text summarization [53]. One approach to achieving this would be to extend existing summarization datasets with explanations related to fact selection and omission, as well as the reasons behind them.

## 10.5 Conclusion

This chapter proposed a novel algorithm, the Fact Omission Explanation (FOE) framework, designed to explain why transformer-based Abstractive Summarization (AS) models omit certain key facts during the summarization process. The framework leverages insights from feature attribution methods and incorporates a large language model (LLM) to generate human-interpretable explanations that reflect the decision-making process of AS models.

The issue of key information omission is particularly prevalent in the summarization of long documents. To evaluate the effectiveness of the proposed framework, experiments were conducted using the PubMed dataset, which contains medical research articles, and the BigBird-Pegasus model, a transformer architecture optimized for processing long input sequences.

This study presents two main contributions. First, it identifies effective feature attribution techniques that can provide insight into the internal mechanisms of AS models. Second, it introduces the FOE framework, which generates user-friendly explanations that reveal the rationale behind the omission of key facts in generated summaries. Experimental results demonstrate that the FOE framework produces explanations that align with the model's internal reasoning while maintaining relevance and usability for end-users.

Future research directions include the development of dedicated evaluation metrics and benchmark datasets to systematically assess the quality of natural language explanations in the context of abstractive summarization.

## List of Abbreviations

| | |
|---|---|
| AS | Abstractive summarization |
| CNN | Convolutional Neural Network |
| DNNs | Deep Neural Networks |
| ES | Extractive Summarization |
| FOE | Fact Omission Explanation |
| GI | Gradient $\times$ Input |
| Grad-CAM | Gradient-weighted Class Activation Mapping |
| LLM | Large Language Model |
| NLP | Natural Language Processing |
| ROUGE | Recall-Oriented Understudy for Gisting Evaluation |
| SP | Summarization Programs |
| XAI | EXplainable Artificial Intelligence |

## References


1. Wang G, Wu W (Apr.2024) Surveying the landscape of text summarization with deep learning: a comprehensive review. Discret Math Algorithms Appl 16(03):2330004. https://doi.org/10.1142/S1793830923300047
2. Wang M, Wang M, Yu F, Yang Y, Walker J, Mostafa J (Sep.2021) A systematic review of automatic text summarization for biomedical literature and EHRs. J Am Med Inform Assoc 28(10):2287–2297. https://doi.org/10.1093/jamia/ocab143
3. Jain R, Jangra A, Saha S, Jatowt A (2022) A survey on medical document summarization. https://doi.org/10.48550/ARXIV.2212.01669
4. Norkute M, Herger N, Michalak L, Mulder A, Gao S (2021) Towards explainable AI: assessing the usefulness and impact of added explainability features in legal document summarization. In: Extended abstracts of the 2021 CHI conference on human factors in computing systems, Yokohama Japan: ACM, pp 1–7. https://doi.org/10.1145/3411763.3443441
5. Shukla A et al (2022) Legal case document summarization: extractive and abstractive methods and their evaluation. In: Proceedings of the 2nd conference of the Asia-Pacific chapter of the association for computational linguistics and the 12th international joint conference on natural language processing (volume 1: long papers), online only: association for computational linguistics, pp 1048–1064. https://doi.org/10.18653/v1/2022.aacl-main.77
6. Zmandar N, Singh AK, El-Haj M, Rayson P (2021) Joint abstractive and extractive method for long financial document summarization. FNP. https://api.semanticscholar.org/CorpusID:239890015
7. Foroutan N, Romanou A, Massonnet S, Lebret R, Aberer K (2022) Multilingual text summarization on financial documents. In: El-Haj M, Rayson P, Zmandar N (eds) Proceedings of the 4th financial narrative processing workshop @LREC2022, Marseille, France: European Language Resources Association, pp 53–58. https://aclanthology.org/2022.fnp-1.7. Accessed 30 Dec 2024
8. Vaswani A et al (2017) Attention is all you need. In: Guyon I, Luxburg UV, Bengio S, Wallach H, Fergus R, Vishwanathan S, Garnett R (eds) Advances in neural information processing systems, Curran Associates, Inc.,. https://proceedings.neurips.cc/paper_files/paper/2017/file/3f5ee243547dee91fbd053c1c4a845aa-Paper.pdf
9. Baan J, ter Hoeve M, van der Wees M, Schuth A, de Rijke M (2019) Do transformer attention heads provide transparency in abstractive summarization? arXiv:1907.00570. https://doi.org/10.48550/arXiv.1907.00570
10. Vig J (2019) A multiscale visualization of attention in the transformer model. In: Proceedings of the 57th annual meeting of the association for computational linguistics: system demonstrations, Florence, Italy: Association for Computational Linguistics, pp 37–42. https://doi.org/10.18653/v1/P19-3007
11. Yeh C, Chen Y, Wu A, Chen C, Viégas F, Wattenberg M (2023) AttentionViz: a global view of transformer attention. IEEE Trans Vis Comput Graph 1–11. https://doi.org/10.1109/TVCG.2023.3327163
12. Shrikumar A, Greenside P, Shcherbina A, Kundaje A (2016) Not just a black box: learning important features through propagating activation differences. https://doi.org/10.48550/ARXIV.1605.01713
13. Selvaraju RR, Cogswell M, Das A, Vedantam R, Parikh D, Batra D (2017) Grad-CAM: visual explanations from deep networks via gradient-based localization. In: 2017 IEEE international conference on computer vision (ICCV), Venice: IEEE, pp 618–626. https://doi.org/10.1109/ICCV.2017.74
14. "KeyBERT." Accessed 13 Dec 2024. https://maartengr.github.io/KeyBERT/index.html
15. "ccdv/pubmed-summarization Datasets at Hugging Face." Accessed 01 Dec 2024. https://huggingface.co/datasets/ccdv/pubmed-summarization
16. "BigBirdPegasus." Accessed 26 Nov 2024. https://huggingface.co/docs/transformers/en/model_doc/bigbird_pegasus

17. Brown T et al (2020) Language models are few-shot learners. In: Advances in neural information processing systems, curran associates, Inc., pp 1877–1901. Accessed 26 Nov 2024. https://papers.nips.cc/paper_files/paper/2020/hash/1457c0d6bfcb4967418bfb8ac142f64a-Abstract.html
18. Sutskever I, Vinyals O, Le QV (2014) Sequence to sequence learning with neural networks. In: Ghahramani Z, Welling M, Cortes C, Lawrence N, Weinberger KQ (eds) Advances in neural information processing systems, Curran Associates, Inc.,. https://proceedings.neurips.cc/paper_files/paper/2014/file/a14ac55a4f27472c5d894ec1c3c743d2-Paper.pdf
19. Zhou Q, Yang N, Wei F, Zhou M (2017) Selective encoding for abstractive sentence summarization. In: Proceedings of the 55th annual meeting of the association for computational linguistics (volume 1: long papers), Vancouver, Canada: Association for Computational Linguistics, pp 1095–1104. https://doi.org/10.18653/v1/P17-1101
20. Wang Q, Ren J (2021) Summary-aware attention for social media short text abstractive summarization. Neurocomputing 425:290–299. https://doi.org/10.1016/j.neucom.2020.04.136
21. Liang Z, Du J, Li C (2020) Abstractive social media text summarization using selective reinforced Seq2Seq attention model. Neurocomputing 410:432–440. https://doi.org/10.1016/j.neucom.2020.04.137
22. Cohan A et al (2018) A discourse-aware attention model for abstractive summarization of long documents. In: Proceedings of the 2018 conference of the North American chapter of the association for computational linguistics: human language technologies, vol 2 (Short Papers), New Orleans, Louisiana: Association for Computational Linguistics, pp 615–621. https://doi.org/10.18653/v1/N18-2097
23. Nallapati R, Zhou B, Santos CD, Gulcehre C, Xiang B (2016) Abstractive text summarization using sequence-to-sequence RNNs and beyond. In: Proceedings of the 20th SIGNLL conference on computational natural language learning, Berlin, Germany: Association for Computational Linguistics, pp 280–290. https://doi.org/10.18653/v1/K16-1028
24. See A, Liu PJ, Manning CD (2017) Get to the point: summarization with pointer-generator networks. In: Proceedings of the 55th annual meeting of the association for computational linguistics (volume 1: long papers), Vancouver, Canada: Association for Computational Linguistics, pp 1073–1083. https://doi.org/10.18653/v1/P17-1099
25. Chen Y-C, Bansal M (2018) Fast abstractive summarization with reinforce-selected sentence rewriting. In: Proceedings of the 56th annual meeting of the association for computational linguistics (volume 1: long papers), Melbourne, Australia: Association for Computational Linguistics, pp 675–686. https://doi.org/10.18653/v1/P18-1063
26. Li W, Xiao X, Lyu Y, Wang Y (2018) Improving neural abstractive document summarization with explicit information selection modeling. In: Proceedings of the 2018 conference on empirical methods in natural language processing, Brussels, Belgium: Association for Computational Linguistics, pp 1787–1796. https://doi.org/10.18653/v1/D18-1205
27. Narayan S, Cohen SB, Lapata M (2018) Don't give me the details, just the summary! topic-aware convolutional neural networks for extreme summarization. In: Proceedings of the 2018 conference on empirical methods in natural language processing, Brussels, Belgium: Association for Computational Linguistics, pp 1797–1807. https://doi.org/10.18653/v1/D18-1206
28. Li W, Xiao X, Lyu Y, Wang Y (2018) Improving neural abstractive document summarization with structural regularization. In: Proceedings of the 2018 conference on empirical methods in natural language processing, Brussels, Belgium: Association for Computational Linguistics, pp 4078–4087. https://doi.org/10.18653/v1/D18-1441
29. Devlin J, Chang M-W, Lee K, Toutanova K (2019) BERT: pre-training of deep bidirectional transformers for language understanding. arXiv:1810.04805. https://doi.org/10.48550/arXiv.1810.04805
30. Zhang H, Cai J, Xu J, Wang J (2019) Pretraining-based natural language generation for text summarization. In: Proceedings of the 23rd conference on computational natural language learning (CoNLL), Hong Kong, China: Association for Computational Linguistics, pp 789–797. https://doi.org/10.18653/v1/K19-1074

31. Lewis M et al (2020) BART: denoising sequence-to-sequence pre-training for natural language generation, translation, and comprehension. In: Proceedings of the 58th annual meeting of the association for computational linguistics, online: association for computational linguistics, pp 7871–7880. https://doi.org/10.18653/v1/2020.acl-main.703
32. Zhang J, Zhao Y, Saleh M, Liu P (2020) PEGASUS: pre-training with extracted gap-sentences for abstractive summarization. In: III HD, Singh A (eds) Proceedings of the 37th international conference on machine learning, in proceedings of machine learning research, vol 119. PMLR, pp 11328–11339. https://proceedings.mlr.press/v119/zhang20ae.html
33. Raffel C et al (2020) Exploring the limits of transfer learning with a unified text-to-text transformer. J Mach Learn Res 21(1)
34. Liu Y, Liu P (2021) SimCLS: a simple framework for contrastive learning of abstractive summarization. In: Proceedings of the 59th annual meeting of the association for computational linguistics and the 11th international joint conference on natural language processing (volume 2: short papers), online: association for computational linguistics, pp 1065–1072. https://doi.org/10.18653/v1/2021.acl-short.135
35. Liu Y, Liu P, Radev D, Neubig G (2022) BRIO: bringing order to abstractive summarization. In: Proceedings of the 60th annual meeting of the association for computational linguistics (volume 1: long papers), Dublin, Ireland: Association for Computational Linguistics, pp 2890–2903. https://doi.org/10.18653/v1/2022.acl-long.207
36. Ravaut M, Joty S, Chen N (2022) SummaReranker: a multi-task mixture-of-experts re-ranking framework for abstractive summarization. In: Proceedings of the 60th annual meeting of the association for computational linguistics (volume 1: long papers), Dublin, Ireland: Association for Computational Linguistics, pp 4504–4524. https://doi.org/10.18653/v1/2022.acl-long.309
37. Gurrapu S, Kulkarni A, Huang L, Lourentzou I, Batarseh FA (2023) Rationalization for explainable NLP: a survey. Front Artif Intell 6:1225093. https://doi.org/10.3389/frai.2023.1225093
38. Han Z et al (2019) SeqViews2SeqLabels: learning 3D global features via aggregating sequential views by RNN with attention. IEEE Trans Image Process 28(2):658–672. https://doi.org/10.1109/TIP.2018.2868426
39. Fukui H, Hirakawa T, Yamashita T, Fujiyoshi H (2019) Attention branch network: learning of attention mechanism for visual explanation. In: Proceedings of the IEEE/CVF conference on computer vision and pattern recognition, pp 10705–10714. https://openaccess.thecvf.com/content_CVPR_2019/html/Fukui_Attention_Branch_Network_Learning_of_Attention_Mechanism_for_Visual_Explanation_CVPR_2019_paper.html. Accessed 06 Dec 2024
40. Zhao B, Wu X, Feng J, Peng Q, Yan S (Jun.2017) Diversified visual attention networks for fine-grained object classification. IEEE Trans Multimed 19(6):1245–1256. https://doi.org/10.1109/TMM.2017.2648498
41. Saha S, Zhang S, Hase P, Bansal M (2023) Summarization programs: interpretable abstractive summarization with neural modular trees. arXiv:2209.10492. https://doi.org/10.48550/arXiv.2209.10492
42. BBC Sport (2016) Rio Olympics 2016: Novak Djokovic and Nenad Zimonjic beaten in men's doubles. BBC. https://www.bbc.com/sport/olympics/37018844
43. BBC News (2017) Floyd Mayweather's van torched in Birmingham. BBC, Birmingham, UK. https://www.bbc.com/news/av/uk-england-birmingham-39182312
44. BBC News (2016) Robber's 'distinctive' grey hair spotted on CCTV. BBC. https://www.bbc.com/news/uk-scotland-glasgow-west-38288921
45. Zhang T, Kishore V, Wu F, Weinberger KQ, Artzi Y (2019) BERTScore: evaluating text generation with BERT. https://doi.org/10.48550/ARXIV.1904.09675
46. Clark K, Khandelwal U, Levy O, Manning CD (2019) What does BERT look at? An analysis of BERT's attention. In: Proceedings of the 2019 ACL workshop BlackboxNLP: analyzing and interpreting neural networks for NLP, Florence, Italy: Association for Computational Linguistics, pp 276–286. https://doi.org/10.18653/v1/W19-4828
47. Panawenna PH, Wickramanayake S (2024) Understanding omitted facts in transformer-based abstractive summarization. In: 2024 Moratuwa engineering research conference (MERCon),

Moratuwa, Sri Lanka: IEEE, pp 624–629. https://doi.org/10.1109/MERCon63886.2024.10688628
48. "PubMed," PubMed. https://pubmed.ncbi.nlm.nih.gov/. Accessed 26 Nov 2024
49. Zaheer M et al (2024) Big bird: transformers for longer sequences. In: Advances in neural information processing systems, Curran Associates, Inc., pp 17283–17297. https://proceedings.neurips.cc/paper/2020/hash/c8512d142a2d849725f31a9a7a361ab9-Abstract.html. Accessed 26 Nov 2024
50. "google/bigbird-pegasus-large-pubmed Hugging Face." https://huggingface.co/google/bigbird-pegasus-large-pubmed. Accessed 29 Nov 2024
51. Ayadi I, Bezzine A, Hamida E, Marrakchi Z (2017) Fetal cyst reveling retroperitoneal enteric duplication. J Indian Assoc Pediatr Surg 22(1):60. https://doi.org/10.4103/0971-9261.194630
52. Lin C-Y (2024) ROUGE: a package for automatic evaluation of summaries. In: Text summarization branches out, Barcelona, Spain: Association for Computational Linguistics, pp 74–81. https://aclanthology.org/W04-1013. Accessed 29 Nov 2024
53. Dhaini M, Erdogan E, Bakshi S, Kasneci G (2024) Explainability meets text summarization: a survey. In: Mahamood S, Minh NL, Ippolito D (eds) Proceedings of the 17th international natural language generation conference, Tokyo, Japan: Association for Computational Linguistics, pp 631–645. https://aclanthology.org/2024.inlg-main.49. Accessed 29 Nov 2024

# Chapter 11
# Explainable Artificial Intelligence for Trustworthy Decision-Making in Designing Photonic Devices

**Yogendra Swaroop Dwivedi, Rishav Singh, Anuj K. Sharma, Ajay Kumar Sharma, and C. Marques**

**Abstract** This chapter explores the use of Explainable Artificial Intelligence (XAI) in the optimization of design parameters in surface plasmon resonance (SPR) sensors and the analysis of oil–water emulsion stability using tilted fiber Bragg grating (TFBG) sensors. These sensors are critical for applications such as biosensing, gas detection, environmental monitoring, and the petroleum industry etc. The study focuses on: (*i*) optimizing key design parameters for SPR sensors, including wavelength (λ) and metal layer thickness ($d_m$) leading to maximum figure of merit (FOM), and (*ii*) analysing surfactant concentration ($C_s$) and rotational speed (RPM) for oil–water emulsion stability using TFBG sensors. Machine learning (ML) models and XAI methods like SHAP (SHapley Additive exPlanations) and LIME (Local Interpretable Model-Agnostic Explanations) are applied to SPR and TFBG sensors to enhance transparency and interpretability in the sensor design process. In the SPR sensor analysis, our findings demonstrate that $d_m$ has a significantly greater impact on optimizing the FOM than λ. Further, the findings are in corroboration with the fundamental concept of radiation damping in plasmonic structures. XAI techniques provided valuable insights into the contribution of these parameters, facilitating a more efficient and reliable sensor design. For oil–water emulsion stability, the

Y. S. Dwivedi · A. K. Sharma (✉)
Physics Division, Department of Applied Sciences, National Institute of Technology Delhi, Delhi, India
e-mail: anujsharma@nitdelhi.ac.in

R. Singh
Department of Computer Science and Engineering, Indian Institute of Technology Patna, Patna, India

A. K. Sharma
Department of Computer Science and Engineering, National Institute of Technology Delhi, Delhi, India

C. Marques
CICECO – Aveiro Institute of Materials, Department of Physics, University of Aveiro, Aveiro, Portugal

Department of Physics, VSB – Technical University of Ostrava, Ostrava, Czech Republic

 
N. Bizon and B. Appasani (eds.), *Explainable Artificial Intelligence for Trustworthy Decisions in Smart Applications*, https://doi.org/10.1007/978-3-031-97007-8_11

study shows that $C_s$ is the dominant factor influencing stability, with RPM and $\lambda$ having comparatively smaller impacts. By utilizing ML and XAI techniques for SPR sensors and oil–water emulsion analysis, this chapter provides a comprehensive understanding of the key factors driving sensor performance and emulsion stability. The chapter concludes by suggesting future research directions, including the integration of advanced materials and real-time XAI applications to further improve the functionality and performance of these sensors. By implementing XAI methods such as SHAP and LIME alongside ML models like Gaussian Process Regression (GPR) and Random Forest, we gain deeper insights into the contributions of design parameters. The findings demonstrate that XAI enhances the reliability and interpretability of ML models, leading to more informed and trustworthy decision-making in photonic device design. This approach not only optimizes device performance but also aligns with the physical principles governing photonic systems, thereby bridging the gap between complex ML models and practical application.

**Keywords** Plasmon · Fiber Bragg grating · Sensor · Explainable Artificial Intelligence (XAI) · SHAP · LIME · Machine learning · Figure of Merit (FOM) · Oil–water emulsion

## 11.1 Introduction

### *11.1.1 Overview of Photonic Sensors*

Photonic sensors have become indispensable tools in modern technology, finding applications across a diverse range of fields such as biosensing, environmental monitoring [1], drug discovery [2], industrial processing, clinical diagnostics [3], and telecommunications. These sensors leverage the interaction of light with matter to detect changes in physical, chemical, or biological parameters with high sensitivity and specificity. Two prominent types of photonic sensors are SPR sensors and TFBG sensors. SPR sensors exploit the resonant oscillations of conduction electrons, known as surface plasmons, at the interface between a metal and a dielectric material. This phenomenon is highly sensitive to refractive index (RI) changes near the sensor surface. Figure 11.1 illustrates the working mechanism of a fiber-optic SPR sensor, where surface plasmon resonance occurs at the metal–dielectric interface, enabling detection of changes in the refractive index. This makes the sensor effective for monitoring biomolecular interactions, gas concentrations, and pollutants. TFBG sensors, on the other hand, are a special class of fiber optic sensors that combine the properties of fiber Bragg gratings (FBGs) [4] and long-period gratings (LPGs) [5]. By introducing a tilt in the grating structure, TFBG sensors enable coupling between core modes and cladding modes within the optical fiber. This coupling makes them highly sensitive to changes in the surrounding RI, temperature, and strain, extending their applicability to fields like structural health monitoring and the petroleum industry.

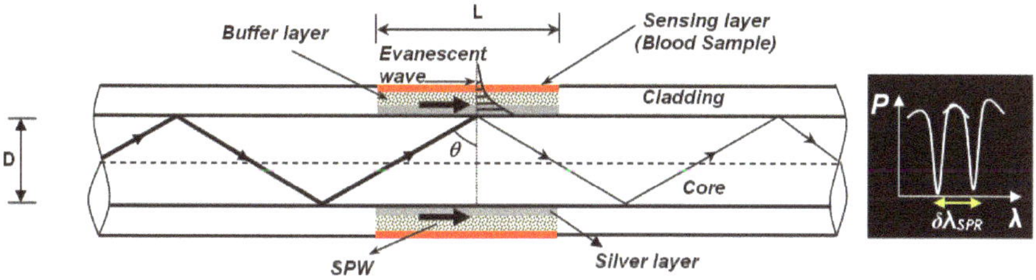


**Fig. 11.1** Schematic of fiber-optic SPR sensor

### *11.1.2 Challenges in Optimizing Photonic Sensor Design*

Despite the advantages of SPR and TFBG sensors, optimizing their design parameters remains a significant challenge. For SPR sensors, key parameters such as metal layer thickness ($d_m$) and light wavelength ($\lambda$) critically influence the sensor's performance, particularly its FOM, which quantifies the sensor's sensitivity and resolution. The interplay between $d_m$ and $\lambda$ affects the excitation of surface plasmons and the resulting resonance conditions. Traditionally, optimizing these parameters involves extensive simulations and experimental trials to explore the parameter space. This process is not only time-consuming but may also fail to capture the complex, nonlinear relationships between the parameters and the sensor's performance metrics.

Similarly, for TFBG sensors used in analysing oil–water emulsions [6] in the petroleum industry, parameters such as $C_s$, RPM during mixing, and $\lambda$ play crucial roles in determining emulsion stability. The stability of oil–water emulsions directly impacts viscosity, production rates, and operational costs [7] in oil extraction and processing. Accurately characterizing and controlling these emulsions is essential for optimizing the corresponding industrial processes. However, TFBG sensors, like many optical sensors, are susceptible to noise and fluctuations in their measurements. External factors and inherent limitations can affect the reliability of the sensor data, posing challenges in interpreting the results and making informed decisions.

### *11.1.3 Need for Advanced Analytical Techniques*

The complexities involved in optimizing photonic sensor designs and interpreting their outputs necessitate the adoption of advanced analytical techniques. Traditional methods may not adequately address issues such as:

- **High-dimensional parameter spaces**: The multitude of interacting parameters makes exhaustive exploration impractical.
- **Nonlinear relationships**: The sensor responses often exhibit nonlinear dependencies on design parameters, complicating analysis.

- **Data noise and variability**: Sensor measurements can be affected by environmental conditions and instrumental limitations.

To overcome these challenges, ML models have emerged as powerful tools capable of handling complex, high-dimensional data and uncovering hidden patterns within it. ML algorithms can learn from data to predict sensor performance metrics based on input parameters [8], significantly reducing the time and resources required for optimization [9].

### *11.1.4 Role of Machine Learning in Sensor Optimization*

Machine learning models, such as GPR and Random Forests, are particularly well-suited for modelling the intricate relationships in photonic sensor data.

- **Gaussian Process Regression (GPR)**: GPR is a non-parametric, probabilistic model that provides not only predictions but also uncertainty estimates. It is effective in modelling complex, nonlinear functions and is valuable in situations with limited data availability, as often encountered in sensor experiments.
- **Random Forests**: This ensemble learning method constructs multiple decision trees during training and outputs the mean prediction. Random Forests can capture nonlinear interactions between variables and are robust against overfitting, making them suitable for analysing data with noise and variability.

By applying ML models to sensor data, we can predict outcomes like the FOM for SPR sensors or emulsion stability for TFBG sensors based on various input parameters. This predictive capability accelerates the optimization process and enables more efficient exploration of the parameter space [10]. However, a significant limitation of traditional ML models is their lack of interpretability. They often function as "black boxes," providing predictions without insights into how individual input features influence the output. This opacity can hinder trust and acceptance among practitioners who require understanding the underlying mechanisms to make informed design decisions.

Additionally, Federated Learning (FL) offers a promising approach for optimizing fiber optic sensor design while preserving data privacy. FL enables collaborative model training across multiple institutions or devices without exchanging raw sensor data, thus addressing concerns related to data security and regulatory constraints [11].

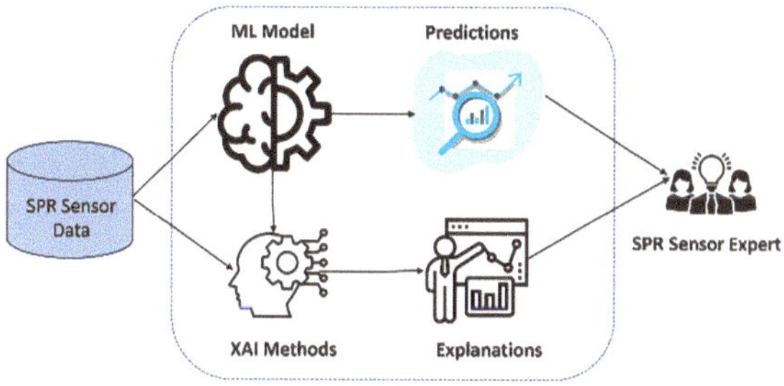


**Fig. 11.2** XAI model approach for fiber optic SPR sensor design under consideration

### *11.1.5 Enhancing Interpretability with Explainable Artificial Intelligence (XAI)*

XAI addresses the transparency issue by offering methods to interpret and understand ML model predictions (Fig. 11.2). XAI techniques elucidate how input features contribute to the output, providing both global and local explanations [12].

Two widely used XAI methods are:

- **SHapley Additive exPlanations (SHAP)**: SHAP assigns each feature an importance value based on Shapley values from cooperative game theory. It quantifies the contribution of each feature to the prediction, enabling a comprehensive understanding of feature importance across the entire dataset [13].
- **Local Interpretable Model-Agnostic Explanations (LIME)**: LIME approximates the ML model locally around a specific prediction using an interpretable model, such as a linear regression. It identifies the features that majorly influence that particular prediction, offering insights at the individual instance level [14].

By integrating XAI with ML models, one can achieve:

- **Transparency**: Understanding how input parameters affect sensor performance.
- **Trustworthiness**: Building confidence in model predictions by aligning them with physical principles.
- **Efficiency**: Identifying key parameters for optimization, reducing unnecessary experimentation.

### *11.1.6 Objectives of the Chapter*

The primary objectives of this chapter are:

1. **For SPR Sensors**:

    - To optimize the design parameters, specifically $d_m$ and $\lambda$, using ML models.
    - To apply XAI techniques to interpret the ML model's predictions, determining the relative importance of $d_m$ and $\lambda$ on the FOM.

- To validate the ML and XAI findings with established physical principles, such as radiation damping [15], enhancing the credibility of the results.

2. **For TFBG Sensors**:

   - To analyse the stability of oil–water emulsions using TFBG sensor data, focusing on parameters like $C_s$, RPM, and λ.
   - To employ ML models to predict emulsion stability and use XAI methods to interpret the contributions of each parameter.
   - To provide actionable insights for the petroleum industry to optimize processes based on the identified key factors influencing emulsion stability.

By the end of this chapter, readers will be able to gain insights into how integrating ML and XAI can revolutionize the design and optimization of photonic sensors, leading to more efficient, reliable, and trustworthy decision-making processes in various technological domains.

## 11.2 Background

### *11.2.1 Surface Plasmon Resonance (SPR) Sensors*

SPR is a well-established optical phenomenon that occurs at the interface between a metal and a dielectric material. When polarized light strikes this interface at a specific angle and wavelength, it can excite surface plasmons—coherent oscillations of free electrons at the metal surface. This excitation leads to a resonance condition characterized by a sharp decrease in the reflected light intensity. The resonance is highly sensitive to changes in the refractive index of the surrounding medium, making SPR sensors exceptionally valuable for real-time detection of molecular interactions. SPR sensors have found widespread applications in various fields due to their high sensitivity [16] and label-free detection capabilities. In biosensing [17], they are employed to monitor antigen–antibody interactions, DNA hybridization, and protein binding events [18], providing crucial insights in medical diagnostics and biological research [19, 20]. Environmental monitoring benefits [21] from SPR sensors' ability to detect pollutants and hazardous substances with high specificity. Additionally, these sensors are utilized in gas detection, where they monitor industrial gases and environmental pollutants. Among the available sensing techniques, SPR stands out for its widespread adoption and superior performance across various applications [22]. In particular, fiber optic SPR sensors have gained prominence due to their flexibility, durability, and ability to enable remote sensing [23], making them highly suitable for challenging environments.

Advancements in SPR technology have led to the incorporation of novel materials, such as two-dimensional (2D) materials like graphene and molybdenum disulfide ($MoS_2$) [24], and plasmonic nanocomposites [25]. These materials enhance the performance of SPR sensors by increasing sensitivity and specificity [26–28] which is critical for applications requiring precise detection. Optimizing the performance of SPR sensors involves careful consideration of key design parameters that significantly influence the sensor's FOM. The FOM is a crucial metric representing the sensor's sensitivity and resolution, defined as the ratio of sensitivity to the full width at half maximum (FWHM) of the resonance curve. Two primary parameters affecting the FOM are the $d_m$ and the operating $\lambda$ of the incident light.

$$\boldsymbol{FOM = Sensitivity \times Accuracy} \tag{11.1}$$

The value of $d_m$ affects the resonance condition by influencing the coupling efficiency between the light and surface plasmons. An optimal thickness ensures maximum sensitivity and minimal damping losses. The operating wavelength determines the excitation of surface plasmons and the penetration depth of the evanescent field into the dielectric medium. Adjusting wavelength allows for tailoring the sensor's sensitivity to specific applications. Traditionally, optimizing these parameters requires extensive simulations and empirical testing, which are time-consuming and resource intensive. As SPR sensor designs become more sophisticated, incorporating multilayer plasmonic structures and advanced materials, the complexity of optimization increases. This complexity poses challenges in balancing sensitivity, stability, and manufacturing constraints.

To address these challenges, ML models have emerged as powerful tools for transforming the optimization process of SPR sensors [29, 30]. ML algorithms, such as GPR, enable researchers to analyse large datasets and predict optimal sensor configurations based on the relationships between $\lambda$, $d_m$, and the FOM. GPR is particularly effective due to its ability to model complex, nonlinear functions and provide uncertainty estimates for predictions.

However, a significant limitation of traditional ML models is their "black box" nature. While they can accurately predict outcomes, they offer little insight into how specific input parameters affect sensor performance. This lack of transparency can hinder trust and acceptance among practitioners who require understanding the underlying mechanisms to make informed design decisions. XAI techniques have been introduced to overcome this limitation, providing transparency by revealing the contributions of individual design parameters to the model's output. Methods like SHAP and Local Interpretable LIME allow for interpreting ML models by quantifying the impact of each feature on the prediction. In the context of SPR sensors, XAI techniques help determine how variations in $d_m$ and $\lambda$ influence the FOM [31], aligning the optimization process with physical principles such as radiation damping [32].

### *11.2.2 Tilted Fiber Bragg Grating (TFBG) Sensors*

A TFBG is created by introducing a periodic modulation of RI within the fiber core, tilted at an angle relative to the fiber axis. This tilt causes coupling between the core mode and backward-propagating cladding modes when light propagates through the fiber. The cladding modes are sensitive to changes in the surrounding refractive index, making TFBG sensors effective for sensing applications. Figure 11.3 shows the schematic of a TFBG sensor, where the tilt in the grating structure enables coupling between core and cladding modes. This design, with the tilt angle facilitating enhanced mode coupling, makes the sensor highly sensitive to changes in the surrounding refractive index, allowing for versatile sensing applications.

In the petroleum industry, TFBG sensors can be used for monitoring the stability of oil–water emulsions [6]. These emulsions are mixtures of oil and water stabilized by surfactants, and their stability significantly impacts production efficiency, viscosity, and operational costs. Stable emulsions can increase viscosity, reducing flow rates and increasing energy consumption, while unstable emulsions may lead to equipment corrosion and fouling.

TFBG sensors enable real-time monitoring of emulsion stability by detecting subtle changes in parameters such as $C_S$, RPM during mixing, and $\lambda$. The value of $C_S$ affects the interfacial tension between oil and water, influencing the formation and stability of emulsions. The RPM determines the mixing intensity and droplet size distribution, which are critical factors in emulsion stability. The operating wavelength impacts the sensor's sensitivity and the resolution of spectral features.

Optimizing TFBG sensors involves addressing the complex interactions between these parameters. Traditional empirical approaches require extensive experimentation, which is inefficient and may not capture the nonlinear and interdependent

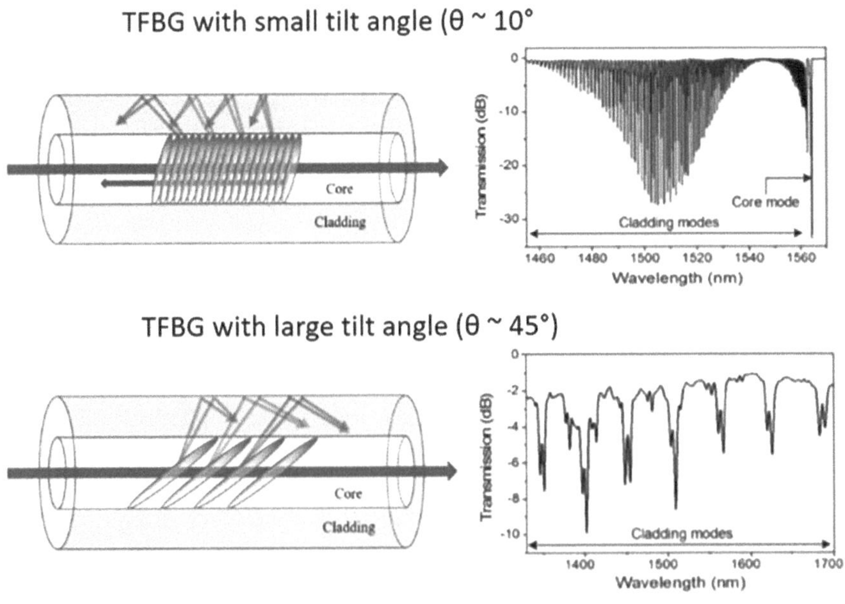


**Fig. 11.3** Schematics of TFBG sensor having different tilt angles along with their corresponding spectra patterns

relationships among the variables. Moreover, TFBG sensors are susceptible to noise and fluctuations in their measurements due to environmental factors and instrumental limitations.

ML models, particularly Random Forests, have been employed to analyse TFBG sensor data, capturing the intricate relationships between input parameters and emulsion stability. Random Forest models are effective in handling nonlinear interactions and are robust against overfitting, making them suitable for datasets with noise and variability. By predicting how different combinations of Cs, RPM, and λ affect emulsion stability, ML models facilitate more precise control over oil processing systems.

To enhance the interpretability of these models, XAI techniques like SHAP and LIME are applied. SHAP analysis identifies the dominant role of surfactant concentration in determining emulsion stability, providing a quantitative measure of each parameter's impact. LIME offers localized explanations for individual predictions, helping to understand the effects of specific parameter values on emulsion stability. This integration of ML and XAI allows for transparent optimization of TFBG sensors, leading to improved performance and reliability in industrial applications.

### 11.2.3 Explainable Artificial Intelligence (XAI) in Sensor Design

The integration of ML and XAI in sensor design addresses the critical need for transparency and trust in model predictions. XAI techniques provide insights into the decision-making processes of ML models by revealing how input features contribute to the output. This transparency is essential in fields like sensor design, where understanding the influence of design parameters on performance metrics is crucial.

In the case of SPR sensors, XAI methods help determine the relative importance of $d_m$ and λ on FOM. SHAP analysis quantifies the contribution of each parameter, revealing that $d_m$ has a more significant impact on the FOM than λ. This finding aligns with the concept of radiation damping, where the metal layer thickness critically influences the resonance condition and coupling efficiency.

For TFBG sensors used in oil–water emulsion stability studies, XAI techniques identify $C_s$ as the most influential parameter affecting emulsion stability. By providing both global and local interpretations, SHAP and LIME analyses help engineers understand the complex interactions between $C_s$, RPM, and λ. This understanding facilitates more efficient optimization of industrial processes, as resources can be focused on controlling the most impactful parameters.

The benefits of integrating ML and XAI in sensor design are multifaceted. It enhances the efficiency of the optimization process by quickly identifying key parameters, reduces reliance on exhaustive simulations and empirical testing, and aligns model insights with physical principles and domain knowledge. Additionally, it improves the performance of sensors by enabling the development of designs with higher sensitivity, specificity, and reliability.

### *11.2.4 Challenges and Opportunities*

While the integration of ML and XAI offers significant advantages, it also presents challenges that need to be addressed. Computational complexity can be a concern, especially when calculating SHAP values for large datasets or complex models. Approximation limitations in methods like LIME may not fully capture global model behaviour. Moreover, interpreting XAI outputs requires a certain level of expertise, and there is a risk of misinterpretation if users are not adequately trained.

Despite these challenges, the opportunities presented by ML and XAI in sensor design are substantial. The methodologies discussed have the potential for broader applications across various fields, including biosensing, environmental monitoring, and industrial processing. By fostering transparency and trust in ML models, XAI facilitates the adoption of advanced analytical techniques in critical applications, ultimately leading to innovations in sensor technology and improved decision-making processes.

## 11.3 Methodology

This section details the methodology used for optimizing SPR and TFBG sensors using ML models integrated with XAI techniques. It includes data collection, dataset augmentation (for TFBG), model training, XAI principles, and steps for SHAP and LIME analysis. The general steps common to both sensors are outlined first, followed by the specific distinctions in data augmentation, feature parameters, and algorithm details for each sensor.

### *11.3.1 Common Methodology*

#### 11.3.1.1 Data Collection and Preprocessing

The data for both sensors were collected from experimental setups (Table 11.1), with different key parameters influencing performance:

a. **SPR Sensor**

**Table 11.1** Few sporadic instances of the used dataset corresponding to fiber optic SPR sensor

| $d_m$ (nm) | λ (nm) | FOM ($RIU^{-1}$) |
|---|---|---|
| 29.74 | 1099 | 524.9882 |
| 29.78 | 1099.29 | 941.0315 |
| 29.80 | 1099.33 | 3050.223 |
| 29.82 | 1099.47 | 907.2284 |
| 29.87 | 1099.5 | 506.9020 |

- **Features**: λ and $d_m$.
- **Target**: FOM, which evaluates the sensor's sensitivity and resolution.
- **Dataset Size**: 714 data points.

The 2D variation offers a color-coded correlation for FOM with λ and $d_m$. It enables the easy identification of how FOM varies with thickness and wavelength, allowing for the pinpointing of the maximum FOM value. This aids in selecting the wavelength and thickness according to specific requirements or scenarios. The reported variation of FOM varying with λ and $d_m$ for the corresponding fiber optic SPR sensor is depicted in Fig. 11.4.

b. **TFBG Sensor**:

- Features: $C_s$, RPM, and λ.
- Target: Emulsion stability, essential for assessing the stability of oil–water emulsions in petroleum applications.

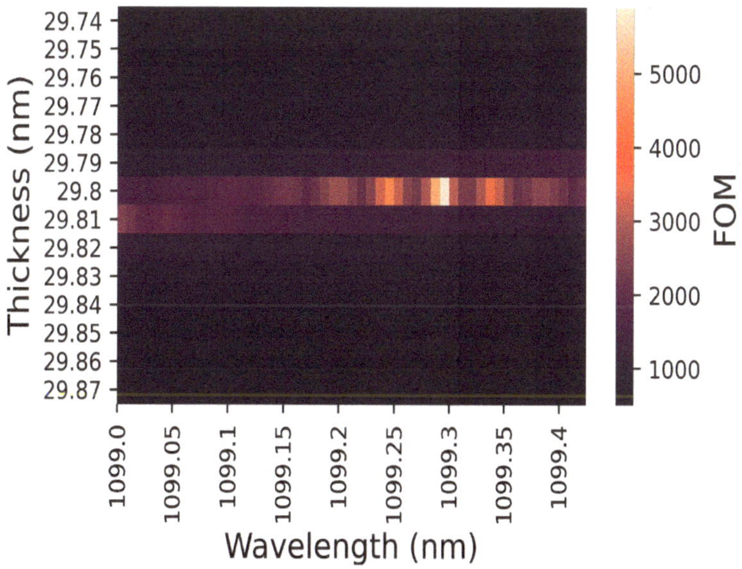


**Fig. 11.4** 2D ($\lambda$-$d_m$) variation of FOM for fiber optic SPR sensor

**Table 11.2** Few instances from TFBG dataset

| Sample | RPM | Cs (%) | Area (Arbitrary units) |
|---|---|---|---|
| 1 | 230 | 0.01 | 294.405 |
| 2 | 520 | 0.1 | 299.8004 |
| 3 | 170 | 0.01 | 294.21 |
| 4 | 690 | 0.1 | 301.4188 |
| 5 | 740 | 1 | 310.8688 |

- Dataset Size: 15 data points initially [6], expanded to 252 through augmentation.

Table 11.2 provides a few selected instances from the TFBG dataset, showcasing variations in RPM, Cs, and the corresponding area (a proxy for emulsion stability). The augmentation of the dataset from 15 to 252 data points allowed for a more robust analysis and improved model performance in predicting emulsion stability [33].

The experimental study used the envelope method that calculates the area between the upper and lower envelopes of the TFBG spectral response at different Cs and RPM values as shown in Fig. 11.5. Large area means greater emulsion stability. The dataset included three levels of Cs (0.01%, 0.10%, and 1.00%) and five RPM (100, 500, 750, 1000, and 1500 rpm). The wavelength (λ) ranged from 1250 nm to 1649.99 nm.

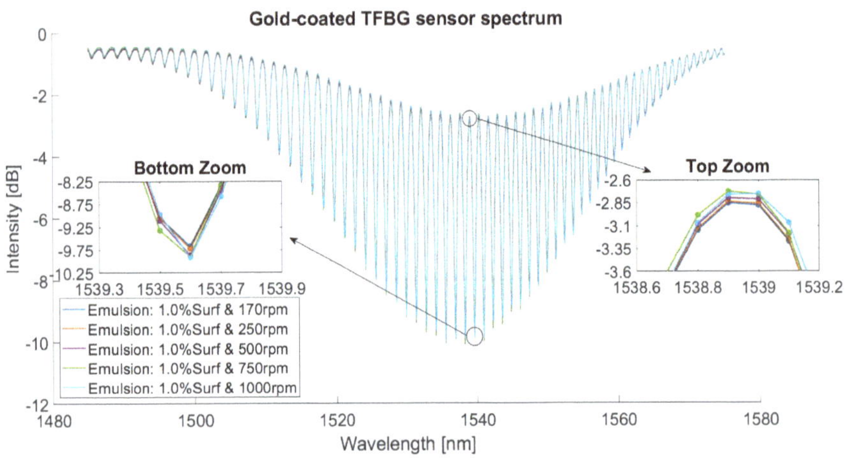


**Fig. 11.5** Spectral analysis of emulsions with different rotation speeds and same surfactant concentrations (adapted from ref. [6])

#### 11.3.1.2 Machine Learning Models

We initially began with less complex models like **Linear** Regression [34], Polynomial Regression [35], Random Forest [36], XGBoost [37], and Gaussian Process Regression [38]. After testing and comparing their performance, we identified the best model for each sensor as follows:

- SPR: Gaussian Process Regression (GPR) was applied due to its capability to model complex non-linear relationships and handle small datasets with noise.
- TFBG: Random Forest Regression (RFR) was used for its ability to handle non-linear interactions between multiple features and prevent overfitting.

#### 11.3.1.3 Model Training and Evaluation

For both sensors, the following general steps were taken:

1. **Train-Test Split**:

   The data was split into training and testing sets [39]. For SPR, an 80:20 split was used; for TFBG, a 70:30 split was applied due to limited data constraint.

2. **Model Training**:

   The models were trained on the training data using cross-validation to optimize model hyperparameters for better performance.

3. **Prediction Generation**:

   Predictions were made using the trained models, and the outputs were evaluated using performance metrics such as $R^2$ [40], MSE [40], and RMSE [41].

4. **Application of XAI Techniques**:

   - SHAP was used to provide global explanations of feature importance, showing how much each feature contributed to the model's predictions.
   - LIME was used for local explanations, offering insight into individual predictions by perturbing input features.

### *11.3.2 Distinctions Between SPR and TFBG Methodologies*

#### 11.3.2.1 Data Augmentation for TFBG Sensors

Unlike SPR data, which did not undergo augmentation, the initial TFBG dataset of 15 data points was expanded to 252 points using both linear and non-linear augmentation techniques:

- Linear Augmentation: New data points were created by interpolating between existing values.
- Non-linear Augmentation: Synthetic data points were generated to simulate more complex behaviours observed in real-world sensor performance.

#### 11.3.2.2 Algorithm for TFBG Sensor Training and Testing

The following algorithm was applied to the TFBG datasets:

This article has computed the results on datasets $\mathcal{D}$, to increase the training sample, both linear and non-linear augmentation techniques have been applied.

**Algorithm 1: Training and Testing of Machine Learning Model**

**Input:** $\mathcal{D}$ – Dataset

**Output:** $\hat{y}$ – Predicted Value (Emulsion Stability)

*/*Training stage*/*

$\mathcal{D}_{train}, \mathcal{D}_{test} = TrainTestSplit(\mathcal{D})$, where $\mathcal{D}_{train} \cap \mathcal{D}_{test} = \emptyset$

*/*Augmentation*/*

$\mathcal{D}_{Aug1} = LinearAug(\mathcal{D}_{train})$

$\mathcal{D}_{Aug2} = NonLinearAug(\mathcal{D}_{train})$

$\mathcal{D}_{train} = \mathcal{D}_{train} \cup \mathcal{D}_{Aug1} \cup \mathcal{D}_{Aug2}$

*/*Model Building*/*

$model = BuildModel()$

$f = model.fit\ (\mathcal{D}_{train})$

*/*Testing stage*/*

$\hat{y} = f(\mathcal{D}_{test})$

### 11.3.3 XAI Analysis for SPR and TFBG Sensors

The XAI analysis was applied similarly to both sensors, with some differences in feature importance and interpretation.

#### 11.3.3.1 SHAP Analysis

- Global Interpretation:
  - SHAP values provided a comprehensive view of the importance of features such as λ, $d_m$, Cs, and RPM across the entire dataset.

- For SPR sensors, $d_m$ had a significantly higher contribution to the prediction of FOM.
- For TFBG sensors, Cs emerged as the dominant feature affecting emulsion stability.

#### 11.3.3.2 LIME Analysis

- Local Interpretation:
  - LIME was applied to individual predictions to explain how changes in specific features (e.g., Cs or RPM) affected the model's output.
  - The localized insights from LIME reinforced the global feature importance results obtained through SHAP.

### *11.3.4 Gaussian Process Regression (GPR) and XAI Techniques: SHAP and LIME*

In this section, we explain the methodologies employed for analysing and interpreting sensor data using GPR and XAI techniques, specifically SHAP and LIME. The GPR model is particularly useful for small datasets with nonlinear characteristics, while XAI techniques provide transparency in machine learning models, revealing the contributions of various features to predictions.

#### 11.3.4.1 Gaussian Process Regression (GPR)

GPR is a nonparametric supervised machine learning model based on Bayesian inference, making it suitable for small datasets where traditional models might struggle. GPR models calculate the probability distribution over all possible functions that fit the data, rather than directly estimating the probability distribution of parameters for a predefined function. This flexibility allows GPR to model complex, nonlinear relationships and incorporate prior knowledge, providing a robust framework for SPR sensor design parameter optimization.

The core principle of GPR is based on assuming a Gaussian Process prior, characterized by a mean function and a covariance function. These functions allow the model to estimate relationships in the data and capture uncertainties in the predictions.

The mathematical foundation of GPR is expressed as:

$$f(x) \sim GP(m(x), k(x, x)) \tag{11.2}$$

Labels are drawn from the Gaussian process with mean function, m, and covariance function, k. More specifically, a Gaussian process is a multivariate Gaussian distribution in infinite-dimensional space, where any collection of the labels of the dataset is joint Gaussian distributed. Within this GP, we can use the prior knowledge about the space of functions through the selection of the mean and covariance statistical functions. Identically distributed (i.i.d) Gaussian noise, N(0,σ2) can also be easily and independently incorporated into the labels by summing the label distribution and noise distribution.

This noise term ensures that GPR can accommodate real-world variations and uncertainties in the data. Due to its flexibility and ability to handle small datasets, GPR was used to predict the FOM in SPR sensors.

#### 11.3.4.2 SHAP (SHapley Additive ExPlanations)

SHAP is an XAI technique rooted in cooperative game theory. It assigns a value (called the Shapley value) to each feature, representing its contribution to the model's prediction. This global interpretability approach allows researchers to quantify how each input parameter affects the model's output.

The mathematical foundation of SHAP is based on the following equation:

$$\phi_i = \sum_{X' \subseteq X \setminus \{i\}} \frac{|X'|!(|X| - |X'| - 1)!}{|X|!}\left[f\left(X' \cup \{i\}\right) - f\left(X'\right)\right] \tag{11.3}$$

In Eq. (11.3), $\phi_i$ is shapley value, F is the set of all features, and (S) is a subset of features not including $(i)$ [42].

This equation calculates the average marginal contribution of each feature across all possible subsets of features. SHAP is particularly useful for providing global feature importance i.e., understanding how much each feature contributes to the overall predictions across the entire dataset. In this work, SHAP was applied to both SPR and TFBG sensor models, revealing the global importance of features such as $\boldsymbol{d_m}$ and Cs in predicting the FOM and emulsion stability, respectively.

#### 11.3.4.3 LIME (Local Interpretable Model-Agnostic Explanations)

**LIME** offers a **local** interpretation of model predictions by approximating the complex model with a simpler, interpretable model for a specific instance. LIME focuses on generating explanations by perturbing input features and observing how the model's output changes. It provides an instance-level view, making it useful for analysing specific predictions in greater detail.

The loss function used in LIME is as follows:

By minimizing this loss function, LIME finds a simple, interpretable model that closely approximates the behaviour of the complex model in the vicinity of a specific instance.

$$L(f, g, \pi_x) = argmin_{g \in G} \sum_{x \in X'} \pi_x(x)\left(f(x) - g(x')\right)^2 + \Omega(g) \quad (11.4)$$

Equation (11.4) represents the loss function ($L$) of LIME, g is the interpretable model, $\Omega$ is the complexity measure, $x'$ is the perturbed dataset generated by LIME around $x$ (features), and $\pi_x$ is the approximity measure [43].

LIME is particularly effective in providing local insights into how variations in Cs, RPM, or λ influence the predictions of emulsion stability for specific data points.

This structured approach ensures that the distinct processes for SPR and TFBG sensors are clearly delineated, while also capturing the common methodology steps that apply to both. Let me know if further changes or refinements are needed!

## 11.4 Results and Discussion

This section presents the findings from the analysis of SPR sensors and TFBG sensors using ML models and XAI techniques. The discussion is centred on model performance, insights derived from SHAP and LIME analyses, and the impact of key features on sensor performance from both global and local perspectives.

### *11.4.1 SPR Sensors: Optimization of Figure of Merit (FOM)*

#### 11.4.1.1 Model Performance

The GPR model was employed to predict the FOM of SPR sensors, using λ and $d_m$ as the input features. The model was trained on a dataset of 714 data points, which covered a wide range of λ and $d_m$ values. The model performance on the test dataset is as follows.

- **Coefficient of Determination ($R^2$)**: The model achieved an $R^2$ score of 0.92, indicating that 92% of the variance in the FOM is explained by the model.
- **Mean Squared Error (MSE)**: The MSE was 0.015, indicating minimal prediction error.

The model predicted a peak FOM of 6,526.23 $RIU^{-1}$ at $\lambda = 1{,}099.34$ nm and dm = 29.8 nm, representing an 11% improvement over the previously recorded FOM of 5,897.08 $RIU^{-1}$. This demonstrates the model's ability to effectively optimize sensor performance. Figure 11.6 illustrates the close match between the predicted and actual FOM values, confirming the model's reliability in optimizing sensor behaviour.

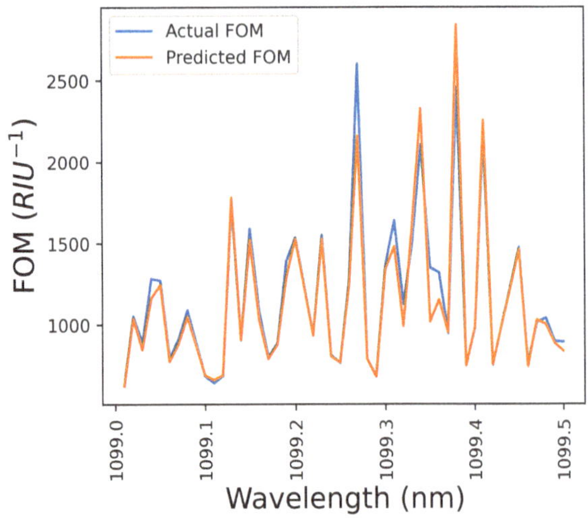


**Fig. 11.6** Pattern similarity between predicted and actual FOM

#### 11.4.1.2 SHAP Global Feature Importance Analysis

SHAP analysis was conducted to interpret the GPR model and provide insights into the influence of the input features on the FOM. The results showed that $d_m$ contributed 70% to the FOM prediction, while λ accounted for 30%, highlighting that $d_m$ plays a significantly larger role in determining sensor performance compared to λ. Figure 11.7 visually displays this distribution of global feature importance, with $d_m$ being the dominant factor.

The SHAP summary plot in Fig. 11.8 further illustrated that variations in dm had a greater effect on the FOM compared to λ. The wider spread of SHAP values for dm demonstrated its critical importance in optimizing the sensor's sensitivity, making dm the key parameter for fine-tuning performance.

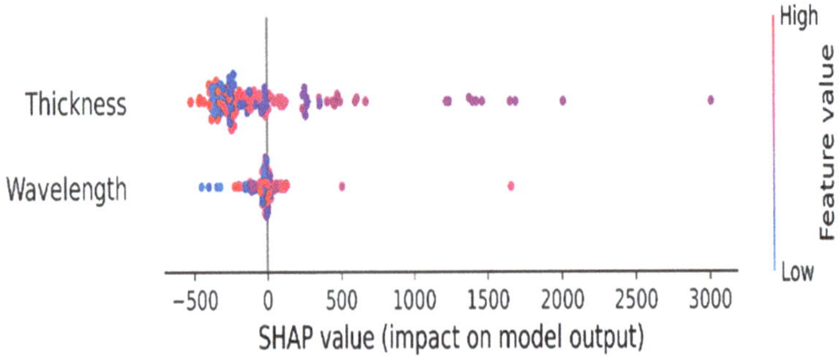


**Fig. 11.7** Global feature importance using SHAP

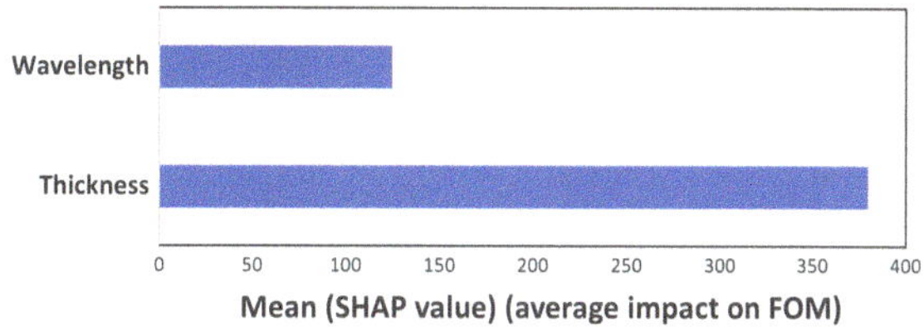


**Fig. 11.8** Global feature importance summary plot using SHAP

These findings align with established principles in SPR sensor design, where $d_m$ has a major impact on plasmonic resonance at the metal–dielectric interface, affecting sensor sensitivity.

#### 11.4.1.3 LIME Local Feature Contribution Analysis

LIME analysis was applied to provide localized insights into individual predictions. The results confirmed that dm consistently contributed more than λ to the FOM in individual cases, with dm accounting for approximately 68% of the FOM, while λ contributed the remaining 32%. The visualized output in Fig. 11.9 illustrates the contribution of λ towards achieving the maximum FOM, showing that λ has a comparatively smaller impact.

The local analysis highlighted that small adjustments in dm could lead to significant improvements in the FOM, suggesting that precise control over dm during sensor

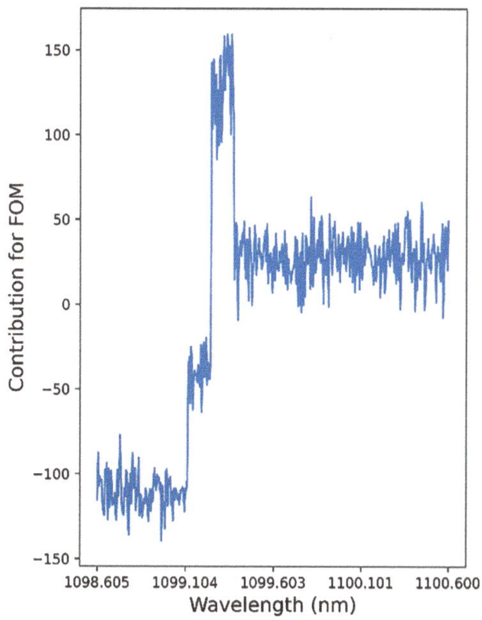


**Fig. 11.9** Contribution of λ towards maximum FOM using LIME

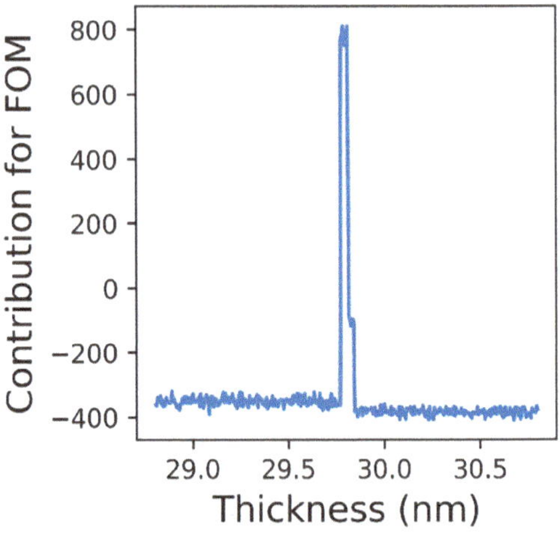


**Fig. 11.10** Contribution of $d_m$ towards maximum FOM using LIME

fabrication is crucial for optimizing performance. This insight is clearly demonstrated in Fig. 11.10, which emphasizes the dominant role of dm in enhancing the FOM.

#### 11.4.1.4 Implications for SPR Sensor Design

- **Optimization of $\mathbf{d_m}$**: The findings suggest that prioritizing the optimization of $d_m$ can result in significant improvements in sensor performance.
- **Efficient Design Process**: Understanding the dominant influence of $d_m$ allows researchers to allocate resources toward optimizing this parameter, reducing the need for extensive variation in other factors like wavelength.

### *11.4.2 TFBG Sensors: Prediction of Emulsion Stability*

#### 11.4.2.1 Model Performance

A **Random Forest Regression (RFR)** model was utilized to predict emulsion stability for TFBG sensors based on Cs, RPM, and λ. The dataset initially consisted of 15 data points, which were augmented to 252 data points through linear and non-linear interpolation techniques to enhance the model's robustness.

- **Coefficient of Determination ($\mathbf{R^2}$)**: The RFR model achieved an $R^2$ score of 0.99, indicating excellent predictive accuracy for emulsion stability.
- **Root Mean Squared Error (RMSE)**: The RMSE was 0.020, confirming the model's ability to predict emulsion stability with high precision.

#### 11.4.2.2 SHAP Global Feature Importance Analysis

SHAP analysis provided insights into the global influence of each feature on the model's predictions:

**Feature Importance Ranking**: SHAP analysis of dataset was carried out with three features (RPM, $C_s$, and λ) focusing on their impact on emulsion stability (or Arca). Figurc 11.11 contains thc corrcsponding rcsult. From Fig. 11.11, it can be observed that while Cs has maximum contribution to emulsion stability, it is noteworthy that the contribution of λ is considerably higher than RPM in all instances.

**SHAP Summary Plot**: More exact quantification of the contribution of all three features is presented in Fig. 11.12, which contains the global feature contribution summary in the form of corresponding bar chart The plot revealed that Cs had the most substantial impact on emulsion stability, with RPM playing a secondary role and λ contributing minimally. SHAP values indicated that surfactant concentration (Cs) is the most important feature, contributing.

65% to emulsion stability predictions. RPM accounted for 25%, while wavelength (λ) contributed 10%.

These findings are consistent with the understanding that surfactant concentration is critical in stabilizing oil–water emulsions by reducing interfacial tension.

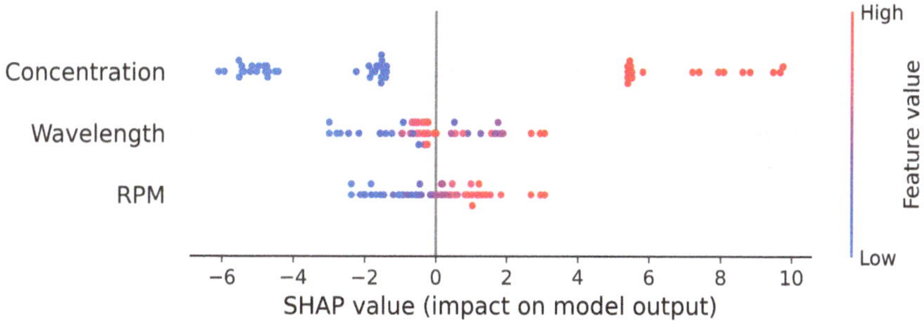


**Fig. 11.11** Global feature importance (RPM, Cs, and λ) using SHAP

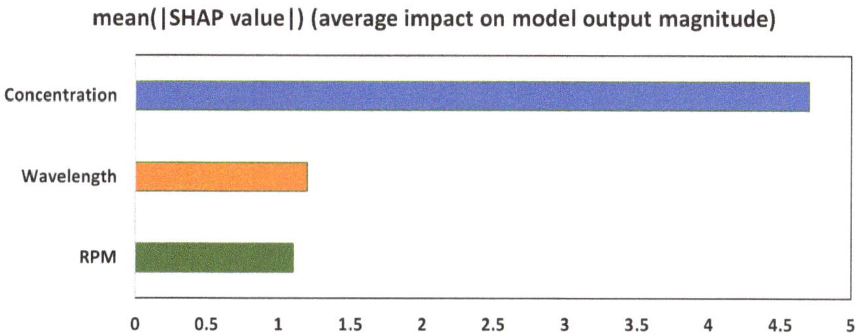


**Fig. 11.12** Global feature (RPM, Cs, and λ) importance summary plot

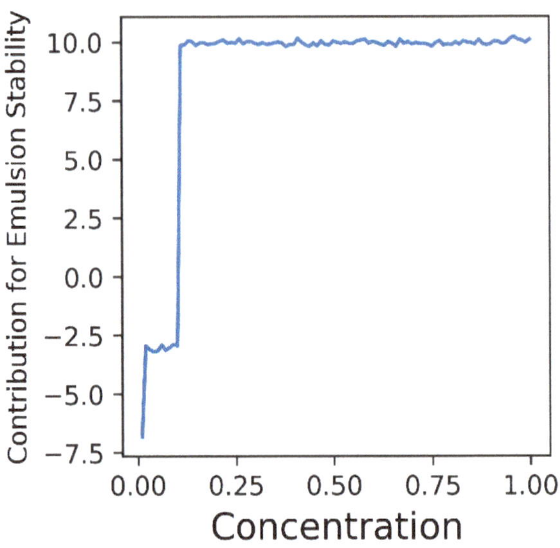


**Fig. 11.13** Contribution of Cs towards maximum emulsion stability using LIME

#### 11.4.2.3 LIME Local Feature Contribution Analysis

For this analysis, three separate datasets (one each for contributions of Cs, λ, and RPM) were generated and used. The corresponding results are shown in Figs. 11.13, 11.14 and 15. As shown in Fig. 11.13, the local contribution of Cs towards emulsion stability ranges from −6.80 to 10.23, resulting in a delta of 17.04, which is significantly larger than the contributions of both λ and RPM. For λ, the contribution ranges between −1.62 and 1.46, as depicted in Fig. 11.14, with a delta of 3.08. Finally, Fig. 11.15 shows the local contribution of RPM, which varies from −2.16 to 2.19, resulting in a delta of 4.35. Cs contributed 1.8 times more to emulsion stability predictions than RPM, reinforcing the global SHAP findings, which indicated Cs as the dominant factor. Localized analysis showed that small adjustments in Cs could significantly affect emulsion stability, underscoring the importance of precise control over surfactant concentration in industrial processes. This highlights the substantial influence Cs has over the stability of emulsions compared to λ and RPM.

#### 11.4.2.4 Practical Implications for the Petroleum Industry

- **Process Optimization**: Focusing on optimizing $C_s$ can improve emulsion stability in oil–water systems, leading to more efficient processing.
- **Cost Efficiency**: Understanding the lesser impact of RPM and λ allows for better resource allocation, potentially reducing operational costs.

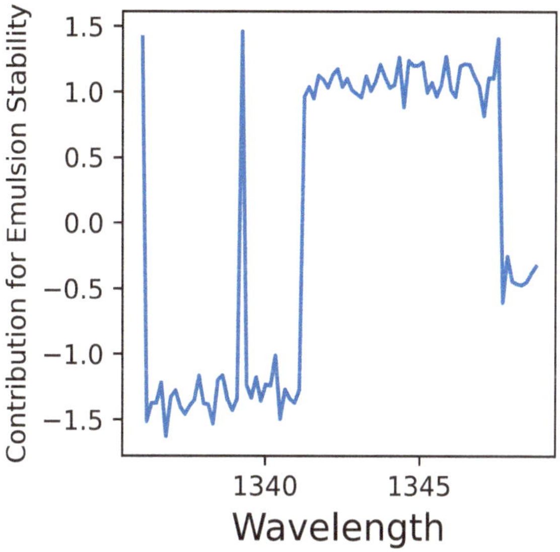


**Fig. 11.14** Contribution of λ towards maximum emulsion stability using LIME

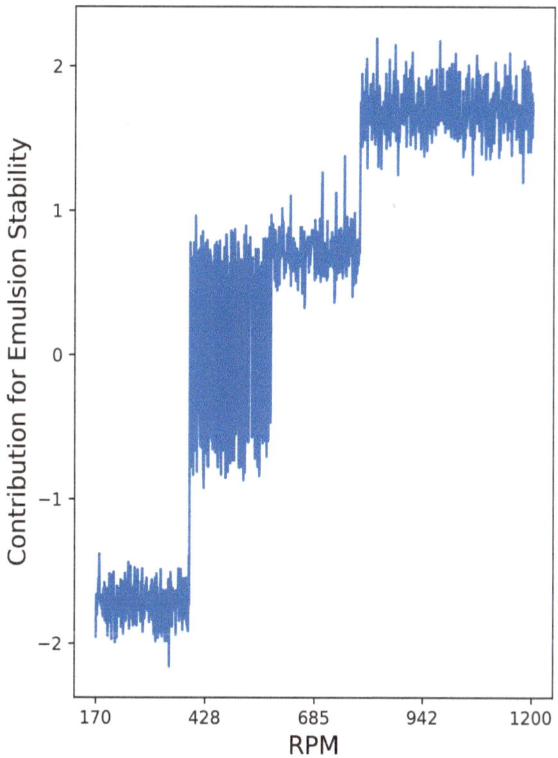


**Fig. 11.15** Contribution of RPM towards maximum Emulsion Stability using LIME

### *11.4.3 Comparative Analysis of SHAP and LIME Results*

Both SHAP and LIME provided valuable insights into the key features impacting sensor performance at global and local levels for both sensor types:

- **SPR Sensors**:
  - SHAP indicated that $d_m$ is the dominant feature influencing FOM.
  - LIME confirmed the local impact of $d_m$, reinforcing its importance in individual predictions.
- **TFBG Sensors**:
  - SHAP revealed that **Cs** is the most influential feature affecting emulsion stability.
  - LIME provided localized validation, demonstrating that Cs had a substantial impact in specific predictions.

The complementary use of SHAP and LIME enhances the interpretability of the models, offering a comprehensive view of feature impacts both globally and locally.

### *11.4.4 Importance of Data Augmentation*

Data augmentation was critical for improving the model performance of TFBG sensors. The dataset was expanded from 15 data points to 252 through linear and non-linear augmentation techniques:

- **Improved Model Training**: The augmented dataset provided a more robust foundation for training the RFR model, contributing to its strong performance.
- **Realistic Representation**: The augmentation techniques ensured that the generated data accurately reflected possible real-world scenarios.

### *11.4.5 Comparative Model Performance*

A comparison of the performance metrics for both the SPR and TFBG sensor models is summarized below (Table 11.3):

**Table 11.3** Summary of performance metrics and key feature impact for SPR and TFBG sensor models

| Model | $R^2$ (Test Set) | Key feature impact (SHAP) | Error metric |
|---|---|---|---|
| GPR (SPR sensor) | 0.92 | $d_m > \lambda$ | MSE: 0.015 |
| RFR (TFBG sensor) | 0.99 | Cs > RPM > $\lambda$ | RMSE: 0.020 |

- **SPR Sensor Findings**: The GPR model effectively optimized FOM, identifying $d_m$ as the most critical feature.
- **TFBG Sensor Findings**: The RFR model highlighted Cs as the most significant factor in emulsion stability, corroborating practical industry insights.

### *11.4.6 Implications for Sensor Design and Future Applications*

- **Enhanced Optimization**: The integration of ML models with XAI techniques allows for efficient optimization of sensor parameters.
- **Broader Applicability**: The methodologies employed can be extended to other sensor types, promoting the use of ML and XAI in various industries.

This section consolidates key findings from both sensors, illustrating the importance of ML and XAI techniques in enhancing sensor design and application. The combination of SHAP and LIME enables deeper insights, improving transparency and decision-making.

## 11.5 Conclusion

This study demonstrated the integration of XAI techniques with ML models to optimize the design and enhance the performance of SPR and TFBG sensors. By employing GPR and RFR models, the study achieved high predictive accuracy in modelling sensor behaviour based on key design parameters, leading to significant improvements in sensor performance and providing valuable insights into the factors influencing their operation.

For SPR sensors, the GPR model effectively predicted the FOM, achieving an $R^2$ score of 0.92 and a MSE of 0.015. The model identified $d_m$ as the most influential parameter, contributing approximately 70% to the prediction of the FOM. This finding aligns with the principle of radiation damping, where $d_m$ critically influences the excitation and propagation of surface plasmons. The optimization process led to a significant improvement in the FOM, achieving a peak value of 6,526.23 $RIU^{-1}$, representing an 11% enhancement over previous values.

For TFBG sensors, the RFR model exhibited exceptional predictive performance, with an $R^2$ score of 0.99 and a RMSE of 0.020. The model determined that Cs is the dominant factor influencing emulsion stability, contributing 65% to the model's predictions, with RPM and $\lambda$ contributing 25% and 10%, respectively. This underscores the crucial role of Cs in stabilizing oil–water emulsions, which is critical for optimizing processes in the petroleum industry.

The application of XAI techniques, specifically SHAP and LIME, enhanced the interpretability of the models:

- SHAP provided global insights into feature importance, confirming the dominant roles of $d_m$ in SPR sensors and Cs in TFBG sensors.
- LIME offered localized explanations, reinforcing SHAP's findings and providing granular insights into individual predictions. This dual approach ensured transparency, built trust in the models, and facilitated informed decision-making in sensor design and application.

The integration of XAI techniques offers significant advantages over traditional methods, including:

- **Efficiency**: Accelerates the optimization process by reducing the need for exhaustive simulations.
- **Interpretability**: Enhances understanding of the models' decision-making processes, aligning predictions with physical principles and domain expertise.
- **Resource Optimization**: Identifies key parameters for targeted improvements, leading to more effective resource allocation.

In conclusion, this work underscores the potential of XAI in revolutionizing sensor design and optimization. By bridging the gap between complex data analysis and practical applications, this approach enables more effective, transparent, and trustworthy decision-making. The methodologies demonstrated here not only enhance sensor performance but also lay the foundation for future advancements in sensor technology, driving innovation across various technological domains.

## 11.6 Future Directions

Building on the promising outcomes of this study, several avenues for future research can further enhance the integration of ML and XAI in photonic sensor design and optimization. These future directions aim to address current limitations, explore advanced techniques, and broaden the applications of SPR and TFBG sensors.

1. **Real-Time XAI Implementation**:

The application of XAI in real-time [44] during sensor operation presents a significant opportunity for dynamic optimization. By integrating XAI techniques into the operational phase, sensor parameters could be adjusted on-the-fly, based on immediate feedback from ML models. This would enable real-time adjustments to improve accuracy and sensitivity in environments where conditions affecting sensor performance fluctuate rapidly.

2. **Development of Generalized Models**:

Extending the current methodologies to develop more generalized ML models that can be applied across a variety of sensor types, such as other fiber optic sensors, optical biosensors, and integrated photonic devices, is a valuable direction for future research. This would promote cross-domain applications and amplify the broader impact of ML and XAI on sensor technology.

3. **Integration with 5G/6G Networks**:

Integrating advanced sensors with upcoming 5G and 6G networks [45] could revolutionize sensor operations by enabling real-time data transmission and processing. This connectivity could enhance sensor performance in remote or challenging environments, offering significant benefits for applications in environmental monitoring, industrial automation, and healthcare.

4. **IoT and Real-Time Inference with 5G/6G**:

The Internet of Things (IoT) [46] presents a significant opportunity for real-time sensor deployment and inference. Future research could focus on designing an architecture that integrates IoT systems with SPR and TFBG sensors, leveraging 5G/6G networks for rapid data transmission and real-time inference. This would allow for continuous monitoring and dynamic sensor adjustment based on real-time feedback, enhancing sensor accuracy and responsiveness in critical applications such as smart cities, industrial automation, and environmental monitoring.

5. **Experimental Validation**:

While the ML models used in this study were validated against experimental data, future work should focus on real-time experimental validation in diverse practical settings. Testing model robustness under dynamic conditions would provide valuable feedback for refining the models and ensuring their reliability across various scenarios. Extensive experimental validation will help strengthen the credibility of these ML and XAI approaches in sensor optimization.

By pursuing these directions, future research can build on the foundational findings of this study, advancing photonic sensor technology and expanding its range of applications. Incorporating real-time XAI applications, developing generalized models, and integrating sensors with IoT systems and 5G/6G architectures can lead to more efficient, accurate, and versatile sensors. These advancements will enhance sensor performance across industries, contributing to technological innovation and societal benefits.

## List of Abbreviations

| | |
|---|---|
| SPR | Surface Plasmon Resonance |
| TFBG | Tilted Fiber Bragg Grating |
| XAI | Explainable Artificial Intelligence |
| SHAP | SHapley Additive exPlanations |
| LIME | Local Interpretable Model-Agnostic Explanations |
| FOM | Figure of Merit |
| GPR | Gaussian Process Regression |
| $\lambda$ | Wavelength |
| $d_m$ | Metal Layer Thickness |

| | |
|---|---|
| $C_s$ | Surfactant Concentration |
| RPM | Rotational Speed |
| MAE | Mean Absolute Error |
| RMSE | Root Mean Square Error |

## References


1. Newman J, Turner A, Rickman A, Harpin A, Shaw M, McKenzie J (1995) Silicon-on-insulator integrated optical biosensors for environmental monitoring. In: IEE colloquium on optical techniques for environmental monitoring, pp 3/1–3/6
2. Tao Y, Chen L, Pan M, Zhu F, Zhu D (2021) Tailored biosensors for drug screening, efficacy assessment, and toxicity evaluation. ACS Sensors 6(9):3146–3162
3. Panesar S, Weng X, Neethirajan S (2017) Toward point-of-care diagnostics of breast cancer: development of an optical biosensor using quantum dots. IEEE Sens Lett 1(4):1–4
4. Zhou B, Zhang AP, He S, Gu B (2010) Cladding-mode-recoupling-based tilted fiber Bragg grating sensor with a core-diameter-mismatched fiber section. IEEE Photonics J 2(2):152–157
5. Laffont G, Ferdinand P (2001) Tilted short-period fibre-Bragg-grating-induced coupling to cladding modes for accurate refractometry. Meas Sci Technol 12(7):765
6. Avellar L, Marques CA, Caucheteur C, Frizera A (2024) Comprehensive evaluation of tilted fiber Bragg grating (TFBG) sensors for characterizing oil-water emulsions: a study on the impact of surfactant concentration and mixing speed. IEEE Sens J
7. Issa RJ, Hunt EM (2015) Rheology of water-in-oil emulsions for a medium crude oil. In: 2015 international mediterranean gas and oil conference (MedGO), pp 1–3. IEEE
8. Avellar L, Frizera A, Rocha H, Silveira M, Díaz C, Blanc W, Leal-Junior A et al (2023) Machine learning-based analysis of multiple simultaneous disturbances applied on a transmission-reflection analysis based distributed sensor using a nanoparticle-doped fiber. Photonics Res 11(3):364–372
9. Tiwari H et al (2023) Exploring deep learning models aimed at favorable optimization and enhancement of fiber optic sensor's performance. IEEE Sens J 23(17):20330–20337
10. Dwivedi YS, Singh R, Sharma AK, Sharma AK (2022) Enhancing the performance of photonic sensor using machine-learning approach. IEEE Sens J 23(3):2320–2327
11. Dwivedi YS, Singh R, Sharma AK et al (2025) Leveraging Federated Learning For Optimizing Fiber Optic Sensor Design For Biodetection. J Opt. https://doi.org/10.1007/s12596-025-02459-3
12. Arrieta AB, Díaz-Rodríguez N, Del Ser J, Bennetot A, Tabik S, Barbado A, Herrera F et al (2020) Explainable artificial intelligence (XAI): concepts, taxonomies, opportunities and challenges toward responsible AI. Inf Fusion 58:82–115
13. Lundberg SM, Lee SI (2017) A unified approach to interpreting model predictions. Adv Neural Inf Process Syst 30
14. Visani G, Bagli E, Chesani F, Poluzzi A, Capuzzo D (2022) Statistical stability indices for LIME: obtaining reliable explanations for machine learning models. J Oper Res Soc 73(1):91–101
15. Sharma AK, Kaur B (2018) Fiber optic SPR sensing enhancement in NIR via optimum radiation damping catalyzed by 2D materials. IEEE Photon Technol Lett 30(23):2021–2024
16. Peveler WJ, Yazdani M, Rotello VM (2016) Selectivity and specificity: pros and cons in sensing. ACS Sens 1(11):1282–1285
17. Mowbray SE, Amiri AM (2019) A brief overview of medical fiber optic biosensors and techniques in the modification for enhanced sensing ability. Diagnostics 9(1):23
18. Bekmurzayeva A, Ashikbayeva Z, Myrkhiyeva Z et al (2021) Label-free fiber-optic spherical tip biosensor to enable picomolar-level detection of CD44 protein. Sci Rep 11:19583

19. Gu R, Duan Y, Li Y, Luo Z (2023) Fiber-optic-based biosensor as an innovative technology for point-of-care testing detection of foodborne pathogenic bacteria to defend food and agricultural product safety. J Agric Food Chem 71(29):10982–10988
20. Murugan D, Bhatia H, Sai VVR et al (2020) P-FAB: a fiber-optic biosensor device for rapid detection of COVID-19. Trans Ind Natl Acad Eng 5:211–215
21. Ni K (2017) Chitosan-coated humidity sensor based on Mach-Zehnder interferometer with waist-enlarged fusion bitapers. Opt Fiber Technol 33:56–69
22. Mahmoud RA, Sagor RH, Khan MZM (2023) Surface plasmon refractive index biosensors: a review of optical fiber, multilayer 2D material and gratings, and MIM configurations. Opt Laser Technol 159:108939
23. Michel D, Xiao F, Alameh K (2017) A compact, flexible fiber-optic surface plasmon resonance sensor with changeable sensor chips. Sens Actuators B Chem 246:258–261
24. Sun H et al (2022) A review of transition metal dichalocogenides-based biosensors. Front Bioeng Biotechnol 10:941135
25. Darmadi I et al (2023) Bulk-processed plasmonic plastic nanocomposite materials for optical hydrogen detection. Acc Chem Res 56(13):1850–1861
26. Chen Y et al (2019) MoS2 nanosheets modified surface plasmon resonance sensors for sensitivity enhancement. Adv Opt Mat 7(13):1900479
27. Mansuri M, Mir A, Farmani A (2022) Sensitivity enhancement of nanoparticle localized surface plasmon resonance pressure sensor based on the MoS2 monolayer for ultra-wide range pressure detection. Mat Sci Eng B 285:115976
28. Thawany P et al (2023) L-cysteine/MoS2 modified robust surface plasmon resonance optical fiber sensor for sensing of Ferritin and IgG. Sci Rep 13:5297
29. Pal D, Kumar A, Gautam A, Thangaraj J (2022) FBG based optical weight measurement system and its performance enhancement using machine learning. IEEE Sens J 22(5):4113–4121
30. Keser S, Hayber ŞE (2021) Fiber optic tactile sensor for surface roughness recognition by machine learning algorithms. Sens Actuators A Phys 332:113071
31. Dwivedi YS, Singh R, Sharma AK, Sharma AK (2024) On the application of explainable AI in optimizing the performance and design of fiber optic SPR sensor. Opt Fiber Technol 85:103801
32. Sharma AK, Gupta J (2019) Fluoride fiber plasmonic sensor with multilayer variants of tungsten disulfide (WS2): seeking enhanced figure-of-merit via thermo-optic tuning of radiation damping. Opt Fiber Technol 53:102037
33. Dwivedi YS, Singh R, Sharma AK, Sharma AK, Marques C (2025) Enhanced prediction and optimization of oil–water emulsion stability through application of machine learning and explainable artificial intelligence on TFBG sensor data. IEEE Sens Lett 9(1):1–4, Art no 3500104
34. James G, Witten D, Hastie T, Tibshirani R, Taylor J (2023) Linear regression. An introduction to statistical learning: with applications in python. Springer International Publishing, Cham, pp 69–134
35. Ostertagová E (2012) Modelling using polynomial regression. Procedia Eng 48:500–506
36. Rodriguez-Galiano V, Sanchez-Castillo M, Chica-Olmo M, Chica-Rivas MJ (2015) Machine learning predictive models for mineral prospectivity: an evaluation of neural networks, random forest, regression trees and support vector machines. Ore Geol Rev 71:804–818
37. Duan J, Asteris PG, Nguyen H, Bui X-N, Moayedi H (2021) A novel artificial intelligence technique to predict compressive strength of recycled aggregate concrete using ica-xgboost model. Eng Comput 37(4):3329–3346
38. Rasmussen CE (2003) Gaussian processes in machine learning. In: Summer school on machine learning. Springer, pp 63–71
39. Salazar JJ, Garland L, Ochoa J, Pyrcz MJ (2022) Fair train-test split in machine learning: mitigating spatial autocorrelation for improved prediction accuracy. J Petrol Sci Eng 209:109885
40. Kang M-C, Yoo D-Y, Gupta R (2021) Machine learning-based prediction for compressive and flexural strengths of steel fiber-reinforced concrete. Constr Build Mater 266:121117
41. Singh R, Srivastava S (2017) Stock prediction using deep learning. Multimed Tools Appl 76(18):18569–18584

42. Hamilton RI, Papadopoulos PN (2023) Using SHAP values and machine learning to understand trends in the transient stability limit. IEEE Trans Power Syst 39(1):1384–1397
43. Slack D, Hilgard S, Jia E, Singh S, Lakkaraju H (2020) Fooling lime and shap: adversarial attacks on post hoc explanation methods. In: Proceedings of the AAAI/ACM conference on AI, ethics, and society, pp 180–186
44. Mill ER, Garn W, Ryman-Tubb NF, Turner C (2023) Opportunities in real time fraud detection: an explainable artificial intelligence (XAI) research agenda. Int J Adv Comput Sci Appl 14(5):1172–1186
45. Wild T, Braun V, Viswanathan H (2021) Joint design of communication and sensing for beyond 5G and 6G systems. IEEE Access 9:30845–30857
46. Ullo SL, Sinha GR (2020) Advances in smart environment monitoring systems using IoT and sensors. Sensors 20(11):3113

# Chapter 12
# Medical Diagnosis System Based on Explainable Artificial Intelligence and Blockchain

**Cosmin-George Nicolăescu, Florentina Magda Enescu, and Nicu Bizon**

**Abstract** An artificial intelligence system represents a set of technologies and algorithms designed to imitate the cognitive processes of human beings, such as language understanding, pattern recognition, reasoning and learning. These systems are capable of processing vast amounts of data, learning from past experiences and improving their performance over time, adapting to new situations and requirements. The purpose of an artificial intelligence system is to perform tasks that traditionally require human intelligence, from providing simple answers to questions to complex problem-solving and decision-making. Blockchain is an emerging technology widely adopted across various sectors that facilitates the secure transfer of information between different participants. This ensures data integrity and confidentiality, eliminating the risk of breach or interception by unauthorised external entities. The implementation of blockchain technology in the medical field paves the way for revolutionary solutions, facilitating the efficient and secure exchange of patient data between different entities such as hospitals, testing laboratories, pharmaceutical companies and healthcare professionals. This paper explores the integration of artificial intelligence technology with blockchain to develop a system that offers significant advantages in medical diagnostics. The proposed system minimises the risk of incorrect diagnosis by providing the possibility to seek additional consultations from various specialists in the medical sector. The system also allows access to a prompt diagnosis through an integrated chat service, supported by artificial intelligence, which provides quick and accurate answers. Data storage security ensures

C.-G. Nicolăescu
Doctoral School, National University of Science and Technology POLITEHNICA Bucharest, Bucharest, Romania
e-mail: cosmingeorge470@yahoo.com

F. M. Enescu (✉)
Associate University Professor, National University of Science and Technology POLITEHNICA Bucharest, Pitesti University Centre, Pitesti, Romania
e-mail: florentina.enescu@upb.ro

N. Bizon
National University of Science and Technology POLITEHNICA Bucharest, Pitesti University Centre, Pitesti, Romania
e-mail: nicu.bizon1402@upb.ro



N. Bizon and B. Appasani (eds.), *Explainable Artificial Intelligence for Trustworthy Decisions in Smart Applications*, https://doi.org/10.1007/978-3-031-97007-8_12

that only authorised doctors can access patient information regardless of geographic restrictions. This guarantees efficient interconnectivity, flexibility, and robust data protection. The system provides an intuitive user interface accessible through a web platform. Here, patients can register by filling out a simple form, and the data entered is collected and rigorously protected by blockchain technology. Once registered in the system, users benefit from the platform's facilities, having the opportunity to communicate with the medical staff via email and chat assisted by specialised medical personnel. Alternatively, they can opt for a quick and accurate diagnosis through dialogue with the artificial intelligence assistant, available in a dedicated chat. This multi-functional approach underscores the system's commitment to patient accessibility and security, while offering flexibility in choosing the consultation and diagnostic method. In conclusion, the integration of this innovative technology into the existing medical infrastructure is not only possible, but also advisable. This would not only enable secure and rapid access to patient data by healthcare professionals irrespective of location, but also improve diagnostic accuracy through enhanced collaboration between specialists and advanced artificial intelligence support. In addition, the proposed system offers the possibility of a prompt diagnosis through an artificial intelligence-assisted chat service, making a significant contribution to the efficiency and personalisation of medical care.

**Keywords** Artificial intelligence · Blockchain technology · Blockchain · Medical diagnosis · Data security · Web platform · Interconnectivity · Smart contracts · Identification · Medical services · Analyse · Diagnosis

## Acronyms

| | |
|---|---|
| AI | Artificial Intelligence |
| XAI | Explainable Artificial Intelligence |
| LIME | Local Interpretable Model-agnostic Explanations |
| SHAP | SHapley Additive exPlanations |
| HER | Electronic health record |
| IBM | International Business Machines |
| EMR | Electronic medical record |
| Ref | Internet of Medical Things |
| NLP | Natural Language Processing |
| DNN | Deep Neural Networks |
| DApps | Decentralised applications |
| ETH | Ether |
| SVC | Support Vector Classifier |
| SVM | Support Vector Machine |
| ReLU | Rectified Linear Unit |

| | |
|---|---|
| ROC | Receiver Operating Characteristic |
| FPR | False Positive Rate |
| TPR | True Positive Rate |

## 12.1 Introduction

Artificial intelligence (AI) is a field of computer science that focuses on creating systems capable of performing tasks that traditionally require human intelligence. These tasks include pattern recognition, learning from experience, interpreting language, and solving complex problems. AI uses advanced algorithms, machine learning, and neural networks to analyse data, learn from it, and make predictions or decisions based on that information, without direct human intervention.

On the other hand, explainable artificial intelligence (XAI) is a branch of AI that focuses on developing intelligent models which are transparent to users. XAI aims to explain how and why an AI system makes particular decisions. The system presented in this work combines machine learning techniques and natural language processing models for medical diagnosis.

Blockchain is a digital ledger system, distributed over a wide network of computers, that securely records transactions. This decentralised structure prevents any retroactive modification of recorded data unless all subsequent blocks are also altered in sequence.

AI systems in medical diagnostics represent a significant evolution in the way patients interact with healthcare services. These systems use advanced AI technologies to provide preliminary medical advice, guidance, and support to patients before they meet a healthcare professional, either in person or online. By analysing symptoms described by patients and comparing them to a vast medical database, AI chatbots can suggest possible diagnoses, offer medical recommendations for care, and even arrange a consultation with a specialist. These systems can also provide immediate answers to patients' frequently asked questions, thereby reducing waiting time and improving the overall efficiency of medical services. The importance of these systems in medical diagnosis is increasingly recognised, with the potential to transform access to healthcare and make the health system more accessible and responsive to patient needs [1].

In [1], the researchers reviewed in detail the existing studies and publications related to the use of blockchain technology in various fields of study. Their findings highlight widespread optimism about the potential and high expectations surrounding this technology. This study categorised the fields in which this technology can be applied into three distinct segments. In addition, the research contributed to the development of a conceptual framework for the advancement of blockchain technology, aiming to identify and remove barriers to its adoption. The study also explored AI integration to enhance efficiency and drive innovation across these fields.

In the study mentioned in [2], the researchers address the difficulties related to the transfer of essential information for health services between different medical institutions. These challenges become even more pressing in the context of the global movement of people, which requires access to the medical history of patients treated in various locations. The study places particular emphasis on how patients can effectively manage their digital identity, ensuring access to their medical data wherever they may be. Current systems generally face significant challenges and are limited in various ways, including efficiency and processing capacity. In contrast, our proposed system delivers superior performance, providing a high degree of data security and fast information transmission, enabling us to process a large volume of records without constraints. This improvement ensures an extended data processing capacity and enhances user experience, increasing the overall effectiveness of the system.

The benefits offered by this system, including the protection of patient data, the simplification of user identity verification, the efficient organisation of information from medical records, and the facilitation of inter-hospital diagnosis, are emphasised by Dr. Jung-San Lee in [3], where these aspects are explored and discussed for the public. The article recommends the adoption of blockchain technology, complemented by smart contracts, as a method to share medical information in a secure and cryptographic manner.

In [4], a blockchain-based framework integrating XAI in medical diagnostics is proposed to enhance transparency and trust in AI-driven healthcare. The study explores the use of machine learning algorithms for medical data analysis, XAI techniques such as Local Interpretable Model-agnostic Explanations (LIME) and SHapley Additive exPlanations (SHAP) for interpretability, and blockchain technology for ensuring data integrity and traceability. Additionally, smart contracts are utilised to automate data validation and sharing while maintaining security and confidentiality. This integration aims to provide accurate, interpretable, and secure AI-driven medical diagnostics, increasing trust among healthcare professionals and patients.

Reference [5] presents studies that focus on protecting extensive databases in the medical field by applying blockchain technology. A model called Secure Block Chain for organising and distributing medical information is proposed. The Cryptographic Hashing Method applies the Discrete Shearlet Transform to encrypt information, integrating blockchain technology within the Hadoop distributed file system. The transaction validation process will be performed using public key cryptography. This approach is developed using Java and Hadoop platforms, and data security is strengthened by using deep learning algorithms within the Hyperledger Blockchain.

The exploration of blockchain technology in the medical field is presented in [6], where it discusses how integrating various techniques into a hybrid framework offers significant advantages.

According to [6], the individual elements of this system collaborate synergistically rather than competing, using each technology to address specific issues. This synergy creates high-performance hybrid solutions, characterised by increased complexity and efficiency compared to the isolated application of a single method.

Blockchain and AI are innovations in the health sector, collecting health-related data from various sources. The integration of these technologies can have a significant impact on healthcare by supporting the adoption of generalisable analytical technologies that can be incorporated into a broader approach to risk management. This article highlights various possibilities for creating reliable AI models in eHealth using blockchain, which is an open network for sharing and authorising information [7].

Another study explores the use of blockchain technology in improving electronic health record (EHR) management for Health Care 4.0 applications, noting that blockchain can enhance the interoperability of medical databases by providing increased access to patient medical records and securing data through a distributed database of transactions [8].

The study presented in [9] addresses the process of patient diagnosis by involving medical specialists, while also adopting innovative diagnostic methods based on blockchain technology. Study [10] presents how blockchain protects user data while simultaneously providing access to essential information, such as geographic location, mobility, and other relevant characteristics.

Current systems that combine AI and blockchain technology in the medical field are still in their infancy, but there are some outstanding initiatives and platforms that explore this innovative approach: IBM Watson Health, AiCure, MediLedger, Medical Sieve, BurstIQ, Moly, Patientory and Guardtime.

- IBM Watson Health is an advanced platform that uses the power of computing to assist in clinical decision-making. Working with various cancer care centres, IBM developed and trained the Watson supercomputer to help clinicians make informed treatment decisions. Watson can process vast amounts of clinical data and suggest treatment options in line with the latest evidence and guidelines. This ability to standardise care and speed up the approval process represents significant benefits for both healthcare providers and patients [11]. In addition, Watson for Oncology, a cognitive computing system developed by IBM, provides evidence-based and confidence-rated cancer treatment recommendations, supporting medical decisions in oncology. A study examined the level of agreement in lung and colorectal cancer treatments between a multidisciplinary tumour board at the Manipal Comprehensive Cancer Centre and Watson for Oncology, highlighting the system's ability to provide therapeutic suggestions [12];
- AiCure, as mentioned in [13], is a system designed to assist patients in the medication administration process. Using the mobile phone camera, the system verifies that patients are following the instructions for administering the treatment. The camera monitors whether the drugs are taken correctly by the user;
- MediLedger uses blockchain technology to secure and decentralise the tracking of the pharmaceutical supply chain, preventing counterfeit medicines. The study [14] proposes Medledger, a system based on Hyperledger Fabric, which utilises smart

contracts (chaincodes) for managing transactions within a private and permissioned network in the pharmaceutical industry. This system ensures data integrity and reliability through automatic recording and verification of transactions;
- Medical Sieve, described in [15], is a device that analyses medical data, applying specific algorithms to optimise the diagnostic process in the fields of radiology and cardiology, providing support to medical personnel;
- BurstIQ, as described in [16], provides a blockchain-based platform that enables large-scale integration, sharing, and securing of health data. The platform uses AI to analyse data and provide personalized insights, thereby improving patient care and operational efficiency;
- Moly, as mentioned in [17], is a virtual healthcare system developed by Sense.ly, designed to assist with treatment management. It is notably useful for medical consultations, acting as a virtual medical assistant for patients;
- Patientory is a platform that uses blockchain to give patients more control over their medical records. The platform enables users to securely store, manage, and transfer health information while facilitating communication between different healthcare providers [18];
- Guardtime is a company that uses blockchain technology to secure data in healthcare and other industries. In the medical sector, Guardtime enhances data management and ensures adherence to security regulations [19].

This paper highlights the development of an innovative architecture in the field of medical diagnostics by integrating intelligent hybrid systems based on XAI, alongside the robust security provided by blockchain technology. The XAI domain offers significant benefits to both patients and doctors, facilitating fast, accurate, and personalised diagnoses, while blockchain provides a secure platform for storing and sharing medical data. This advanced implementation of XAI incorporates a decision tree classifier, a Support Vector Machine (SVM) method, and neural networks with PyTorch, ensuring precise and reliable diagnostics. The decision tree ensures high interpretability, clearly explaining the factors influencing the diagnosis, while SVM optimises accuracy by effectively distinguishing medical cases. Meanwhile, the PyTorch-based neural network enhances learning capabilities, capturing complex relationships between symptoms and diseases. Additionally, security and personalisation of medical data and verification mechanisms are ensured by using CSV files for patient records. On the other hand, the implementation of a private blockchain allows interaction with smart contracts in both a simulated environment and the main Ethereum network. Virtual ether (ETH) is used, and transaction management is handled via each patient's MetaMask digital wallet. The system facilitates transaction execution, ETH balance management, and direct connection to the decentralised application via the browser. Transaction monitoring is carried out in Ganache, which displays the blocks within the network, while Truffle generates the smart contracts DataMigration.sol, DataPatient.sol and DataSave.sol for data migration, management and updates.

By merging these technologies, this work opens new horizons in specialised medicine, providing patients with easier access to diagnostics and improving the quality of medical care.

The structure of this work is organised as follows:

- The paper begins with a detailed introduction to the research area;
- Sect. 12.2 illustrates the examination of recent research that aligns with the objectives of the proposed study;
- Sect. 12.3 is dedicated to the presentation and design of the proposed system's architecture for medical diagnosis;
- Sect. 12.4 focuses on the implementation structure of the medical platform;
- Sect. 12.5 presents the results obtained after data processing;
- The document concludes with Sect. 12.6, which highlights the future directions of development, the system's limitations, and the final observations.

## 12.2 Analysis of Existing Works

XAI plays an essential role in generating new knowledge and advancing modern society. It proves its utility across a wide range of domains, including industry, medicine, and social interaction. XAI algorithms have a profound influence on the way we interact and make decisions, providing transparency and interpretability, adapting to the context around us, and enhancing our ability to solve complex problems in various situations.

Fundamentally, these emerging technologies, such as AI and blockchain, are reshaping the landscape of contemporary medicine by providing innovative solutions to traditional challenges. XAI, through its advanced algorithms and ability to provide transparency in the decision-making process, significantly contributes to improving diagnostic accuracy and personalising treatments. In parallel, blockchain technology ensures a high level of security and transparency in medical data management, strengthening trust in digital processes within the healthcare sector.

AI is proving to be a revolutionary tool in the rapid and accurate interpretation of medical images, often surpassing the performance of human radiologists in detecting subtle abnormalities. AI algorithms are also able to analyse patients' comprehensive medical histories and propose personalised treatment options based on complex data patterns. This ability to deeply personalise care can lead to improved patient outcomes and increased efficiency in the healthcare system.

On the other hand, blockchain introduces a new level of security and efficiency in the exchange of healthcare information. Through its decentralised nature, blockchain ensures that data is immutable and transparent, reducing the risks of fraud and error. Implementing blockchain technology in electronic medical record (EMR) management gives patients greater control over their own data and facilitates the secure exchange of information between different healthcare providers [20].

The integration of AI and blockchain in medicine paves the way for more intelligent and more adaptive healthcare systems that can quickly and efficiently address individual patient needs. These technologies have the potential to transform disease prevention, diagnosis and treatment, while ensuring the integrity and privacy of medical data. Therefore, the integration of AI and blockchain technologies in healthcare is a crucial evolution towards a future where healthcare is more accessible, personalised and safe for all patients.

The field addressed in this research consists of the use of AI and blockchain to create a system in the healthcare sector that offers improvements to the current system. Our aim is to achieve significant results in the development process of the medical system through the explanation and interpretation of XAI models. It is important to be able to draw some conclusions regarding the performance of the proposed system in the analysis of the results. Thus, AI and blockchain systems play a crucial role in ensuring transparency, understanding and acceptance in the field of modern medicine.

Currently, there are several systems and implementations that demonstrate the effective integration of AI and blockchain technologies in the healthcare sector, bringing significant improvements in medical data management, security, operational efficiency, and personalisation of medical care.

The study presented in [21] suggests a detailed analysis of XAI methods applied in the healthcare domain, focusing on tabular and time series data. It emphasises the necessity of clinical validation, consistency evaluation, objective and standardised quality assessment, as well as user-centred evaluation to ensure effective explanations for end users.

Another study conducted in [22] presents a blockchain-based EHR system for the Healthcare 4.0 applications. This system uses blockchain technology to improve medical record transactions, insurance billing and the use of smart contracts, providing permanent and secure access to data as well as a distributed database of transactions. The benefits of using blockchain technology in the healthcare sector include improved interoperability of medical databases and increased access to patient medical history.

Another approach is presented in [23] where an Internet-based Medical Assistance System was implemented, using blockchain technology to integrate the medical data of patients from over 350 hospitals with a decentralised storage system, ensuring the highest level of security possible and addressing security and data management issues at the local hospital level.

Another remarkable study is presented in [24], utilising XAI and blockchain technology in the medical metaverse. An innovative architecture is proposed, incorporating XAI techniques such as GradCAM and LIME to provide transparency and interpretability in disease prediction.

This research, cited in [25], examines the role of AI, Internet of Medical Things (IoMT) and blockchain technology in public health, highlighting current applications, challenges encountered, and future research directions. The study emphasises the benefits these technologies offer to the IoT ecosystem while also underscoring the complexity and difficulties associated with their effective integration into medical systems.

In study [26], a systematic analysis of the use of XAI in clinical decision support systems is provided. The authors examine various XAI methods, including techniques based on interpretable models, local and global explanations, as well as hybrid methods used to enhance the understanding of decisions made by machine learning algorithms.

In the study [27], the XAI methods used in medical image analysis are explored, highlighting the challenges related to the interpretability of deep learning models. The authors explore XAI techniques, including visualisation based on attention maps, surrogate models, and rule-based methods, to make the models more transparent and more comprehensible for medical professionals.

The systematic review of current medical research using new technologies is presented in Table 12.1.

The medical platform we have developed aims to facilitate more effective and faster communication between specialised medical personnel and patients. With an intuitive web interface, the platform allows patients to create an account by filling out a simple form when registering. It also makes the interaction with the doctors straightforward and user-friendly. Patients can quickly create an account on the platform by filling out the form available at the login page. The platform is also accessible without creating an account, but in order to interact with the doctors and to view the medical history of the visitor, respectively the patient in this case, they must create an account. This system approach is also proposed in [28].

In the proposed system, the patient has the option to choose the doctor, assistant, or specialised medical staff they wish. The patient can also choose the medical centre of their choice, such as a hospital, a medical clinic, or any medical centre available on the platform. A study was also conducted that shows patients' tendency to choose a medical centre based on health issues [29].

The medical system provided by this platform is available globally and can be accessed by any future patient, allowing them to choose to be treated by medical professionals from any location or corner of the world.

Patients can communicate with the medical team via email, chat with a live specialist or through an advanced AI chatbot. The following article presents the importance of the digitalisation of the medical system and the virtualisation of the contact environment with doctors [30].

The AI chatbot has a very high accuracy in disease identification and provides quick and accurate diagnoses, thanks to an extensive training database. The AI chatbot has the ability to discuss various topics, but specialises in medical diagnosis. Any visitor to the platform has the opportunity to talk with the implemented AI chatbot to find out more information about the medical system implemented within the platform or about any topic the interlocutor wishes to discuss. However, there is a possibility that the AI chatbot does not know the answer if the question addressed does not refer to the integrated medical system, because the AI chatbot specialises in the medical diagnosis of patients. In [31] an approach is presented capturing the effectiveness of chatbots in the medical field through different methods of diagnosing patients.

**Table 12.1** Synthesis of the latest trends in blockchain, XAI and AI respectively

| Ref | Technology used | Purpose | Technical design method | Result |
|---|---|---|---|---|
| [1] | Blockchain | It analyses the challenges, barriers and potential benefits of blockchain in the context of digital transformation | Processes and systematic approaches | Identifying the current state of research in this area, clarifying existing research gaps and highlighting agendas for future research |
| [2] | Blockchain, Shibboleth identity management system | It facilitates access to the history of a patient's treatments performed by different doctors in one location and secures medical data spread across various locations | Chain-like Structure, Patient Involvement in Authorization | Improves performance compared to other record sharing systems by reducing the time required to read, write, delete, and revoke a record |
| [3] | Blockchain, smart contract | Data sharing | Cryptography | The accuracy of medical data and patient confidentiality through the implementation of electronic health records |
| [4] | Blockchain, explainable artificial intelligence | To enhance transparency and trust in AI-driven medical diagnostics by integrating explainability and data security | Machine learning algorithms for data analysis, LIME and SHAP for interpretability, blockchain for data integrity, and smart contracts for secure data validation and sharing | A framework that ensures AI-based medical diagnostics are accurate, interpretable, and secure, increasing trust among healthcare professionals and patients |
| [5] | Blockchain | Improve the security and privacy of medical records by using blockchain technology | Hash code generation technique for cryptographic implementation | Data security is enhanced through the implementation of a system based on blockchain technology for the protection of medical data using Big Data |

(continued)

**Table 12.1** (continued)

| Ref | Technology used | Purpose | Technical design method | Result |
|---|---|---|---|---|
| [6] | Blockchain | Blockchain technology in the decentralised medical system and the creation of a solid database | Implementation of smart contracts for the management of medical data, diagnostics, laboratory analysis, medical imaging and drugs | The results suggest that blockchain technology offers efficiency and reduced costs in healthcare, enabling global population health monitoring through encryption programs for patient database storage |
| [7] | Blockchain | Examines blockchain technology in the health sector, with a focus on improving the efficiency of the medical system through decentralization and rapid access to information | The study explores the implementation of smart contracts for the management of medical data and evaluates the blockchain architecture in the context of healthcare systems | The results ensure the efficiency of medical data security, the improvement of interoperability between the health system and the facilitation of better access to medical information for patients and medical service providers |
| [8] | Blockchain | The designed system aims to improve the efficiency and reliability of electricity networks using Blockchain technology | The system approaches a technical design and includes the proposal of blockchain architectures for various applications within the smart electric grid | The system has directions for future research efforts aimed at decentralised applications and security in smart electric grids using blockchain technology |
| [9] | Artificial intelligence | Informative—data processing | Description of platforms that aim to help modernize medical systems | Collecting data to determine the performance of modern medical systems |
| [10] | Artificial intelligence | A system that uses artificial intelligence to help doctors diagnose and treat cancer | It combines AI and machine learning to analyse medical data, including journals, records, and genetics, to provide personalised cancer treatment suggestions | The system can identify cancer treatments with an accuracy close to 90% of human oncologists, demonstrating AI's significant potential in advancing medical care |

(continued)

**Table 12.1** (continued)

| Ref | Technology used | Purpose | Technical design method | Result |
|---|---|---|---|---|
| [11] | Artificial intelligence | Describes the use of the IBM Watson supercomputer in clinical decision support for treatments | Watson is a supercomputer that integrates advanced AI and machine learning technologies, with the ability to process and analyse large amounts of data, including natural language, to provide evidence-based suggestions for clinicians' treatment decisions | It contributes to the improvement of decision-making in the treatment of patients, providing a possible benefit for both healthcare providers and patients |
| [12] | Artificial intelligence | The aim of the study was to examine the level of agreement between lung and colorectal cancer therapies recommended by Watson for Oncology and those of a multidisciplinary tumor committee at the Manipal Comprehensive Cancer Centre in Bangalore, India | Watson for Oncology utilises structured and unstructured medical data, including literature, guidelines, records, medical images, and pathology reports, along with expert insights, to develop personalised treatment plans | The study compares treatment suggestions provided by Watson for Oncology with multidisciplinary committee decisions, highlighting the ability of this technology to support clinical decisions in cancer therapy |
| [13] | Artificial intelligence | The goal of the AiCure system is to improve adherence to drug treatment, which is crucial in the effective management of chronic conditions and the success of clinical trials | The technical design method involves the use of artificial intelligence algorithms to analyse videos recorded by patients while taking the drugs | The results of using AiCure include significantly improving adherence to drug treatment, reducing the risk of disease relapse, and improving overall treatment outcomes |
| [14] | Blockchain, hyperledger fabric | The purpose of MediLedger is to provide a drug tracking and authentication system to prevent counterfeit drugs from entering the supply chain | The technical design of the MediLedger system includes the use of chaincodes to regulate interactions between supply chain participants | The implementation of the MediLedger system results in increased efficiency and security in the pharmaceutical supply chain, reducing the risk of intrusion of counterfeit drugs |

(continued)

**Table 12.1** (continued)

| Ref | Technology used | Purpose | Technical design method | Result |
|---|---|---|---|---|
| [15] | Artificial intelligence | Medical Sieve's goal is to help radiologists and cardiologists make more informed clinical decisions by analysing medical imaging data and other relevant clinical information, thereby reducing eye strain and improving diagnostic accuracy | Medical Sieve collects and analyses multimodal data to detect abnormalities and summarizes patient records, collecting all relevant information related to a chief complaint | Clinical trials in the fields of cardiology and breast radiology have demonstrated the accuracy of the system in differential diagnosis and imaging studies, providing comprehensive summaries of the patient while using raw data for detailed review when necessary |
| [16] | Blockchain | BurstIQ is a platform that uses blockchain technology to enable the secure exchange of health data | Medical data collection platform | The aim is to streamline and organize medical data |
| [17] | Artificial intelligence | Molly uses AI, natural language processing and machine learning algorithms to interact with patients. The primary goal is to assist healthcare providers by monitoring patients, managing their questions, and ensuring they follow care plans outside of traditional healthcare settings | The system is designed to analyse patient responses, medical data, and behavior to provide personalized care and suggestions, using a conversational interface for ease of use | The result is improved patient engagement, adherence to treatment plans, and a possible reduction in hospital readmissions or in-person visits by providing a virtual and personalized healthcare assistant |
| [18] | Blockchain, artificial intelligence, mobile app | Patientory aims to democratise the ownership of health data by providing comprehensive medical data, analytics, and clinical trial management through the use of AI. The Patientory mobile app and Neith software include blockchain for secure data management | Their method integrates decentralised applications to collect and analyse healthcare data for personalized care plans aimed at improving population health outcomes | The result is improved interoperability, secure data management and personalized health information, contributing to better health outcomes and lower costs of care |

(continued)

**Table 12.1** (continued)

| Ref | Technology used | Purpose | Technical design method | Result |
|---|---|---|---|---|
| [19] | Blockchain | Guardtime generally focuses on cyber security and data integrity, using blockchain technology. The aim is to provide a tamper-proof infrastructure for digital assets, thereby ensuring trust and compliance of information | The technical design likely involves advanced cryptographic methods and blockchain architecture to secure data across various platforms | The result is increased data integrity, security and trust in digital transactions and information storage |
| [20] | Explainable artificial intelligence, blockchain, immersive technology | To enhance intelligent healthcare by integrating XAI, blockchain, immersive technologies, and EMR in the metaverse for secure and transparent patient data management | Utilisation of XAI for transparent AI-driven decisions, blockchain for secure and tamper-proof EMR storage, and immersive technology for enhanced patient engagement | Improved trust, security, and efficiency in managing and using EMRs within a digital healthcare ecosystem |
| [21] | Explainable artificial intelligence | To analyse and evaluate XAI methods applied in healthcare, focusing on tabular and time series data for improved decision-making transparency | Review of XAI techniques for interpreting machine learning models in healthcare, focusing on validation, consistency and user-centred evaluation | Identified challenges and limitations in current XAI methods, emphasising the need for clinical validation and standardised quality assessment |
| [22] | Blockchain | The paper outlines a blockchain-based system for managing electronic healthcare records in Healthcare 4.0 applications, aiming to improve data security and interoperability | The system uses Hyperledger Fabric in blockchain, as it enables secure access to data and updates. It integrates an access control policy algorithm for data sharing between healthcare providers, improving efficiency and confidentiality | The results indicate significant improvements in healthcare data management, highlighting increased security and efficiency in patient care |

(continued)

**Table 12.1** (continued)

| Ref | Technology used | Purpose | Technical design method | Result |
|---|---|---|---|---|
| [23] | Blockchain | The development of an Internet-of-Healthcare system using blockchain aims to improve the security and management of patient data in several hospitals | Decentralised storage through blockchain ensures secure data access. By using the Message Queuing Telemetry Transport protocol and mobile applications, efficient and secure management of patient data is supported | This approach has been implemented in more than 350 hospitals, reflecting improvements in security and user satisfaction regarding system use and maintenance |
| [24] | Explainable artificial intelligence, blockchain, metaverse | To integrate XAI and blockchain in the medical metaverse for enhanced transparency and security in disease prediction | Utilisation of XAI techniques such as GradCAM and LIME for interpretability, with blockchain ensuring secure and immutable medical data storage | Improved trust, security, and accuracy in AI-driven medical diagnostics within the metaverse |
| [25] | Blockchain, explainable artificial intelligence, IoMT | The article explores the combination of AI, the IoMT and blockchain technology in healthcare | These technologies aim to personalize care, improve outcomes and reduce costs | The study covers their applications, challenges and future research directions, highlighting the potential and obstacles in implementing these innovations in public health systems |
| [26] | Explainable artificial intelligence | The study provides a systematic review of XAI methods in CDSS, aiming to enhance transparency and trust in medical decision-making | Analysis of interpretable models, local and global explanations, hybrid methods for improving decision understanding | Highlights challenges in integrating XAI into clinical practice and suggests future research directions for better adoption |
| [27] | Explainable artificial intelligence, medical image analysis | To review and evaluate XAI methods used in medical imaging for improved transparency and interpretability | Analysis of various XAI techniques, including attention maps, surrogate models, and rule-based methods, applied to medical image analysis | Identified challenges and future directions for integrating XAI into medical imaging to enhance trust and diagnostic accuracy |

The AI chatbot, integrated into the platform, is able to determine a patient's condition by analysing the symptoms described by the patient, thus providing an interactive and personalised experience in medical diagnosis.

The AI chatbot analyses the conversation with the patient and, following the discussion, identifies the disease and recommends methods for treating the disease and precautionary measures to the patient. If the AI chatbot identifies a more serious case, it recommends the patient to consult an emergency doctor, as the problem is more complex and requires urgent human medical intervention. In [32], a study is presented that discusses diagnostic applications using symptomatology to render analysis results.

The confidentiality of patient information as well as of any users is rigorously protected by adopting Blockchain technology, which significantly strengthens the security of personal data.

The platform benefits from the advanced integration of Ethereum blockchain technology, ensuring a higher level of security for all users, whether they are doctors, platform operators or patients. This systems approach is also presented in work [33], which illustrates an efficient way of working in this field and offers extensive explanations regarding its applicability.

In addition, the designed medical platform offers the opportunity for patients to view, in the 'Profile' section, the details of their medical history, loyalty points earned through their activity on the platform and other benefits.

Patients have the flexibility to pay for medical services using their bank card, PayPal or virtual currencies, the latter being elegantly integrated within the platform. These virtual models can be earned through the patient's loyalty points or purchased directly by the patient. A service payment system using blockchain technology is also discussed in the document [34], which specifies the benefits and safety of currency transfers.

The patient has the option to select the video mode for the conversation with the doctor and to provide the doctor with visible information about their health condition. This functionality facilitates more efficient identification of comorbidities or other relevant conditions by the specialist. If the doctor considers that the patient requires a physical consultation, the patient will be redirected to the nearest medical centre where they can be treated by the same doctor or by a colleague of the respective doctor, who has been informed of the patient's history and health condition.

Widespread implementation of this medical system has the potential to significantly reduce waiting times in hospitals and other medical centres, potentially reducing these intervals by as much as 50%. In [35], the benefits of such a system across different sectors in the medical field are illustrated. Thus, the system helps to reduce the number of patients waiting in physical queues to be treated. Moreover, car trips are drastically reduced, leading to a decrease in greenhouse gas emissions, air pollution, noise pollution, heat pollution, and space consumption, as well as the impact on habitats.

## 12.3 The Proposed System for Determining Medical Diagnosis

Today's healthcare system faces many challenges, including managing patient data effectively, protecting the privacy and security of that data, and delivering personalised and efficient healthcare. In many cases, traditional infrastructure and processes are overburdened and inefficient, and in certain situations, patient information may be compromised or inaccessible when needed.

Implementing a system based on AI and blockchain technology could bring many significant benefits to the healthcare system:

- Efficient data management: AI systems can be used to process and analyse massive amounts of medical data, extracting valuable information and providing insights that can be used to improve patient diagnosis and treatment;
- Data Privacy Protection: Blockchain technology provides a high level of security and privacy, enabling secure and immutable recording of medical data. By using blockchain, access to this data can be precisely controlled and information can only be shared with authorised parties;
- Personalised healthcare: AI systems can analyse patient data to generate personalised profiles and provide individualised recommendations for diagnosis and treatment. This could lead to a significant improvement in the quality of medical care and patient outcomes;
- Operational efficiency: Automating certain processes through AI and blockchain can lead to reduced overhead costs and improved efficiency in data management and healthcare system workflow.

However, implementing an AI and blockchain-based system in the healthcare system also comes with certain challenges and considerations:

- The initial costs and investment in infrastructure and training can be substantial;
- Ensuring compliance with data protection and medical information privacy regulations is critical and requires special attention;
- System integration and interoperability with existing medical infrastructure and other data recording systems can be difficult and requires careful planning and coordination.

Implementing an AI and blockchain-based system in healthcare could bring significant benefits, including improving the quality of healthcare, protecting data privacy and streamlining operations.

### *12.3.1 Methodology of the Main Objectives of the Medical System*

In this paper, we address the subject of creating a medical system for diagnosing patients based on AI and Blockchain.

The stages involved in the implementation of the proposed system are as follows:

1. **Definition of requirements and objectives**: The detailed objectives of the project are outlined, determining the variety of diagnostics managed by the system, as well as the strategy of merging AI and Blockchain technologies to increase security, effectiveness and accuracy in the diagnostic process;
2. **Technology research and selection**: An analysis of the diversity of existing AI algorithms and Blockchain solutions is carried out to identify the most efficient and compatible technological tools with the project specifications;
3. **Elaboration of the system structure**: A detailed diagram of the system architecture is developed, emphasising the collaboration between AI elements and medical information to generate diagnoses, as well as the strategy of applying Blockchain technology to guarantee the security and confidentiality of data;
4. **AI development and integration**: Implementing selected AI algorithms, training models with relevant medical datasets, and integrating them into the system to analyse data and provide diagnostic recommendations;
5. **Blockchain implementation**: Configuring the Blockchain infrastructure, including choosing the appropriate protocol, and integrating it into the system for secure data and transaction management;
6. **User interface and integration of system components**: Designing an accessible and intuitive interface for both doctors and patients and orchestrating the smooth integration of all system elements, from AI and Blockchain to medical databases to ensure flawless functionality;
7. **Testing and validation**: Conducting extensive tests to verify the accuracy of AI diagnostics, the security provided by Blockchain, and the overall functionality of the system;
8. **Continuous monitoring and optimisation**: Continuous evaluation of system performance in a real-world context is carried out, users' impressions are gathered and regular improvements are made, thus ensuring that the system remains up to date with the latest advances in AI and Blockchain technology.

The system is designed to ensure security and optimise the medical diagnosis process, offering an efficient and reliable structure.

In the next section, we will explore the architecture of the system, exposing its components and structure to provide detailed insight into its organisation and operation.

### 12.3.2 The Functional Architecture of the Medical Diagnostic System

The functional architecture of the system is designed to facilitate the medical diagnosis process through a secure digital platform. The foundation of this architecture is based on blockchain technology to protect users' personal data and manage financial transactions securely using cryptocurrencies such as Ethereum. AI is also a component of the system architecture foundation, facilitating patient diagnosis through an advanced and secure digital platform.

Users can navigate through various stages of the platform, starting with registration and log in, followed by uploading medical data and interacting with available healthcare options.

This integrated architecture aims to improve diagnostic efficiency, reduce waiting time and facilitate access to qualified healthcare for patients worldwide, all in a secure and user-friendly digital environment.

The development of a system architecture for diagnosing patients, using AI and Blockchain technology, aims at the following work stages:

- **Stage 1: Access and registration**

In the first stage, the patient will choose the device to use to access the platform. If the patient has an account created, they can log in directly by entering the username and password in the relevant fields. If the patient has not created an account, they can access the webpage directly or create an account quickly by filling in the appropriate forms. The interface facilitates easy navigation and is designed to be accessible on various mobile and desktop devices (see Fig. 12.1).

- **Stage 2: Interface and use of the platform**

After registration and authentication, users are introduced to an interface with a wide range of options, where they can update their profile, access information about departments and services, contact medical staff via email or chat, and make payments. This step brings together all the actions required to manage the interaction with the healthcare system (see Fig. 12.2).

- **Stage 3: Diagnostics and data security**

The system combines two Blockchain and AI technologies required in this context to ensure high performance.

The architecture of the medical system integrates Blockchain technology to secure users' personal information and enable payments with cryptocurrencies, including Ethereum, doubling its functionality to secure data and simplify financial transactions. Users can purchase virtual coins or obtain loyalty points in the medical platform by using it frequently. Later, the obtained loyalty points can be converted into Ethereum cryptocurrency. These virtual currencies are used to pay for medical services.

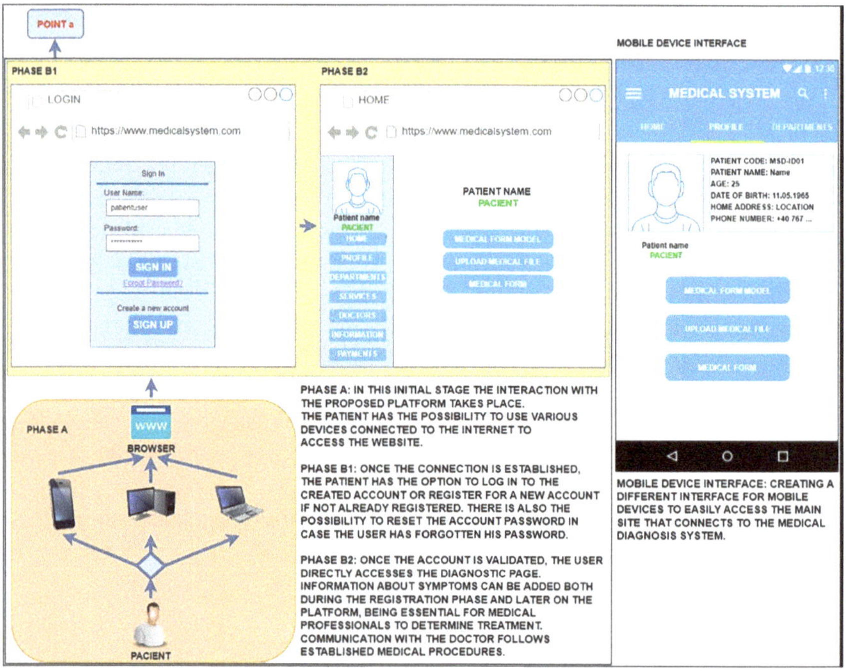


**Fig. 12.1** The architecture of the access stage of the medical diagnostic platform

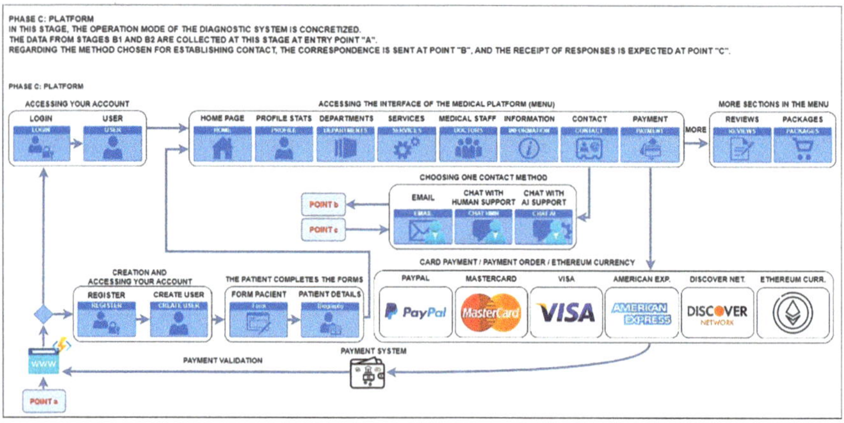


**Fig. 12.2** Platform interface and functionality architecture

Three main methods are integrated through which users can communicate within the medical diagnostic system: e-mail, chat with a human specialist and chat with an AI specialist. The AI system specialises in chat interaction. In the chat with AI support, the system has a conversation with the patient, collects information about

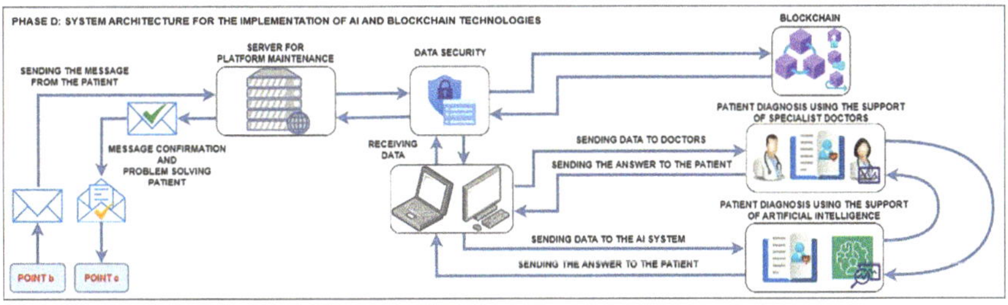


**Fig. 12.3** Blockchain and chat system architecture with AI support

symptoms and assigns a diagnosis. The AI system, trained on an extensive dataset, achieves a high level of accuracy in identifying ailments. In addition, this AI chat provides prevention tips and treatment recommendations, and in more serious cases, urges the patient to urgently consult a specialist doctor.

Medical staff can also interact with the AI system to verify or seek second opinions on patient diagnoses, which increases the accuracy of assessing patients' health conditions.

Due to the large volume of training data, the accuracy of AI diagnosis is very high, and the security of the data and the diagnosis process is strengthened by using Blockchain technology, making this system an advanced support for medical diagnosis (see Figs. 12.2 and 12.3).

The complete architecture, which serves as the starting point for the construction of this system, is developed further (see Fig. 12.4).

### *12.3.3 Payment Security Methods and Systems*

The method of payment for medical services is varied, giving patients flexibility in paying the costs.

Users of the medical diagnosis platform can pay for medical services in the following ways:

- Payment by bank card;
- Payment by means of a payment order;
- Payment via virtual models.

Regardless of the choice made, the security of personal data is a priority, with access to this information being strictly controlled and allowed only with the explicit consent of the owner, when necessary.

When creating an account, patients enter their personal data, symptoms and request a diagnosis. They can select their preferred country, medical facility and doctor for consultation. After this process is completed, the patient is quickly contacted for a virtual assessment. Following the investigation, they are presented

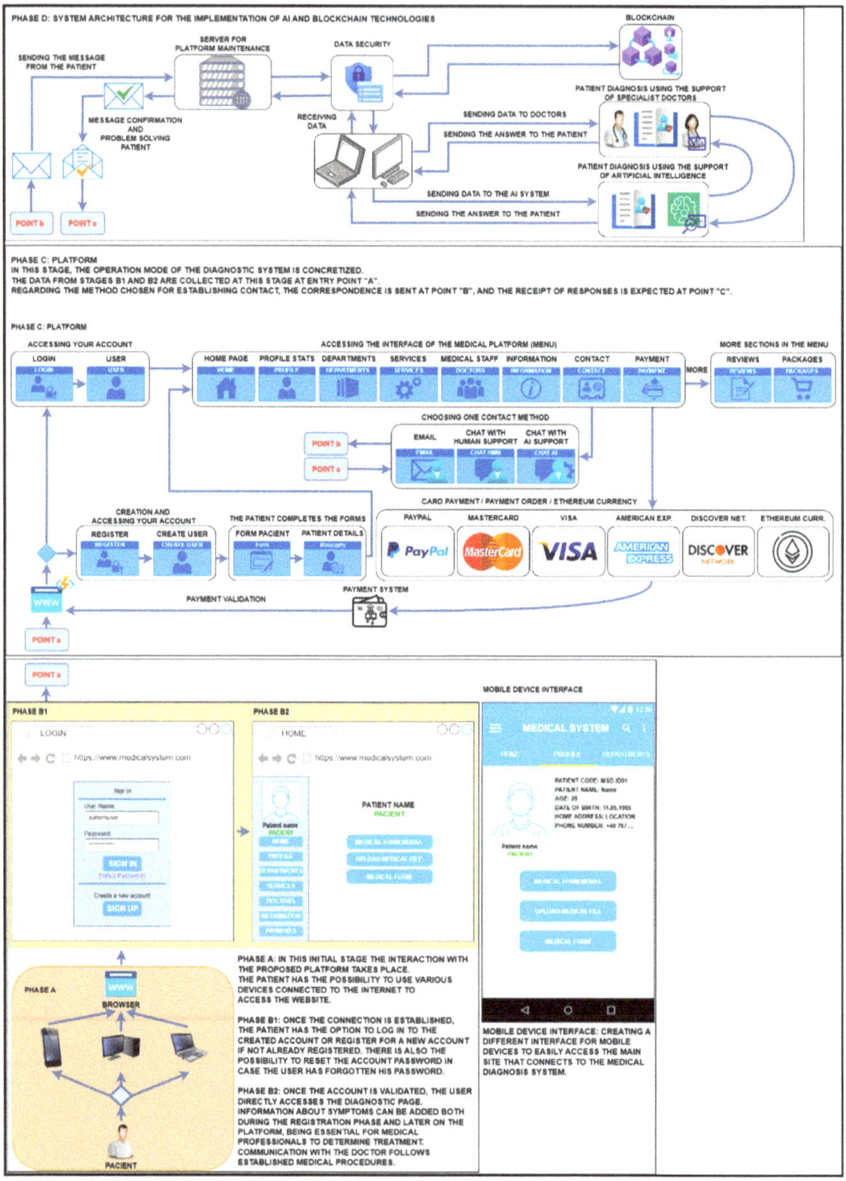


**Fig. 12.4** The complete architecture of the medical diagnosis system

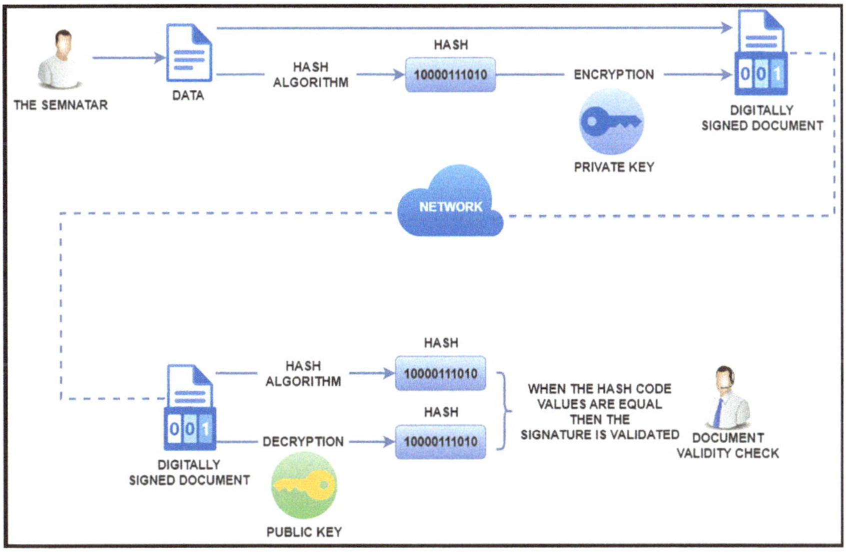


**Fig. 12.5** Mode of operation for the Merkle tree

with a medical quote, which can be paid according to their preferences. Transactions are secured using a blockchain system, employing the Merkle Tree structure, which ensures data integrity through chains of hash codes. The Merkle tree is shown in the Fig. 12.5.

The blockchain is a technology that stores data in blocks linked together by unique hash values. Each block contains a hash of the previous block, creating a cryptographic bond that ensures the security and immutability of the entire network. This structure prevents retroactive changes, thus guaranteeing the authenticity and integrity of the data. Also, the following image illustrates this (see Fig. 12.6).

In a Merkle tree, data is divided into small blocks, each with its own hash value. These blocks are then grouped into pairs, and each pair is hashed by a new hash value. The process continues hierarchically until a single root hash, called the Merkle root, is reached, which is the summary of the entire dataset. This structure facilitates efficient data verification in blockchain networks, allowing data integrity to be verified without requiring access to the entire dataset (see Fig. 12.7).

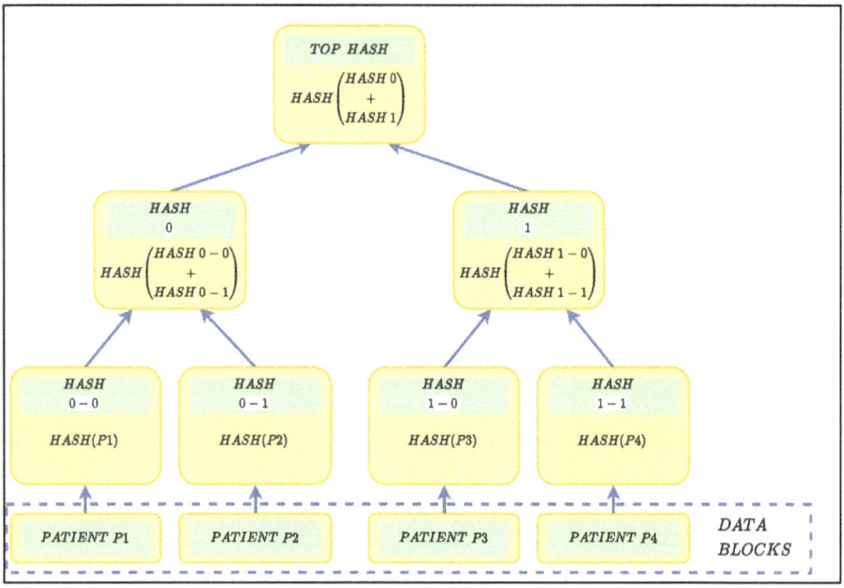


**Fig. 12.6** Mathematical data structure of chained blocks linked by hash code

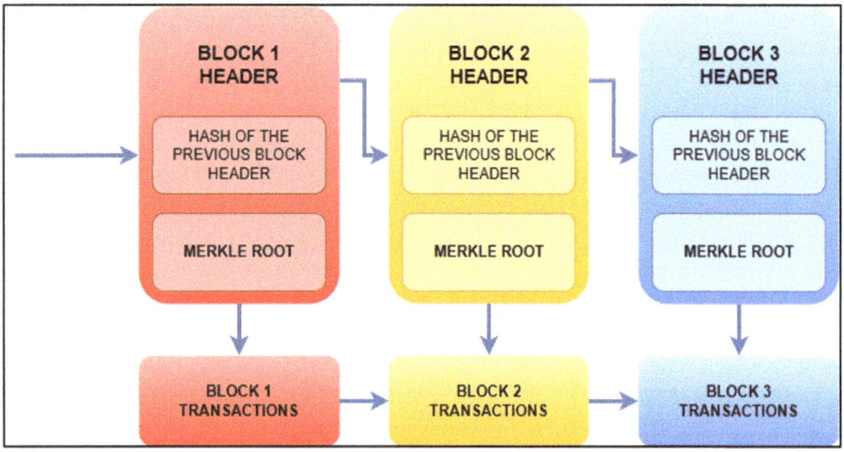


**Fig. 12.7** Splitting the data into parts for the Merkle abore

## 12.4 The Visual Structure of the Medical Diagnostic System

The design of this medical patient diagnosis system is designed to provide an integrated experience, seamlessly blending the user interface with cutting-edge technologies. At its core is the use of AI to interpret symptoms and suggest diagnoses, increasing the accuracy and efficiency of the diagnostic process. Data security is ensured through the implementation of blockchain technology, which adds an additional layer of protection and privacy for patients' personal information.

The system allows users to register and interact with medical staff through various modalities, including AI-powered chat, ensuring constant and accessible communication. The platform also includes versatile payment options, accepting both traditional methods and cryptocurrencies, thus offering maximum flexibility in financial management. At the same time, the system is designed to be accessible from various devices, facilitating access to medical services regardless of the patient's location.

### *12.4.1 Conceptualisation and Implementation of the Interface System Design for User Access Through Login or Registration*

This stage outlines the login process of a user who already has an account registered in the medical diagnostic system.

There are three fields on this web page:

- **Patient ID**: The user must enter the patient identification code received during the registration process. This ID code is generated by the system during the registration process in the medical diagnosis system. The identification code is generated based on previous registrations in the system and serves as an index for all registered users;
- **Name**: This field is intended for entering the full name of the user of the medical diagnosis platform;
- **Password**: The user must enter the password they set during registration in this field.

For increased security, each user must complete these three fields with the data they used during registration. Once identified and validated, the user is granted access to the medical diagnosis platform and can use the system.

The system provides an interface for the user. The image below shows the login stage in the medical diagnosis platform (see Fig. 12.8).

During the registration process of the medical diagnostic system, users are welcomed by an accessible web page where they can create a new account using their existing login credentials. This initial phase is critical to personalising each patient's experience. By filling in basic information and medical details, the system can begin to tailor the diagnostic process to the specific needs of patients.

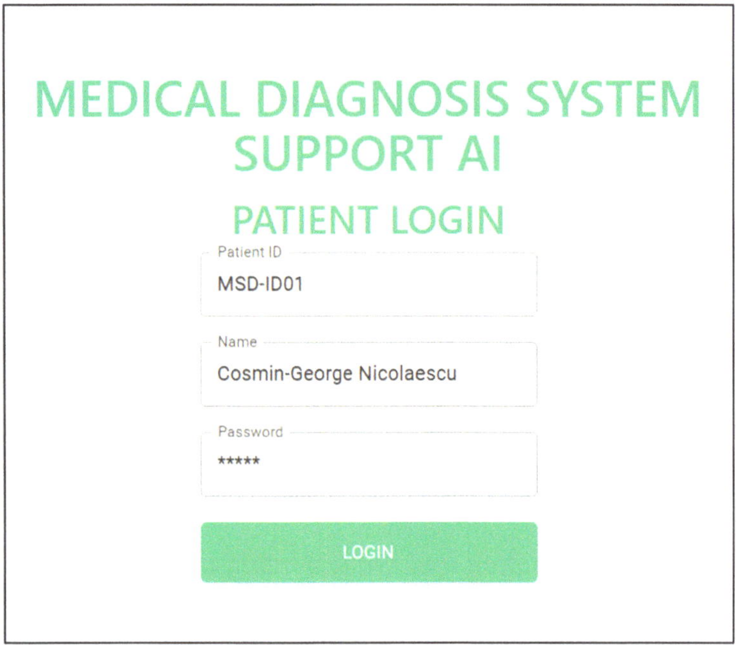


**Fig. 12.8** Logging stage in the medical diagnosis platform

The registration process on the platform is designed to be as simple and intuitive as possible, asking the patient to provide the necessary information to create their profile. This process includes security steps to ensure that personal information is protected and managed to the highest standards of privacy. Once the account is created and verified, patients can access medical history, schedule appointments and communicate with medical specialists, all of which contribute to efficient and centralised management of their healthcare.

At this stage, we carry out the registration process for the user in the medical diagnostic system.

There are six fields dedicated to this web page:

- **Patient ID**: The user does not have to fill in this field, because the identification code is generated by the system and constitutes an indexed order of all user records within the medical diagnostic system. The text field is disabled so that the patient can view the ID code, but cannot alter its integrity. The user must remember this unique patient identification code, as it is used to log in to the medical system;
- **Name**: This field is intended for adding the full name of the user. The purpose of this text field is to facilitate the identification of the patient within the system. It is also used as a security measure, as it is required during the login stage, and an incorrect entry on the login page causes an error in the validation of the patient's identity;

- **Password**: To enhance security, the user must fill in the text field intended for their password. The password is a step in validating the user's identity. It is recommended that the user sets a more complex password to prevent unauthorised access by third parties;
- **Birth-date**: The date of birth will be written manually in the text field or selected by pressing the button located on the right side of the text field. The button within the field is intended for selecting the user's full date of birth. It also provides a friendlier and more interactive way of configuring personal data (see Fig. 12.10);
- **PhoneNumber**: This field is intended for adding the user's phone number, and after creating the account, a confirmation message will be sent to the phone number to complete the account creation process. The code "+40" corresponds to the country of the telephone number. This code cannot be manually modified. To change this code, it must be activated, and a small interface for selecting the desired country code will open (see Fig. 12.10);
- **Address**: This text field is for entering the user's home address. If the patient requires a physical medical intervention, the operators of the medical diagnostic platform have access to the patient's information and can recommend the nearest partner medical centres within the system for a physical examination by the specialist medical personnel present in the system.

At this stage of registration, it is necessary to fill in all the fields to successfully validate the provided information. If the user attempts to create an account with the same data, this action is not allowed and a warning message will be displayed. To change the existing data, it must be updated within the medical diagnostic platform, and if the login details have been forgotten, the user must contact the operators of the technical support to generate and send a link to reset the patient's account data via email (see Fig. 12.9).

By activating the calendar icon, and the index placed before the phone number, as suggested by the arrows in the displayed image, the data configuration interface will open.

The configuration of text boxes that require selection from preset data are illustrated in this image (see Fig. 12.10).

On the second page of the registration form, the platform invites patients to provide additional information about their symptoms. Although this step is optional, providing this data helps speed up the diagnostic process and contributes to the efficiency of medical scheduling. The registration form is designed intuitively to facilitate the identification and selection of symptoms by the patient, giving them an easy opportunity to describe their medical conditions.

This pre-collection of data is valuable because it can point the AI system in the right direction even before a detailed interaction with a doctor. It helps prepare the system for patient interaction by providing an initial picture of health issues that can be refined and supplemented later. Through this mechanism, the platform aims to minimise the time required for patients to reach the later stages of diagnosis, generally improving the user experience and the workflow of healthcare professionals.

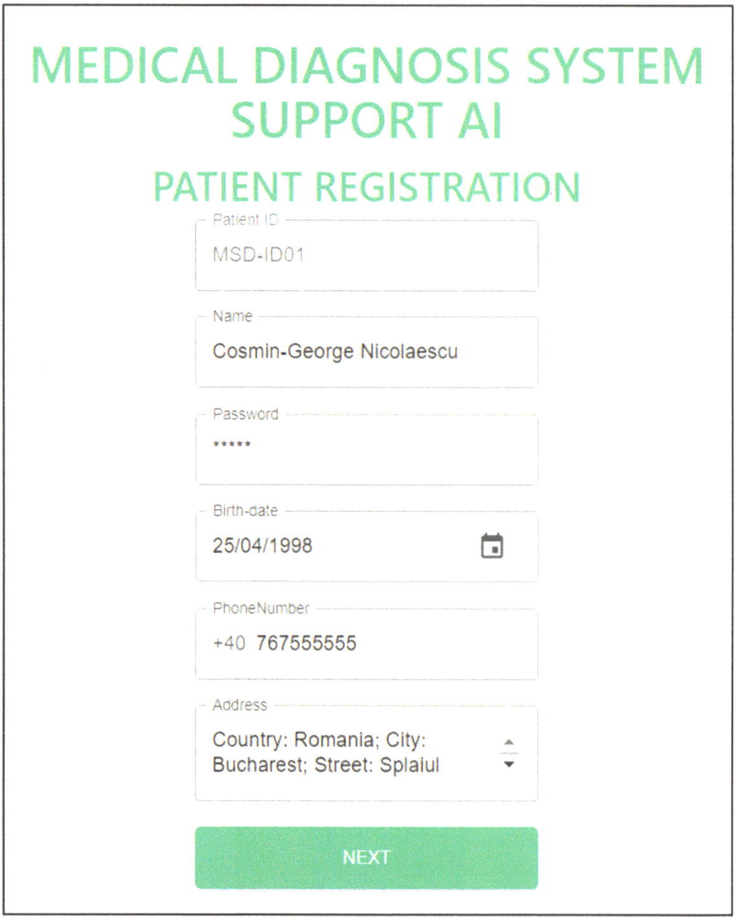


**Fig. 12.9** Registration stage in the medical diagnosis platform

This web page represents the continuation of the user registration stage in the medical diagnosis system.

There are seven fields on this webpage:

- **Medical Report ID**: The user does not need to fill in this field, as the medical report's identification code is generated by the system. This code illustrates the ordered and indexed creation of all medical reports recorded in the system. The text field is disabled, as the user can only view the identification code but cannot alter its integrity. The user does not need to remember this unique identification code, as it helps to identify the patient's medical issues more quickly. The generated reports can be viewed within the user's profile;

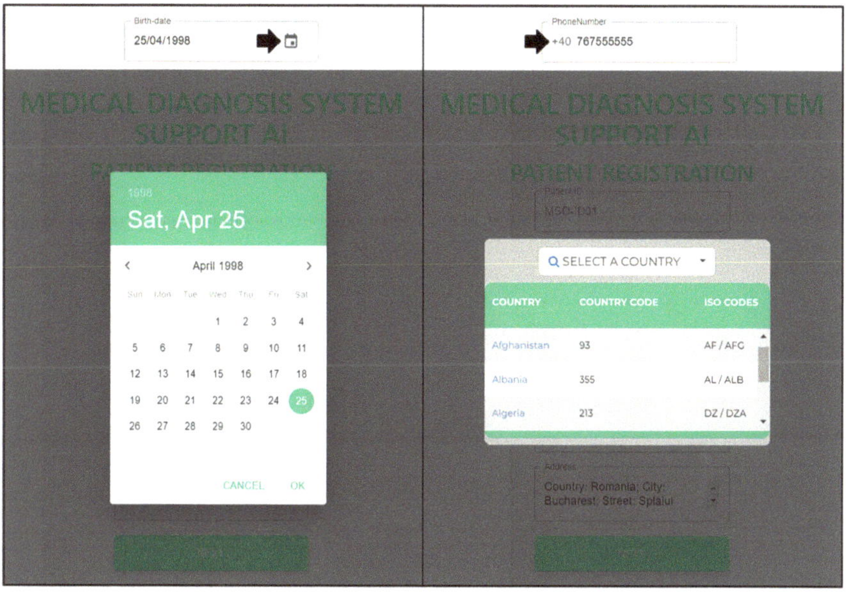


**Fig. 12.10** Setting the data in the dedicated fields

- **Weight**: This field is for entering the user's weight in kilograms. The recorded data can also assist medical personnel in prescribing medication in relevant medical situations;
- **Height**: his text field is for entering the patient's height in centimetres. The recorded data can help inform the medical staff and contribute to the patient's treatment in certain situations;
- **Disease/Diseases Name (Only if you have)**: This text field aims to provide information about the patient's current diseases. Here are written the names of the patient's diseases if he still has them. If the person is not sick and only wants to be registered in the medical diagnosis system, then this field is not completed;
- **Blood Group**: The text field for the patient's blood group is intended to be filled in here. Also, if the user has no symptoms of the disease, this field does not have to be completed, because it can be edited in the user's profile;
- **Disease Started From**: If the user has symptoms of the disease, then he must fill in the approximate date when he started having symptoms. This field does not need to be completed if the user only wants to create the account;
- **Description**: In this section, the patient must detail as concisely as possible the symptoms of the diseases he has or only the symptoms of the disease if the patient does not know the diagnosis of the disease. Also, the symptoms of the patient's existing diseases are filled in or not filled in if the patient does not have any disease and only wants to be registered in the medical diagnosis system.

In the second stage of the registration process, the user is asked to complete the fields as intended. The second page of the registration process details the patient's health status by completing the intended sections.

This stage contributes to a deeper understanding of the patient's needs from the very beginning of the interaction with the system (see Fig. 12.11).

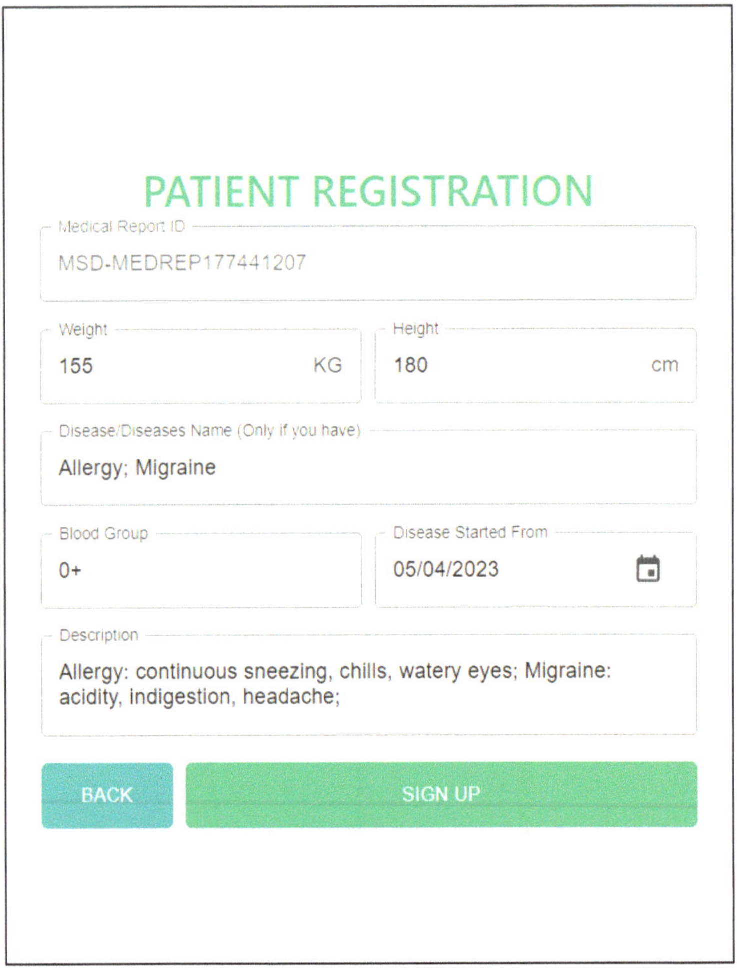


**Fig. 12.11** Continuation of the registration stage in the medical diagnosis platform

### *12.4.2 Conceptualisation and Implementation of the Design of the Interface System for User Access to the Medical Diagnosis Platform*

In this stage, the medical diagnosis platform is introduced. The Medical Diagnostic System is designed to provide quick and accurate medical assessments using advanced technology.

The first page of this stage intended for the medical diagnosis platform is presented in this column.

The menu bar within the webpage is made up of the following elements:

- **Home**: Its role is to navigate to this first home page;
- **Profiles**: This section presents the user's current information and health history;
- **Departments**: It is a general overview of the departments so that all the services within all the specialised departments are included. This category presents the facilities of the medical diagnosis system;
- **Services**: Illustrates a stage of presentation of several medical services within the medical diagnostic system. It represents a short and concise presentation of the approach to various medical problems;
- **Doctors**: In this category, all the medical personnel within the medical system are presented. For easier classification, the user has the freedom to search for the desired doctor, assistant or department;
- **Information**: This category has the role of presenting the necessary information to a user and to facilitate the mode of operation between the patient and the platform;
- **Contact**: Illustrates how a user of the medical platform can get in touch with specialised medical personnel to solve health problems. The system has three categories of contact: e-mail, chat with a specialised operator and chat with a specialised AI operator. Also, the platform has a video call option for providing more visual details to doctors;
- **Payment**: The purpose of this section is to pay for the medical services provided. The medical diagnostic platform has categories intended for payment by bank card, payment by payment order and payment by means of virtual currencies. Thus, ensuring versatility in the choice process;
- **More**: This is a selection section and includes three categories: Reviews, Packages, Download All Stats;
- **Reviews**: This section illustrates the opinions and scores provided by the users of the platform. It is a way for new users to check if already registered patients were satisfied with the services offered by the proposed medical platform;
- **Packages**: This section illustrates the promotional packages that a patient can purchase. The platform is not limited only to the presented medical packages and can be configured to the client's needs;
- **Download All Stats**: In this category, a patient can download with a single click all his medical data and his medical history so that the client can have it in physical format throughout his medical treatment.

This webpage acts as the home page for a medical diagnostic platform, providing a personalised dashboard for users. The date and time when the user logged in to the platform is presented at the top, followed by the patient's identification code and the amount of virtual coins held by the client in their account. The patient's name and a unique identification code are highlighted centrally on the page. On the left, the advantages of the platform are highlighted, while on the right, the possibility of quick scheduling is offered by filling out a simple form, without the need for registration (see Fig. 12.12).

The second page, called 'Profile', is intended to serve as a dashboard and provides data and information about the patient's health status, medical history, and statistics. On the left-hand side are illustrated metrics such as years of loyalty, the score of the treatment method, the hours spent on the platform since account registration, and the loyalty points collected by the user. The patient's name and home address are displayed on the right-hand side. In the lower section of the profile, the patient can view various evaluations related to their medical course, including the health assessment at the last medical consultation, the time remaining until the completion of the medical treatment, the effectiveness yield of the ongoing treatment, the initial health status at the first consultation, the current health status, and the recovery progress of the patient. These statistics are calculated based on the patient's health history. This information is essential for tracking patient progress and personalising medical care (see Fig. 12.13).

The third page is titled 'Departments' and offers information about different sectors and services within the medical diagnosis platform. It also has a range of options that include medical institutions and clinics, medical staff, diagnostic systems, the most in-demand staff, and contact information. The page is aimed

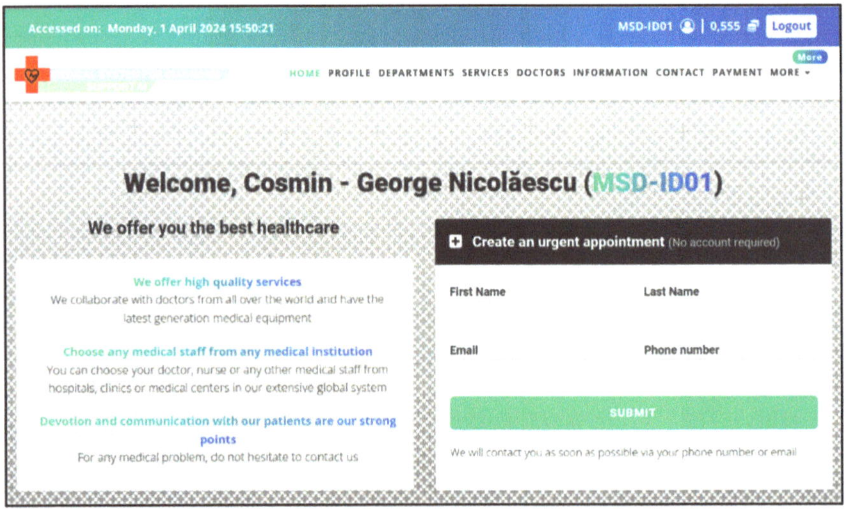


**Fig. 12.12** Home page interface of the medical diagnosis system

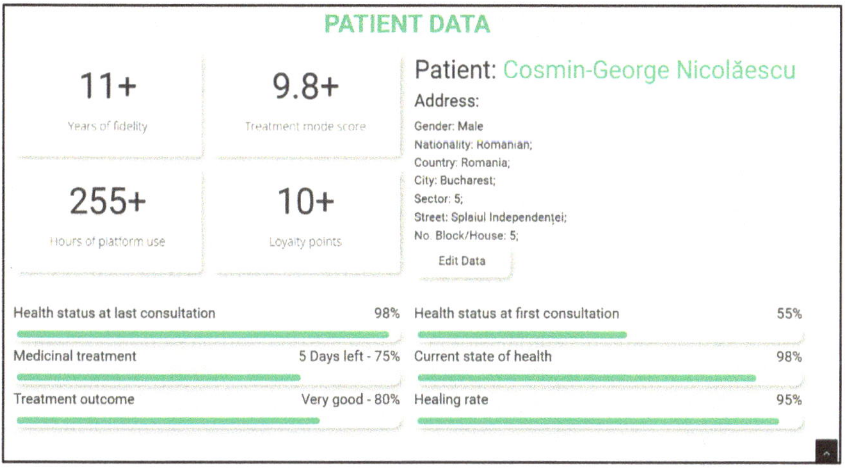


Fig. 12.13 Interface of the profile page of the medical diagnosis system

at departments and is designed to facilitate users' access to information about the services offered, available staff, and support for questions or concerns. Each section is intended to guide the patient to the most appropriate service for their needs, while also providing support for any questions. It is a useful resource for quick and efficient navigation within the medical system on the platform (see Fig. 12.14).

The fourth page is titled 'Services' and aims to outline several types of services for patients. These various medical services are highlighted for informational purposes, providing an overview of the available care options. The items shown in the illustration represent only a part of the services, suggesting the working approach. These

Fig. 12.14 The interface of the departments page of the medical diagnosis system

services include regular medical checks, gynecology services focusing on prevention and education, medical assistance, neurology for nervous system conditions, a pharmacy providing prescriptions and pharmaceutical advice, and a sleep centre specialising in the diagnosis and treatment of sleep disorders. This section is essential for users who want a concise understanding of the types and modes of care available (see Fig. 12.15).

The fifth page of the medical diagnosis platform called ‘Doctors’ has the role of presenting the members of the medical team who are distributed among different work departments. Each medical staff is assigned to its specialization and the user has the possibility to search in the search field for the desired department and view the available doctors or directly search for the name of the doctor he wants. The medical diagnosis platform is in a continuous expansion, because any medical framework or medical institution can be a partner in this globally spread medical system. Thus, the sphere of influence is very large and users have the opportunity to choose any doctor they want and at the same time patients can check the history of the doctor and the success rate of his treatments. The easy-to-navigate interface fosters connections between patients and physicians, making it easier to match health needs with the right expertise. This is an outstanding example of how technology can improve access to health care (see Fig. 12.16).

The sixth page in the patient diagnosis medical platform is called ‘Informations’ and presents information about the available services, including details on making medical appointments, choosing promotional packages to benefit from the reduction of medical investigation costs, specialised assistance and the generation of the complex medical diagnosis report supported by AI and based on the patient’s symptoms. The focus is on accessibility, medical expertise and advanced diagnostic support powered by AI. The page highlights the importance of quick access to high-tech medical services and efficient emergency response, providing easy appointments

**Fig. 12.15** The service page interface of the medical diagnosis system

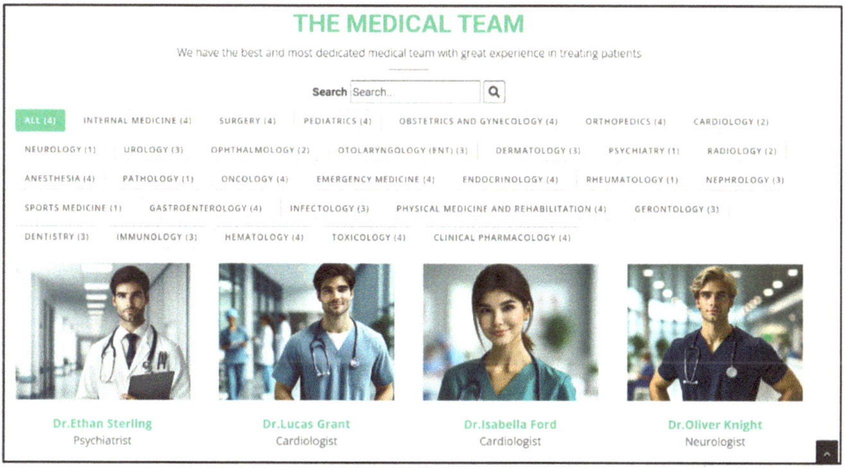


**Fig. 12.16** The interface of the medical staff page of the medical diagnosis system

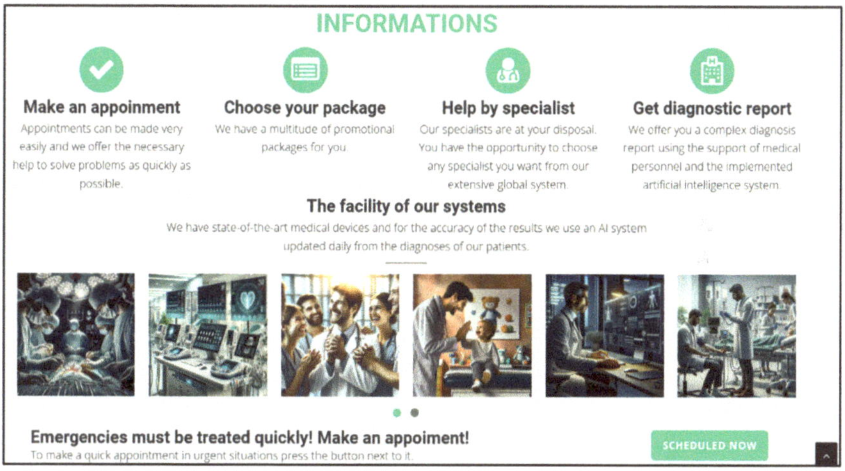


**Fig. 12.17** The interface of the information page of the medical diagnosis system

and advice to ensure the best patient care outcomes. Also, within the web page there is a prominent button that is dedicated to making quick medical appointments (see Fig. 12.17).

The seventh page of the medical diagnosis system is called the 'Contact Form' and illustrates the contact capabilities of the medical platform. On this page, visitors are provided with various methods of contact to interact with the medical team or for technical support.

There are three categories of connection via correspondence:

- **E-mail**: By filling in a form, a response is received promptly in the patient's email inbox. The user will describe the medical problem they are facing in as much detail as possible;
- **Chat with human support**: The purpose of this is to simplify interaction with the operators of the medical diagnosis platform. Through this category, the patient can get in touch with the doctor they wish to consult and communicate their health concerns. If the doctor believes a more detailed visual examination is necessary, they can use the video call integrated into the platform or schedule a physical appointment at the nearest partner medical centre, where the patient can be treated by the doctor they spoke with, or another specialist in the field;
- **Chat with AI support**: A complex AI system integrated in a chat was created to facilitate the interaction between the patient and the medical AI diagnostic system. The patient can have a conversation with the AI bot to find out the answers to any confusion in the context of the medical platform or the medical diagnosis process. This system has the ability to identify the patient's disease by collecting data about the patient's symptoms. The patient has to answer some questions asked by the AI bot to identify the disease that causes the medical problems. The accuracy in identifying diseases is very high due to the large volume of training data of the AI system.

The medical diagnostic platform provides advanced features and contact methods designed to offer users a simple and efficient experience. The platform is equipped with intuitive communication interfaces, facilitating interaction between patients and healthcare professionals. The diversity of contact options and technological adaptability underscore the commitment of this platform to respond promptly and efficiently to the needs of its users, ensuring constant support in their health management (see Fig. 12.18).

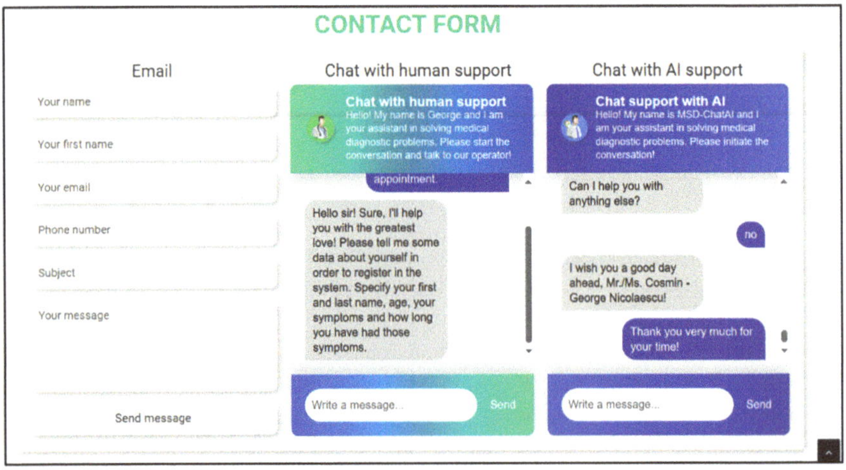


**Fig. 12.18** Contact page interface of the medical diagnosis system

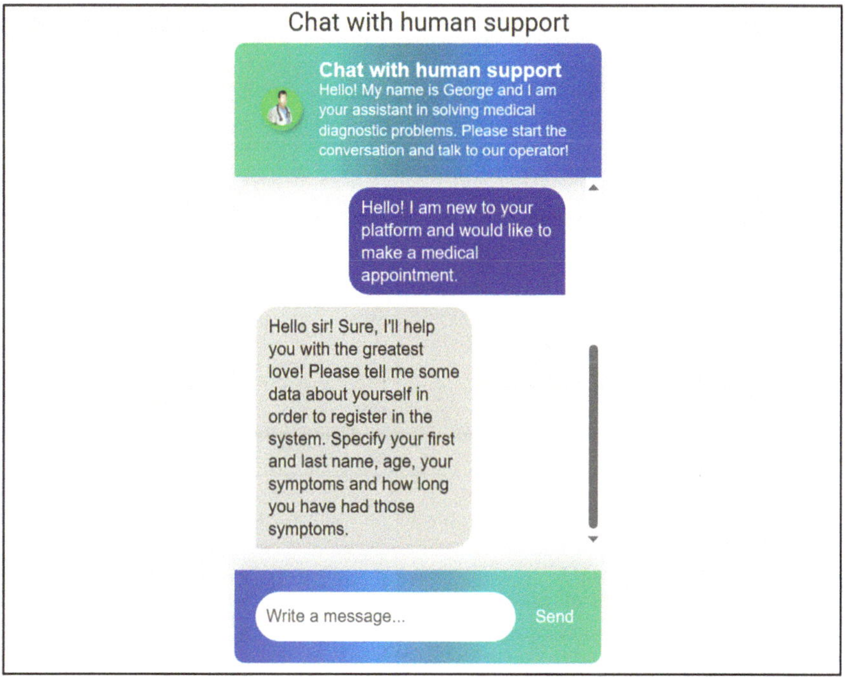


**Fig. 12.19** The chat discussion with the human operator for medical diagnosis

Next, the discussion via chat with human support between the patient and the authorised medical staff is illustrated (see Fig. 12.19).

The presentation of the conversation through the chat with AI support between the patient and the bot, which has synthetic intelligence, is illustrated below (see Fig. 12.20).

The chatbot has concluded that the patient may have Cervical Spondylosis based on a decision-making process involving a decision tree and SVM. The user entered the symptoms into the system, specified that the manifestations had been present for one day and then the chatbot utilised two classification methods, Decision Tree and SVM, to identify the most likely disease based on the stated symptoms. The Decision Tree and SVM are models trained on a medical dataset contained in the Training.csv file, which includes symptoms and their associated diseases.

In this examination, the decision tree has an accuracy of 0.9766 (97.66%) and the SVM achieves 1.0 (100%) on the test set. Thus, the values indicate that the models are rigorously trained and capable of making precise predictions.

The decision-making process of the decision tree works as follows:

1. The main symptoms that define each disease are checked;
2. It is asked whether the patient presents certain additional symptoms;
3. The patient's set of symptoms is compared with those in the database;
4. The most probable disease is estimated.

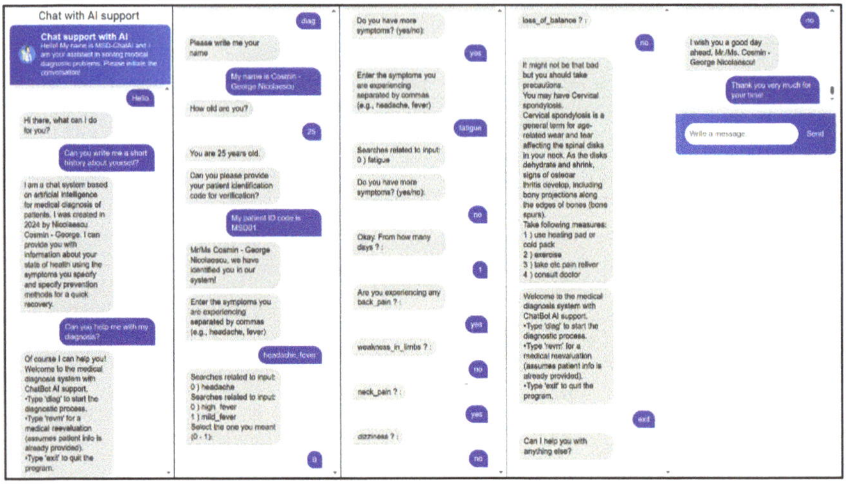


**Fig. 12.20** The chat discussion with the AI operator for medical diagnosis

After analysing these symptoms, the decision tree determined that Cervical Spondylosis is the most likely condition.

The process is continued after the decision tree has made a prediction, and the SVM model has been used for confirmation.

The SVM creates a multidimensional space in which each disease is represented as a point, and the patient's symptoms are placed within this space. This model then determines which disease is closest to the given set of symptoms. The result confirmed that the most probable disease is Cervical Spondylosis.

The calculation process is continued by verifying the severity of the disease based on Eq. (12.1):

$$Severity = \frac{\sum_i severity(symptom_i) \times days}{number\_of\_symptoms + 1} \tag{12.1}$$

where:

- *severity(symptom_i)*—is a predefined score in the symptom_severity.csv file;
- *days*—is the number of days reported by the patient;
- *number_of_symptoms*—is the number of identified confirmed symptoms.

The obtained result was below the threshold of 13, which means that it is not a medical emergency, but preventive measures are recommended.

The workflow of the chatbot is presented in the diagram below (see Fig. 12.21).

The eighth page is titled 'Payment' in the medical patient diagnosis system. The payment form for medical services is illustrated on this web page. Users can enter their details in the dedicated payment fields. Several payment methods are also indicated and accepted for the user's convenience. Patients have the freedom to make

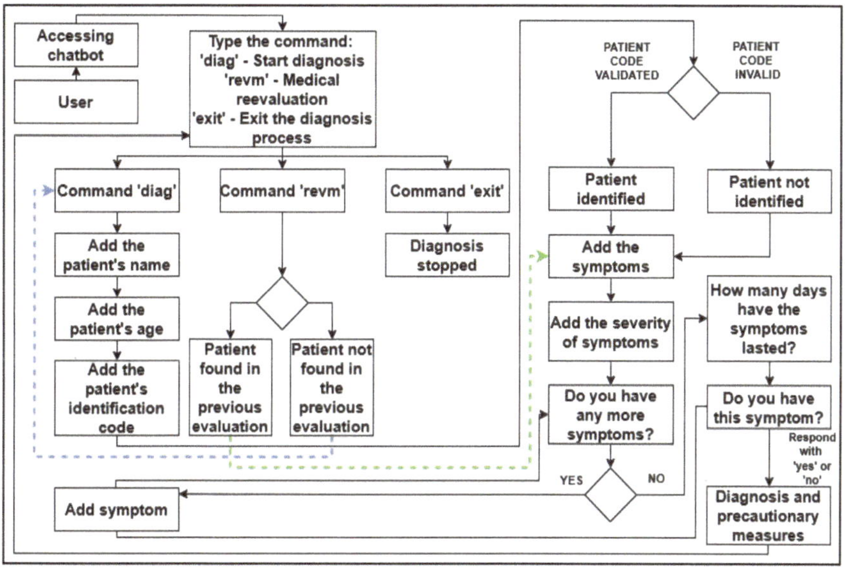


**Fig. 12.21** Chatbot usage diagram

payments via credit cards, payment orders, or cryptocurrencies, with the system supporting Ethereum (see Fig. 12.22).

The ninth page, titled 'Reviews', refers to the evaluation of patients who used the medical diagnosis platform. It is a way for new users to check if the patients already registered in the system were satisfied with the quality of medical services. Also, in this category, you can search for information about the medical personnel who treated the patients. Or for more details about specialist doctors, look for the doctor's profile on the system's web page.

On this page, positive reviews cover a wide range of services, from medical consultations to complex surgical procedures, and praise the professionalism and efficiency of the medical staff. Each review is accompanied by the patient's name and location, giving personal insight and authenticity to the comments (see Fig. 12.23).

The tenth page is titled 'Packages' and it presents various medical promotional packages offered by the platform, each with its own included services, such as consultations, anthropometric measurements, various blood tests, and specialised examinations. Prices are clearly displayed, and users can purchase the desired package directly from the platform. These offerings are created to meet varied healthcare needs and promote a healthy lifestyle. It should be noted that users can opt for personalised packages according to individual requirements (see Fig. 12.24).

The last page of the interface of the medical system for diagnosing patients is dedicated to thanking users for using the platform. Insights into the medical system and the medical centre are given, emphasising data security through blockchain architecture and the availability of human and AI assistance. Contact details and

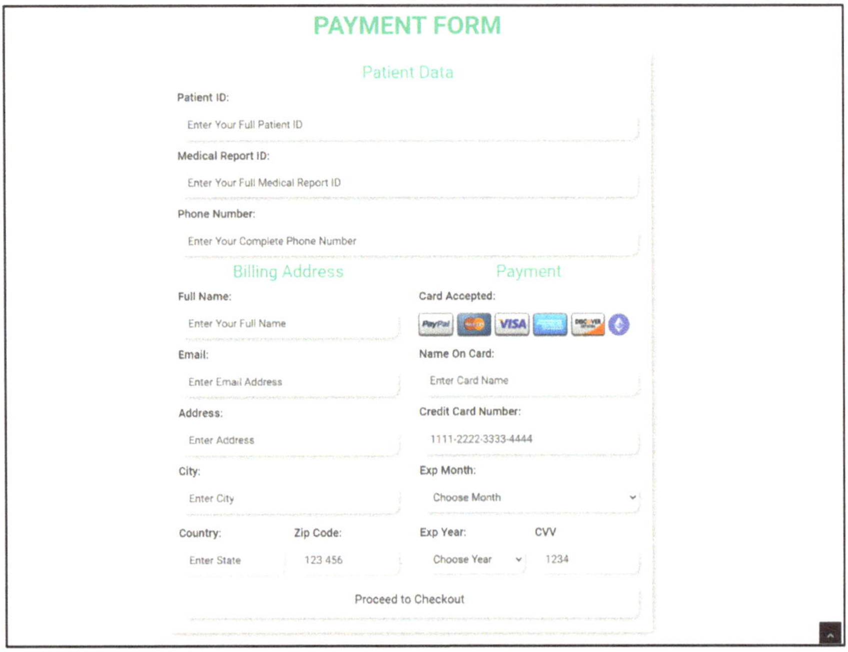


**Fig. 12.22** Interface of the payment page of the medical diagnosis system

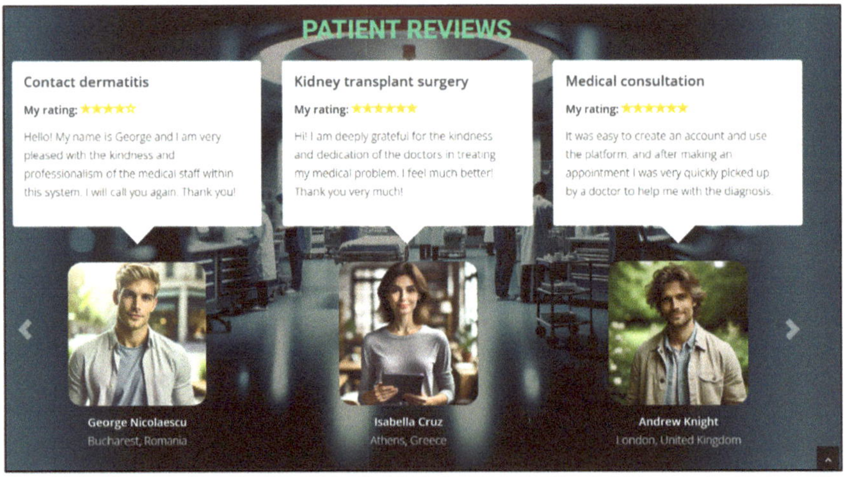


**Fig. 12.23** Interface of the reviews page of the medical diagnosis system

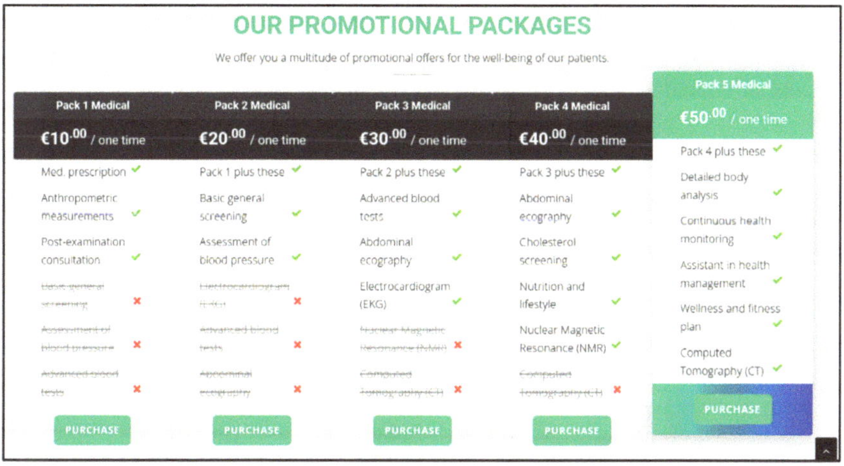


**Fig. 12.24** The interface of the promotional packages page of the medical diagnosis system

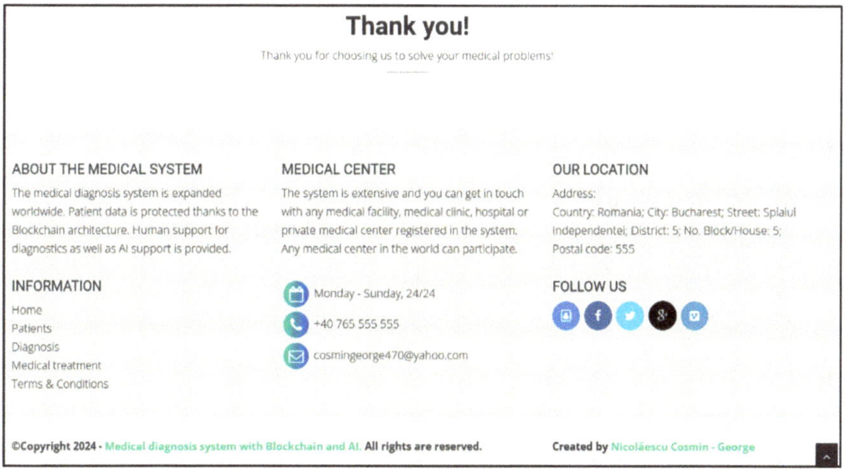


**Fig. 12.25** The interface of the promotional footer page of the medical diagnostic system

location are clearly provided, along with options to follow on social media (see Fig. 12.25).

## 12.5 Analysis of the Obtained Results

The proposed medical system combines secure registration through blockchain technology with AI diagnostic support.

The methodology for obtaining the results is based on, on one hand, the meticulous process of collecting registration information or completing the form regarding the patient's medical problems, with this procedure attributed to the blockchain technology integrated in the proposed medical system.

On the other hand, in the conversation process through the patient's use of an AI-assisted chat, the system collects the data received from the user, analyses it contextually, and extracts relevant information to provide a correct answer.

AI chat provides technical and medical support, including preliminary diagnoses and prescriptions, with the option to arrange emergency physical examinations.

Once the data enters the system, it is analysed to improve accuracy and efficiency, and, in the final phase, the result is generated. It can later be viewed as a direct response in the system or found in the patient's medical history. Various test modalities were implemented to generate a wide spectrum of output values, thus facilitating an exhaustive comparison between the obtained data pairs. Through the forms used for registration, manual entry of medical issues, or chat conversations with the AI system, the data is stored in the system, and the answer received is determined based on them.

### *12.5.1 The Results of Implementing Blockchain Technology in the Medical Diagnosis System*

The first step in realising the functionality of the system is the use of the Ganache program, which creates a private blockchain, allowing developers to deploy, test, and interact with smart contracts in a simulated environment before launching them in a live environment. It is a crucial tool for developing decentralised applications (DApps) before they are deployed on the main Ethereum network.

In the following illustration, I have chosen an address for setting up a test account that has a certain amount of virtual ETH, to carry out transactions and tests within the private network.

I also copied the code for the private key and later imported the account into my virtual wallet (see Fig. 12.26).

The next step is to use the MetaMask program which is used as a digital wallet for interacting with the Ethereum network. In the development of DApps, MetaMask allows users to perform transactions, manage Ethereum balances and connect to decentralised applications directly from the browser, facilitating the testing and deployment of smart contracts.

The following illustration shows the stages of configuring the MetaMask program to add the account generated with Ganache. At the beginning, the program is logged in to, and then, afterwards, the account is imported. As seen in the third section, the account has been accessed, and the amount of ETH is visible. Finally, in the last section, the transaction received after a user tried to create an account within the medical diagnosis system is shown (see Fig. 12.27).

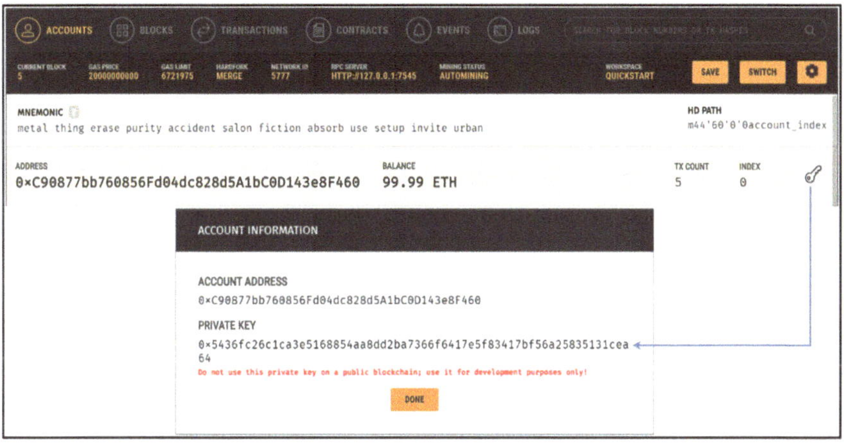


**Fig. 12.26** Private key generation process for test account setup

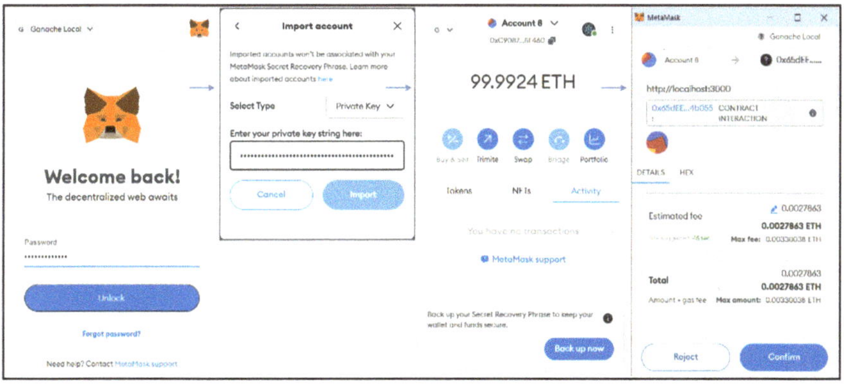


**Fig. 12.27** Configuring the virtual wallet and adding the generated private key

After registering a new account and validating it using the MetaMask application, we can also view the information in the Ganache application. In Ganache, a list of blocks created within the test blockchain network is displayed. One can see a specific transaction selected for block 5, with additional details such as the hash of the transaction, sender and recipient addresses, as well as the gas used for the contract's call.

This figure illustrates that test transactions were performed to verify the correct operation of smart contracts in the local network (see Fig. 12.28).

For the distribution of smart contracts, we used Truffle, which is part of the standard workflow for developing decentralised applications. Three smart contacts were compiled within the program:

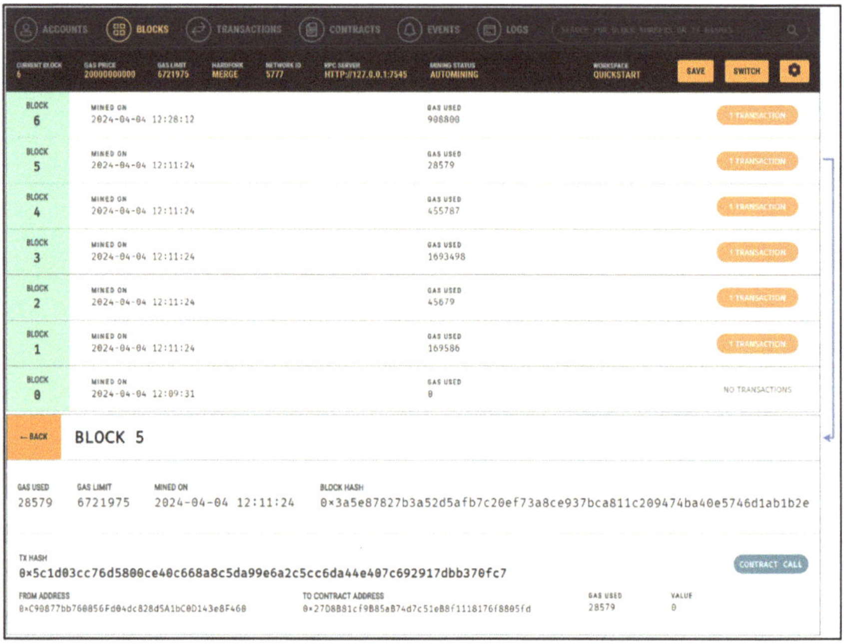


**Fig. 12.28** Display of transactions made in the account generation program

- DataMigration.sol;
- DataPatient.sol;
- DataSave.sol.

These three smart contracts, with the same specified address, have been redeployed.

After that, the contract data and the total amount of ETH consumed for the deployment of this contact are displayed (see Fig. 12.29).

Also, within this illustration, an attempt is made to replace the previous versions of the contracts and substitute them with a new instance on the blockchain. The data for the transaction and for the executor account are displayed. Finally, the total number of contracts and the total cost in ETH for all deployed deployments are displayed (see Fig. 12.30).

At this stage, the records from the system, along with only the data visible in the patients' medical diagnosis platform, are presented.

During the registration process, the data were collected and are available for viewing by the platform administrator or by authorised users who can manage the data within the system.

The figure below shows the patients registered in the medical diagnosis system, with the data presented in tabular form for easier identification (see Fig. 12.31).

```
$ truffle deploy

Compiling your contracts...
===========================
> Compiling .\contracts\DataMigration.sol
> Compiling .\contracts\DataPatient.sol
> Compiling .\contracts\DataSave.sol
> Artifacts written to C:\Users\Cosmin\Desktop\blockchain-master\...
> Compiled successfully using:
   - solc: 0.5.16+commit.9c3226ce.Emscripten.clang


Starting migrations...
======================
> Network name:    'development'
> Network id:      5777
> Block gas limit: 6721975 (0x6691b7)

1_initial_migration.js
======================

   Replacing 'DataMigrations'
   ----------------------
   > transaction hash:    0x65f1af02c76a1791c116bda3d430d4bd05a6c291edd54741...
   > Blocks: 0            Seconds: 0
   > contract address:    0x27D8B81cf9B85aB74d7c51eB8f1118176f8805fd
   > block number:        1
   > block timestamp:     1712221884
   > account:             0xC90877bb760856Fd04dc828d5A1bC0D143e8F460
   > balance:             99.99942764725
   > gas used:            169586 (0x29672)
   > gas price:           3.375 gwei
   > value sent:          0 ETH
   > total cost:          0.00057235275 ETH

   > Saving migration to chain.
   > Saving artifacts
   -------------------------------------
   > Total cost:       0.00057235275 ETH
```

Fig. 12.29 Generation of the first stage of smart contracts within the blockchain program

```
2_deploy_contracts.js
=====================

   Replacing 'DataPatient'
   -----------------------
   > transaction hash:    0xb3c5aefea035ddea75095db67d0dd1da0b369...
   > Blocks: 0            Seconds: 0
   > contract address:    0x08233bA0E79C98551D68c63E6379320860ACD72c
   > block number:        3
   > block timestamp:     1712221884
   > account:             0xC90877bb760856Fd04dc828d5A1bC0D143e8F460
   > balance:             99.993899571960035717
   > gas used:            1693498 (0x19d73a)
   > gas price:           3.17606086 gwei
   > value sent:          0 ETH
   > total cost:          0.00537865271428828 ETH

   Replacing 'DataSave'
   --------------------
   > transaction hash:    0xaee813e93ef830cfaa27d9248fa5c50472109c6d678b89dc...
   > Blocks: 0            Seconds: 0
   > contract address:    0x65dEEe61302d0a9239703242aEDEBbB867e4b055
   > block number:        4
   > block timestamp:     1712221884
   > account:             0xC90877bb760856Fd04dc828d5A1bC0D143e8F460
   > balance:             99.992471074409257811
   > gas used:            455787 (0x6f46b)
   > gas price:           3.134134038 gwei
   > value sent:          0 ETH
   > total cost:          0.001428497550777906 ETH

   > Saving migration to chain.
   > Saving artifacts
   -------------------------------------
   > Total cost:     0.006807150265066186 ETH

Summary
=======
> Total deployments:   3
> Final cost:          0.007379503015066186 ETH
```

Fig. 12.30 Generation of the second stage of smart contracts within the blockchain program

Patient data is carefully managed in the medical diagnostic system, ensuring both accessibility for authorised personnel and the integrity and confidentiality of information. To increase the level of security and ensure a backup copy of the information, the system stores patient data in an Excel register, in addition to the main database. This method reflects the platform's dedication to meticulous and responsible handling of sensitive user data. And for a faster search for a user, you can write the patient's name in the search section and check the results obtained. Thus, all users are filtered and only those whose names match the correspondence are displayed (see Fig. 12.32).

THE PATIENT'S MEDICAL DATA AFTER REGISTRATION

| No | Name | Birth Date | Phone Number | Address | Weight | Height | Blood type | Disease/Diseases Name | Description | Disease Started From |
|---|---|---|---|---|---|---|---|---|---|---|
| 1 | Cosmin-George Nicolaescu | 25/4/1998 25 years | 767555555 | Country: Romania, City: Bucharest, Street: Splaiul Independentei, | 155 | 180 | 0+ | Allergy, Migraine | Allergy: continuous sneezing, chills, watery eyes; Migraine: acidity, indigestion, headache, | 5/4/2023 0 years |
| 2 | George-Florin Nicolae | 24/4/1998 25 years | 767555555 | Country: Romania, City: Cluj-Napoca, Street: Calea Dorobanților, | 55 | 181 | B+ | Flu, Allergy, | Flu: cough, fever, sore throat, runny nose, muscle pain, headaches, fatigue, Allergy: chills, continuous sneezing, watery eyes, | 15/2/2024 0 years |
| 3 | Isabelle Dubois | 25/4/1999 24 years | 767555555 | Country: France, City: Paris; Street: Rue Dussoubs, | 65 | 168 | AB+ | Allergy; | Allergy: watery eyes, continuous sneezing, chills, | 15/12/2023 0 years |
| 4 | John Smith | 25/1/1998 26 years | 767777555 | Country: United States, City: Washington, D.C., Street: Independence Avenue, | 88 | 185 | 0+ | Chickenpox, | Chickenpox: irritations, skin rashes, itching, mild fever, fatigue, | 22/2/2024 0 years |
| 5 | Avi Cohen | 11/1/1998 26 years | 767555000 | Country: Israel; City: Jerusalem, Street: King George Street: | 59 | 167 | A+ | Asthma, | Asthma: wheezing, coughing (especially at night), rapid breathing, pressing sensation, | 5/4/2022 1 years |

**Fig. 12.31** The interface for viewing patients registered in the medical diagnosis system

| No. | Name | Birth Date | Phone Nam | Address | Weight | Height | Blood Type | Disease/Diseases | Description | Disease Started From |
|---|---|---|---|---|---|---|---|---|---|---|
| 1 | Cosmin-George Nicolaescu | 25/4/1998 25 years | 767555555 | Country: Romania; | 155 | 180 | 0+ | Allergy; Migraine; | Allergy: cont | 5/4/2023 0 years |
| 2 | George-Florin Nicolae | 24/4/1998 25 years | 767555555 | Country: Romania; | 55 | 181 | B+ | Flu; Allergy; | Flu: cough, fe | 15/2/2024 0 years |
| 3 | Isabelle Dubois | 25/4/1999 24 years | 767555555 | Country: France; Ci | 65 | 168 | AB+ | Allergy; | Allergy: wate | 15/12/2023 0 years |
| 4 | John Smith | 25/1/1998 26 years | 767777555 | Country: United St | 88 | 185 | 0+ | Chickenpox; | Chickenpox: | 22/2/2024 0 years |
| 5 | Avi Cohen | 11/1/1998 26 years | 767555000 | Country: Israel; Cit | 59 | 167 | A+ | Asthma; | Asthma: whe | 5/4/2022 1 years |

SEARCH George

| No. | Name | Birth Date | Phone Nam | Address | Weight | Height | Blood Type | Disease/Diseases | Description | Disease Started From |
|---|---|---|---|---|---|---|---|---|---|---|
| 1 | Cosmin-George Nicolaescu | 25/4/1998 25 years | 767555555 | Country: Romania; | 155 | 180 | 0+ | Allergy; Migraine; | Allergy: cont | 5/4/2023 0 years |
| 2 | George-Florin Nicolae | 24/4/1998 25 years | 767555555 | Country: Romania; | 55 | 181 | B+ | Flu; Allergy; | Flu: cough, fe | 15/2/2024 0 years |

**Fig. 12.32** Extracting and copying patient data into a separate file

## 12.5.2 *The Results of the Implementation of AI Technology in the Medical Diagnosis System*

In the second step of system development, we focus on strengthening the interaction between AI and the user through the intuitive AI-supported chat platform. In realising the functionality of the system, we use the chat platform with AI support to create a conversation scenario between the AI system and a patient. The simulated conversation scenario serves as the mechanism through which the AI system collects essential information by interrogating the patient in a structured and accessible way. This interactive dialogue makes it easy to capture the patient's symptoms and medical history with increased accuracy.

Moreover, the process of analysing the AI system's responses highlights the effectiveness of the implemented machine learning algorithms, as reflected in performance scores and metrics. Therefore, we ensure that the system not only meets clinical standards for diagnosis but also continuously improves its ability to understand and evaluate various medical cases, adapting to the individual needs of each patient.

This vital stage of verification and fine-tuning of the AI system emphasises the system's effectiveness in achieving high diagnostic accuracy, patient safety, and improving the user experience, ensuring the system remains at the cutting edge of medical technology innovation.

At the heart of the functionality of the AI medical diagnosis system are powerful machine learning algorithms, including the Decision Tree classifier and the Support Vector Classifier (SVC).

The Decision Tree classifier proves its effectiveness through its ability to model clinical decisions and logically infer reasoning leading to a diagnosis, similar to how a physician might analyse presenting symptoms. On the other hand, SVC contributes its ability to distinguish between conditions with similar symptoms, thus providing increased accuracy in the prediction of complex diagnoses.

These algorithms are trained on large and diverse datasets, allowing them to detect subtle nuances and patterns in clinical data. During the interaction with the patient through the chat platform, the system uses these patterns to evaluate the information and provide a preliminary diagnosis along with the appropriate recommendations. Through continuous analysis of results and feedback, the AI system refines its predictive models, ensuring that each new case enriches its overall understanding and diagnostic capability.

To achieve the implementation of this medical chatbot, the use of the Flask, PyTorch, and Scikit-Learn frameworks was necessary. These frameworks enable users to describe symptoms and receive an estimated diagnosis based on machine learning. The programme utilises both artificial neural networks and a decision tree to make predictions based on symptoms.

All the technologies used within the chat system are:

- Flask—a framework for the backend and web API;
- Flask-CORS—enables interactions with other domains;
- PyTorch—used for training a neural network that understands user queries;
- Scikit-Learn—utilised for the classification model based on decision tree and SVM;
- NLTK—natural language processing, tokenisation, and stemming;
- Pandas & NumPy—for data manipulation;
- Pyttsx3—for text-to-speech voice generation;
- CSV—for reading files containing data on symptoms, severity, and precautions.

The general workflow of the application starts with training the model using the train.py file. Within this file, a dataset containing information about symptoms and diseases from intents.json is processed. Texts are converted into vectors using the bag of words approach. The neural network is then trained with PyTorch based on this data. Finally, at the end of the training process, the model's results are saved in the data.pth file for future use.

The interaction with the user is handled in the app.py file, where the application runs on a Flask server that receives messages through an endpoint. The messages are then processed and interpreted by the neural network, and the responses are sent to the user in JSON format.

In the final stage of this general workflow, the medical diagnosis is carried out in the chat_bot.py file. Based on the symptoms entered by the user, the chatbot applies a decision tree to estimate the illness. An SVM model is also used to improve the accuracy of the diagnosis. Subsequently, recommendations and precautions are assigned to the patient from a medical dataset.

Within the project, various mathematical formulas were used for data processing, model training, and symptom classification, as shown below.

1. The Rectified Linear Unit activation function in the neural network:
   The Rectified Linear Unit (ReLU) is used to introduce non-linearity into the model. It is defined in Eq. (12.2):

   $$ReLU(x) = \max(0, x) \tag{12.2}$$

   This means that:

   - If x is positive or 0, the value remains the same;
   - If x is negative, it is transformed into 0.

   This function is applied in the hidden layers of the neural network.
2. The CrossEntropyLoss loss function:
   For training the model, this loss function is used as it measures how accurately the model performs, penalising incorrect predictions. It is defined in Eq. (12.3):

   $$L = -\sum_i y_i \log(\hat{y}_i) \tag{12.3}$$

   where:

   - $y_i$ is the actual label (1 for the correct class and 0 otherwise);
   - $\hat{y}_i$ is the probability predicted by the model for the respective class.

   This function evaluates how well the model predicts the correct classifications.
3. The Gradient Descent training algorithm:
   The model is optimised using the Adam Optimiser, which is based on gradient descent and helps adjust the model's weights. Thus, it is defined as follows in Eq. (12.4):

$$\theta = \theta - \eta \nabla J(\theta) \quad (12.4)$$

where:

- $\theta$ represents the model parameters (weights, biases);
- $\eta$ is the learning rate (set at 0.001);
- $\nabla J(\theta)$ is the gradient of the loss function.

4. Measuring the impurity of the decision tree using Entropy:
   The decision tree uses entropy to determine how pure a node is. It is presented below in Eq. (12.5):

$$H(S) = -\sum p_i log_2(p_i) \quad (12.5)$$

   where:

   - $p_i$ represents the probability of a class in the set S.

   Thus, if H(S) is low, this means that the node is pure and has a clearer classification.
5. Data normalisation with Label Encoding:
   To convert textual data into numerical values, normalisation is used, as presented in Eq. (12.6):

$$y = LabelEncoder(y) \quad (12.6)$$

   This converts each label into a numerical value to be used in machine learning models.
6. Creation of the Bag of Words vector:
   To convert the user's sentences into a numerical format, this model, presented in Eq. (12.7), is used:

$$vector = [w_1, w_2, w_3, w_4, w_5, \ldots, w_n] \quad (12.7)$$

   where:

   - $w_i = 1$ if the word i appears in the sentence;
   - $w_i = 0$ otherwise.

   This vector is used as input for the neural network.

The presented image shows a dialogue section with the AI-based healthcare system. In this session, the user provides the symptoms they are experiencing and the AI system, by analysing and processing this information, offers a preliminary diagnosis and recommendations for further steps. The interface and dialogue illustrate a practical application of AI in healthcare, demonstrating how the technology can assist in the diagnostic process and guide patients towards appropriate next steps (see Fig. 12.33).

```
0.9735274524649129
for svm:
1.0
-----------------------------------------------------------------------------------------

Welcome to Chat with AI support for medical diagnosis
Type 'diag' to start the diagnostic process.
Type 'revm' for a medical reevaluation (assumes patient info is already provided).
Type 'exit' to quit the program.
Enter your command: -> diag
----------------------Chat with AI support for medical diagnosis-----------------------

Please write me your name -> Cosmin - George Nicolaescu
Hello, Mr/Ms  Cosmin - George Nicolaescu
How old are you? -> 25
You are 25 years old.
Can you please provide your patient identification code for verification? -> MSD01
Mr/Ms Cosmin - George Nicolaescu, we have identified you in our system!

Enter the symptoms you are experiencing separated by commas (e.g., headache, fever) -> chills
Searches related to input:
0 ) chills
Do you have more symptoms? (yes/no): yes

Enter the symptoms you are experiencing separated by commas (e.g., headache, fever) -> vomiting,acidity
Searches related to input:
0 ) vomiting
Searches related to input:
0 ) acidity
Do you have more symptoms? (yes/no): no
Okay. From how many days ? : 1
Are you experiencing any
back_pain ? : no
weakness_in_limbs ? : yes
neck_pain ? : no
dizziness ? : yes
loss_of_balance ? : yes
It might not be that bad but you should take precautions.
You may have Cervical spondylosis
Cervical spondylosis is a general term for age-related wear and tear affecting the spinal disks in your
neck. As the disks dehydrate and shrink, signs of osteoar thritis develop, including bony projections
along the edges of bones (bone spurs).
Take following measures :
1 ) use heating pad or cold pack
2 ) exercise
3 ) take otc pain reliver
4 ) consult doctor

Welcome to Chat with AI support for medical diagnosis
Type 'diag' to start the diagnostic process.
Type 'revm' for a medical reevaluation (assumes patient info is already provided).
Type 'exit' to quit the program.
Enter your command: -> exit
Thank you for using Chat with AI support for medical diagnosis. Goodbye!
-----------------------------------------------------------------------------------------
```

**Fig. 12.33** Simulating a dialogue with the AI system for the medical diagnosis of the patient

The following figures illustrate the responses of the medical diagnosis system in different scenarios for analysing the results.

The first graph from the top left is a confusion matrix and is used to evaluate the performance of the classification model. The numbers on the vertical axis represent the true class (True Class), while the numbers on the horizontal axis represent the predicted classes (Predicted Class). The values along the main diagonal, which are highlighted in blue, indicate the number of correct predictions for each class. The boxes that are not positioned on the diagonal represent the number of wrong classifications that occurred.

In the perfect confusion matrix, all incorrect predictions are zero, and only the values on the diagonal are greater than zero. In this specific case, there is a large number of classes, over 40, which indicates that the model performs multiclass classification.

If the numbers on the diagonal are relatively large compared to those off the diagonal, the model correctly predicts most classes, and the strong diagonal is a good indication of performance.

The second graph from the top right is a Receiver Operating Characteristic (ROC) curve for the multiclass classification model. The ROC curve is a tool for evaluating the performance of binary classification models. On the horizontal axis we have the False Positive Rate (FPR), which represents the proportion of negative examples that are wrongly classified as positive. On the vertical axis we have the True Positive Rate (TPR), also known as sensitivity, which measures the proportion of real positive examples that are correctly identified. The dotted line illustrates a guideline indicating the performance of a completely random classifier. If the ROC curve follows this line, then the model has no ability to discriminate between the positive and negative classes. The solid line in the upper-left corner represents the model's performance. In this case, all the curves of the classes within the model are located in the upper-left corner, overlapping, which represents a very good performance of the model.

The third graph from the bottom left is a horizontal bar chart and displays the relative importance of the first 10 features in the predictive model. This form of visualization is often used to show the importance of each feature in the context of a machine learning model, such as a decision tree-based classifier. The horizontal bars represent different features, with their names displayed on the vertical (Y) axis. The length of each bar indicates the relative importance of that feature in the prediction model. Importance is measured on the horizontal (X) axis and is usually calculated based on the influence a feature has in improving the model's predictions, the more a feature contributes to the model, the more important it is considered to be.

In this particular graph, it can be seen that „ulcers_on_tongue" appears to be the most important feature, having the largest bar and the highest relative importance.

The fourth graph from the bottom right is of the vertical bar type and aims to show the Average Precision Score per class, which represents the average precision score for each class in a classification model. Each bar corresponds to a different class, labeled on the horizontal (X) axis, and the height of the bar indicates the average accuracy score for that class, represented on the vertical (Y) axis.

The average accuracy score is an indicator of the classification quality of the model for each class. It reflects the ability of the model not to classify negative labels as positive. A score of 1.0 indicates perfect accuracy, where all positive predictions are correct, while a lower score reflects a higher rate of false positives.

In the graph shown, all classes have an accuracy score very close to 1.0, suggesting that the model performs exceptionally well for all individual classes (see Fig. 12.34).

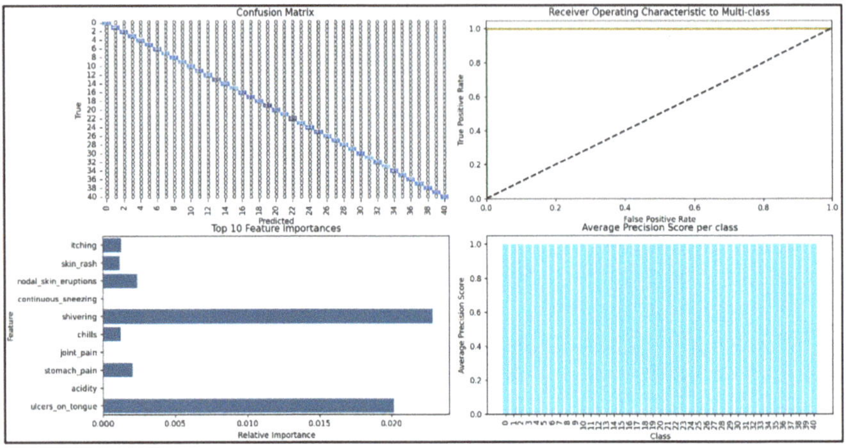


**Fig. 12.34** The first stage of the analysis graphs of the system response in different scenarios

In this figure, other outputs of the medical diagnosis system are illustrated in other scenarios for results analysis.

The first graph from the top left represents a validation curve, which illustrates how the performance of the automatic learning model varies with the change of a parameter of the model, in this case "Max Depth", for a decision tree. The horizontal axis, named "Max Depth", shows different values for the maximum depth of the decision tree, which is a hyperparameter that controls the complexity of the model. A higher maximum depth allows the model to learn finer details from the data, but may also lead to overfitting. The vertical axis represents the accuracy of the model, with values that vary between 0 and 1. An accuracy of 1 indicates maximum performance. The blue line represents the accuracy score on the training set and changes as the maximum depth of the decision tree increases. The orange line represents the accuracy score obtained by cross-validation and indicates the performance of the model on datasets that the model has not seen during training, providing a better estimate of performance on new data. In this image, the lines are very close, indicating a consistency between the model's performance on the training and validation sets.

The second graph in the top-right corner illustrates a learning curve that represents how the machine learning model's accuracy score changes as the number of training examples increases. The horizontal axis indicates the number of training examples, while the vertical axis shows the score of the model, usually measured by accuracies. There are two lines that represent two types of scores. The red line with circle markers shows the score on the training set and indicates how well the model fits the training data. The green line with circle markers represents the cross-validation score and shows indications of the model's performance on data it did not see during training, which may be a better indicator of the model's generalization ability. In the given graph, it is observed that both scores are quite high and converge to a value close to 1, which indicates good performance and good generalization ability of the model.

The third graph from the left middle represents a correlation matrix for the first 10 characteristics, which is a common technique in statistical analysis to evaluate linear relationships between variables. The correlation values are between $-1$ and $+1$. A value of $+1$ indicates a perfect positive correlation and a value of $-1$ indicates a perfect negative correlation, and 0 indicates the absence of correlation. Values on the diagonal are always 1, because each feature has a perfect correlation with itself. Positive values mean that as one characteristic increases, the other characteristic also tends to increase. Negative values indicate that as one characteristic increases, the other tends to decrease. This information is valuable for understanding how different symptoms or clinical signs may be associated and can help in building and interpreting predictive models.

The fourth graph from the middle right illustrates a histogram and is used to visualize the frequency distribution of the probabilities predicted by the classification model. On the horizontal axis (X), we have „Predicted Probability", which represents the probability that a certain observation belongs to a specific class, as predicted by the model. On the vertical axis (Y), we have „Frequency", which indicates how many observations fall in each probability interval. This distribution suggests that the model is likely to be overly cautious or pessimistic about classifying observations as belonging to the positive class. This may be indicative of a very specific pattern in the cases it classifies as positive.

The fifth graph from the bottom left represents a violin plot, which is used to represent the distribution of a characteristic according to different classes. This type of plot combines elements of the distribution box with a probability density estimate. The horizontal axis (X) shows different classes, numbered from 0 to 40 distinct classes. The vertical (Y) axis indicates the value of the „itching" feature, perhaps a score or probability, from 0 to 1. Each „violin" represents the distribution of the „itching" feature values for a specific class. The wider side of the „violin" indicates where the data is more concentrated, in other words, where there is a higher frequency of observations. The blue points represent the medians of the data for each class. The graph shows that for many classes, itching has a distribution with generally small values, but with significant variation for certain classes, indicating that itching may be a more prominent or relevant symptom for some conditions than for others. For example, for some classes, the distribution is wider and the medians are higher, suggesting that itching is a more common symptom for those conditions. While for other classes, the „violin" is very thin or even absent, indicating that „itching" is not a reported symptom or is not relevant for those classes.

The sixth graph from the bottom right illustrates the error band and is used to measure the accuracy of a machine learning model's predictions. The blue line represents the actual or current values of the dataset. The orange line illustrates the model's predictions. The brown area represents the confidence interval around the model's predictions. The horizontal axis (X) indicates the order of the observations in the dataset. The vertical axis (Y) shows the observed values and the predicted values. The error bar plot can be used to evaluate how well the model predictions match the actual data. Areas where the orange prediction line closely aligns with the blue line of actual values indicate a good fit (see Fig. 12.35).

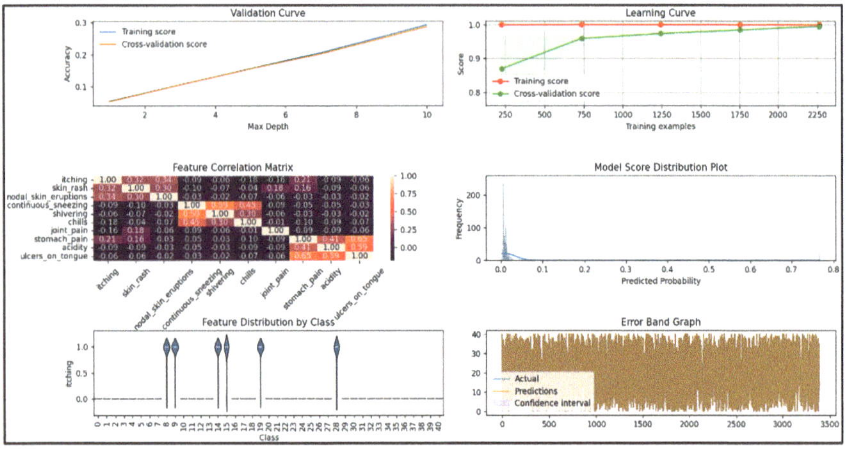


**Fig. 12.35** The second stage of the analysis graphs of the system response in different scenarios

## 12.6 Conclusions and Future Research and Development Directions

The research incorporates blockchain technology and AI in the creation of a medical system for diagnosing patients. The developed medical diagnostic system represents an innovative fusion between blockchain technology and AI, providing a digitisation solution for people's health, prioritising security, accessibility and accuracy. The integration of blockchain technology not only fortifies the privacy and integrity of medical data, but also facilitates a variety of payment options, expanding access to medical services through a modern and convenient method.

Blockchain technology was used to increase security, but also to harness the benefits of Ethereum-based cryptocurrencies. Thus, users have the opportunity to pay for medical services through different methods such as: payment by bank card, payment order, or Ethereum-based cryptocurrencies.

The AI integrated into the system plays a crucial role in facilitating a fast and efficient diagnostic process based on patient symptoms and chat-based interaction. The diagnosis of patients is carried out either by specialised medical staff, by the AI system that identifies the patient's disease based on their symptoms, or by using both methods to increase the accuracy of the diagnosis. The AI utility serves as a first point of triage and consultation, suggesting potential directions for investigation or action and the possibility of escalation to emergency medical services if appropriate. By implementing a continuous feedback cycle, the data collected are analysed to refine the system's accuracy and performance, resulting in accurate medical recommendations that can be accessed directly in the system interface or integrated into the patient's electronic medical record. Extensive and diversified testing ensures the validation and calibration of AI algorithms, in order that the range of diagnoses and medical recommendations is as comprehensive as possible and adapted to the needs of each patient. Data entry through digital forms and interaction with AI help build a detailed medical profile, allowing the system to provide personalized and well-informed answers.

The web interface platform simplifies users' access to medical services and improves their experience through various communication channels, including email and live chat. This technology, combined with the expertise of medical personnel, increases the degree of confidence and accuracy of the diagnosis. The interface has three contact methods: by e-mail, by chat with a human operator and by chat with an AI agent. The system also has a video call to create a connection as close as possible to reality with the medical personnel. However, in the case of a medical emergency, the patient is advised to travel in person to the nearest partner medical centre for a more detailed analysis of their health status.

The data collection and analysis process is meticulous and efficient, allowing the AI system to provide contextual and relevant answers that are essential in formulating a preliminary diagnosis. In addition, the system provides the option to initiate physical consultations for situations that require immediate medical attention.

In the final stages, the data is subjected to in-depth analysis, resulting in continuous improvement of the system and ensuring the accuracy of diagnoses. Various testing allowed the system to be calibrated to respond to a wide spectrum of medical conditions. The results are then presented to the user in an accessible way, contributing to a complete information and strengthening of the medical history of each patient.

Essentially, this platform innovates in the field of medical technology, bridging the gap between human healthcare capability and technological efficiency to deliver better and more affordable healthcare.

Based on a robust and flexible architecture, the medical diagnostic system has a great potential for development. Continued training with expanded datasets will allow the system to encompass a wider medical spectrum and recognise a greater variety of conditions. Although, in its current form, the system demonstrates remarkable performance, it is recognised that it is currently limited to the domains in which it has been trained. To reach a comprehensive coverage of diagnostic possibilities, it is essential to continuously provide AI with updated and detailed medical information.

Expanding the system globally and integrating it into a network of diverse medical partners will not only increase diagnostic accuracy, but also enhance its relevance in different cultural and epidemiological contexts. Collaboration with medical institutions and specialists in various fields will enrich the system's knowledge base, thus strengthening the platform as a global healthcare tool.

The proposed medical diagnostic system sets a high standard in digital health technology and foreshadows a new era where technology and human medical expertise come together to provide more efficient and affordable care services worldwide.

## References

1. Gurzhii A, Islam AKMN, Haque AKMB, Marella V (2022) Blockchain enabled digital transformation: a systematic literature review. IEEE Access 10:79584–79605
2. Butt GQ, Sayed TA, Riaz R, Rizvi SS, Paul A (2022) Secure healthcare record sharing mechanism with blockchain 12(5):2076–3417
3. Lee JS, Chew CJ, Liu JY, Chen YC, Tsai KY (2022) Medical blockchain: data sharing and privacy preserving of EHR based on smart contract. J Inf Secur Appl 65(C)
4. Smith A, Johnson B, Williams C (2023) Enhancing transparency in AI-driven healthcare: a blockchain-based framework for explainable AI (XAI) in medical diagnostics. J Med Inform 145(B):215–230
5. Marichamy VS, Natarajan V (2023) Blockchain based securing medical records in big data analytics. Data Knowl Eng 144
6. Enescu FM, Bizon N, Cirstea A, Stirbu C (2018) Blockchain technology applied in health the study of blockchain application in the health system (I). In: 2018 10th international conference on electronics, computers and artificial intelligence (ECAI), Iasi, Romania, pp 28–30
7. Angraal S, Krumholz HM, Schulz WL (2017) Blockchain technology: applications in health care
8. Tanwar S, Parekh K, Evans R (2020) Blockchain-based electronic healthcare record system for healthcare 4.0 applications. J Inf Secur Appl 50
9. Ghosh PK, Chakraborty A, Hasan M, Rashid K, Siddique AH (2023) Blockchain application in healthcare systems: a review. Systems 11(1):8
10. Haleem A, Javaid M, Singh RP, Suman R, Rab S (2021) Blockchain technology applications in healthcare: an overview. Int J Intell Netw 2:130–139
11. Susan DL (2015) Watson will see you now: a supercomputer to help clinicians make informed treatment decisions. Clin J Oncol Nurs CJON 19:31–32
12. Somashekhar SP, Sepulveda MJ, Norden AD, Rauthan A (2017) Early experience with IBM Watson for oncology (WFO) cognitive computing system for lung and colorectal cancer treatment. J Clin Oncol 35:8527–8527
13. Salcedo J, Rosales M, Kim JS, Nuno D, Suen S, Chang AH (2021) Cost-effectiveness of artificial intelligence monitoring for active tuberculosis treatment: a modeling study
14. Mueen U (2021) Blockchain Medledger: Hyperledger fabric enabled drug traceability system for counterfeit drugs in pharmaceutical industry. Int J Pharm 597:15
15. Syeda-Mahmood T, Walach E, Beymer D, Gilboa-Solomon F, Moradi M, Kisilev P, Kakrania D, Compas C, Wang H, Negahdar R, Cao Y, Baldwin T, Guo Y, Gur Y, Rajan D, Zlotnick A, Rabinovici-Cohen S, Ben-Ari R, Guy A, Prasanna P, Morey J et al (2016) Medical sieve: a cognitive assistant for radiologists and cardiologists. In: SPIE medical imaging, San Diego, California, United States
16. Ricotta F, Jackson B, Henry T, Webb J, Huber B, Hartley A, Haselhorst D, Gionfriddo M, Yablon B (2017) Bringing health to life. BurstIQ. https://www.burstiq.com/
17. Meskó B, Görög M (2020) A short guide for medical professionals in the era of artificial intelligence. NPJ Digit Med
18. Shuaib K, Saleous H, Shuaib K, Zaki N (2019) Blockchains for secure digitized medicine. J Pers Med 9(3):13
19. Yun D, Chen W, Wu X, Ting DSW, Lin H (2020) Blockchain: chaining digital health to a new era. Ann Transl Med
20. Han Y, Zhang Y, Vermund SH (2022) Blockchain technology for electronic health records. Int J Environ Res Public Health
21. Tagde P, Tagde S, Bhattacharya T, Tagde P, Chopra H, Akter R, Kaushik D, Rahman MH (2021) Blockchain and artificial intelligence technology in e-Health 28:52810–52831
22. J. Li, P. Carayon: Health Care 4.0: A Vision for Smart and Connected Health Care. Frontiers in Digital Health, 15 February 2021.
23. Yongjoh S, So-In C, Kompunt P, Muneesawang P, Morien R (2021) Development of an internet-of-healthcare system using blockchain. IEEE Access 9:113017–113031

24. Sikandar Ali A, Armand TPT, Athar A, Hussain A, Ali M, Yaseen M, Joo M-I, Kim H-C (2023) Metaverse in healthcare integrated with explainable AI and blockchain: enabling immersiveness, ensuring trust, and providing patient data security. Sensors 23(2):565
25. Sarker B, Sharif NB, Rahman MA, Parvez AHMS (2023) AI, IoMT and blockchain in healthcare. J Trends Comput Sci Smart Technol 5(1):30–50
26. Kim SY, Kim DH, Kim MJ, Ko HJ, Jeong OR (2024) XAI-based clinical decision support systems: a systematic review. Appl Sci 14(15):6638
27. Muhammad D, Bendechache M (2024) Unveiling the black box: a systematic review of explainable artificial intelligence in medical image analysis. Comput Struct Biotechnol J 24:542–560
28. Kitsios F, Stefanakakis S, Kamariotou M, Dermentzoglou L (2023) Digital service platform and innovation in healthcare: measuring users' satisfaction and implications. Electronics 12(3):28
29. Bahadori M, Teymourzadeh E, Ravangard R, Nasiri A, Raadabadi M, Alimohammadzadeh K (2016) Factors contributing towards patient's choice of a hospital clinic from the patients' and managers' perspective 8(5):2378–2387
30. Paul M, Maglaras L, Ferrag MA, Almomani I (2023) Digitization of healthcare sector: a study on privacy and security concerns. ICT Express 9(4)571–588
31. Jingquan L (2023) Security implications of AI chatbots in health care. J Med Internet Res
32. Müller R, Klemmt M, Koch R, Ehni H-J, Henking T, Langmann E, Wiesing U, Ranisch R (2024) That's just future medicine—a qualitative study on users' experiences of symptom checker apps
33. Enescu FM, Cosmin-George N, Bizon N (2023) Smart system using blockchain for medical diagnosis. In: 2023 15th international conference on electronics, computers and artificial intelligence (ECAI), Bucharest, Romania
34. Kim S-I, Kim S-H (2020) E-commerce payment model using blockchain. J Ambient Intell Humlzed Comput 13:1673–1685
35. Krishnan G, Singh S, Pathania M, Gosavi S, Abhishek S, Parchani A, Dhar M (2023) Artificial intelligence in clinical medicine: catalyzing a sustainable global healthcare paradigm. Front Artif Intell 6:29

# Chapter 13
# Explainable Spatio-Temporal Prioritisation for Mobilising High-Demand and Constrained Resources

**Anupam Khan and Soumya K. Ghosh**

**Abstract** The effective mobilisation of inadequate resources is always a challenging task for the administrators especially when it is in high demand during situations like disasters, medical emergencies etc. They usually try to prioritise the most vulnerable areas in such situations. However, they often face challenges in identifying these areas as multiple factors may define the vulnerability. Interestingly, these factors may change their behaviour over time, thereby adding more complexities into the overall process. Sometimes the lack of transparency makes the beneficiaries impatient in such situations as well. In this work, we have presented an explainable spatio-temporal platform, empowered with the novel wheel based clustering techniques, to address these challenges. As per the availability of resources, it facilitates prioritising a subset of administrative regions based on multiple factors and also explains the reasons thereof. In order to validate the effectiveness of the proposed framework, we have conducted an experiment with Covid-19 dataset of India. The proposed platform is able to prioritise vaccination in the areas with 70% caseload during May 2021, when it was struggling with the peak of second wave. Interestingly, these regions constitute only 40% of the national population.

**Keywords** Spatio-temporal prioritisation · Explainable AI · Random wheel · Resource mobilisation · Disaster management

---

A. Khan (✉) · S. K. Ghosh
Computer Science and Engineering, Indian Institute of Technology Kharagpur, Kharagpur, India
e-mail: anupamkh@iitkgp.ac.in

S. K. Ghosh
e-mail: skg@cse.iitkgp.ac.in

 
N. Bizon and B. Appasani (eds.), *Explainable Artificial Intelligence for Trustworthy Decisions in Smart Applications*, https://doi.org/10.1007/978-3-031-97007-8_13

## 13.1 Introduction

> According to the World Economic Forum, in the past 20 years, 7348 disasters have been recorded. These disasters have taken 1.23 million lives. China alone has reported 577 disasters, followed by the US which has 467 and India which has 321—India Today, December 2021 [1].

Any natural disaster can be a catastrophic event that may cause significant loss or damage of life and property [2–4]. During a flood, drought, cyclone, heavy snowfall, or pandemic, the appropriate relief team usually take the initial responsibility to address the situation and reach out to the maximum number of people requiring assistance. However, they often face challenges in relief work due to the limited availability of life-saving resources such as relief materials, medicines, volunteers, or vaccines [5–8]. A planned distribution can help them optimising the resources even in a constrained condition. The administrators often try to identify the magnitude of the damages in various affected regions and prioritise them accordingly for mobilising life-saving resources [9]. However, measuring the damage is a multivariate problem as different factors can define the vulnerability of the affected regions. The recent Covid19 pandemic, for example, has created a similar situation. The initial supplies were insufficient to cater to the high vaccine demand. At the same time, some regions were going through high mortality rates, may be due to the unavailability of sufficient healthcare facilities, while some others reported a high number of infected cases.

Besides, the dynamics of the affected regions may change over time in some disasters which may further complicate the decision-making process [10]. Therefore, the administrators must continually reassess the situation and change their strategies based on the evolving dynamics at regular intervals. In addition to this, another notable challenge in resource mobilisation is the lack of transparency, which can exacerbate the impatience of beneficiaries [2]. Affected communities expect swift action at the time of crisis, and impatience can naturally arise if there is a lack of clear communication regarding the allocation process or delays in resource distribution [11]. Significantly, transparency could be a crucial factor in such situations that can help in building trust and maintaining open lines of communication between administrators and vulnerable populations. By explaining the decision-making process and engaging with the affected population, administrators can address their concerns and alleviate some of the impatience that may arise [2]. However, they may require a decision support system for this purpose, that can help in enhancing their ability to mobilise resources efficiently [12].

In literature, several works have in fact proposed the utilisation of machine intelligence for effectively carrying out relief works during disasters. A study [8] exemplifies this by ranking affected areas based on their resource requirements identified using crowd-sourced information, and further facilitating multi-criteria decision making. Another work [13] delves into the analysis of spatio-temporal mobility patterns of affected populations, leading to the proposition of a dynamic demand estimation model specifically tailored for flood relief distribution. Similarly, in the context of earthquake relief, a study [14] has introduced a modelling framework that can allocate relief materials efficiently while considering cost-effectiveness parameters as well. During the recent Covid-19 pandemic, various spatio-temporal decision support systems have also been proposed. For instance, a study [15] has utilised a spatio-temporal model to identify the regions and populations with lower vaccination uptake rates for the initial, secondary, and booster doses, and thereby allowing a more targeted distribution strategy. Additionally, similar spatio-temporal models are suggested for establishing an adequate number of intensive care units [16], effective distribution of vaccines [17], comprehending the disease spread by analysing human mobility patterns to guide the reopening of various economic services [18], and for various other purposes. Despite all these efforts, we could not find a third-generation multi-purpose prioritisation platform that can help in mobilising high-demand and constrained resources, and also explain the reasons thereof.

It is worth noting that the initial wave of intelligent systems primarily concentrated on developing expertise in specific domains, such as chess playing, finding optimal transportation routes etc. These systems lacked flexibility and were designed to solve a particular problem only [6]. There was a significant advancement in the second generation due to their capability of learning and applying knowledge to similar tasks, such as predicting unknown observations, grouping objects, finding associations etc. This generation has laid out a strong foundation for artificial intelligence. However, a notable limitation of these second-generation intelligent machines is their inability to provide explanations against the outputs [19, 20]. Consequently, there is a growing need to enhance these systems to make them more explainable and trustworthy. It is important to mention here that efficiency and explainability represent two contrasting aspects, often necessitating a trade-off between them [21]. Any modification in existing intelligent algorithms to extract explanations may potentially reduce their efficiency due to this trade-off. We, therefore, believe that a completely different approach is needed to address this problem.

Significantly, we have introduced a novel wheel-based machine learning approach recently that leverages the concept of wheel rotation in a bicycle race to classify an unknown observation [22–24]. This innovative approach not only demonstrates efficiency in supervised learning but, unlike traditional approaches, also possesses the ability to provide further explanations [23]. We have utilised the wheel-based approach in this work to develop an explainable spatio-temporal platform primarily due to this reason. Instead of using the wheel-based approach for supervised learning, we have employed it for unsupervised clustering of administrative regions into several distinct zones. The higher priority zone includes the regions that require more attention concerning the regions with comparatively lower priority. Notably, the proposed platform.

- regularly updates priority recommendations based on the recent trend of various factors
- explicitly explains the assigned priority of each region, and
- detects the regions that fall under high-priority zone but are not receiving sufficient supplies.

In order to assess the capabilities of this platform, we have conducted a case study focused on the Covid-19 vaccine distribution in India during the first six months while it was also going through the second wave. Intriguingly, the proposed platform demonstrates the ability to prioritise regions based on their substantial caseload, but with a relatively smaller proportion of the population. It is important to mention here that the platform presented in this chapter is an extension of our previous work [22] where we tried to address a specific problem only. The platform proposed here is the generic version that can be used for multiple purposes.

In this chapter, Sect. 13.2 first introduces the proposed spatio-temporal prioritisation platform. We have presented a case study in Sect. 13.3, illustrating how the proposed platform can aid in the mobilisation of Covid-19 vaccines during a period of high demand but limited supply. Section 13.4 thereafter discusses the effectiveness of the platform in the context of Covid-19 vaccine distribution. Finally, Sect. 13.5 provides the concluding remarks for this work.

## 13.2 The Prioritisation Platform

Before introducing the proposed platform, let us briefly discuss the wheelbased approach used here for clustering. It was initially developed for classification where each wheel represents a particular class. In independent trials, these wheels rotate with different angular velocities due to the application of randomly chosen elementary forces, each representing the impact of a deciding factor. The cumulative angular velocity in multiple such trials finally helped us in classification. Besides, the measurable impact of every factor helped in explaining the classification as well. Importantly, we measured these impacts based on the training knowledge while using wheel-based approach for classification. However, clustering is an unsupervised learning technique that does not possess any such knowledge beforehand. In this work, we have refurbished the wheel-based approach for clustering to facilitate the grouping of multiple regions into a few predefined priority zones.

In order to facilitate the grouping of various regions, the proposed platform organises various functionalities into three distinct layers: (i) data, (ii) prioritisation, and (iii) presentation (refer Fig. 13.1). The data layer assumes the primary role of collecting relevant data from various sources, processing it, and storing it in a local repository. The prioritisation layer, empowered with the wheel-based approach, takes the data from the repository at regular intervals and recommends the priority along with the corresponding explanations. The data layer also maintains these recommendations and explanations in its' repository. The presentation layer, on request from the user, shows the priority recommendations and corresponding explanations of each administrative region.

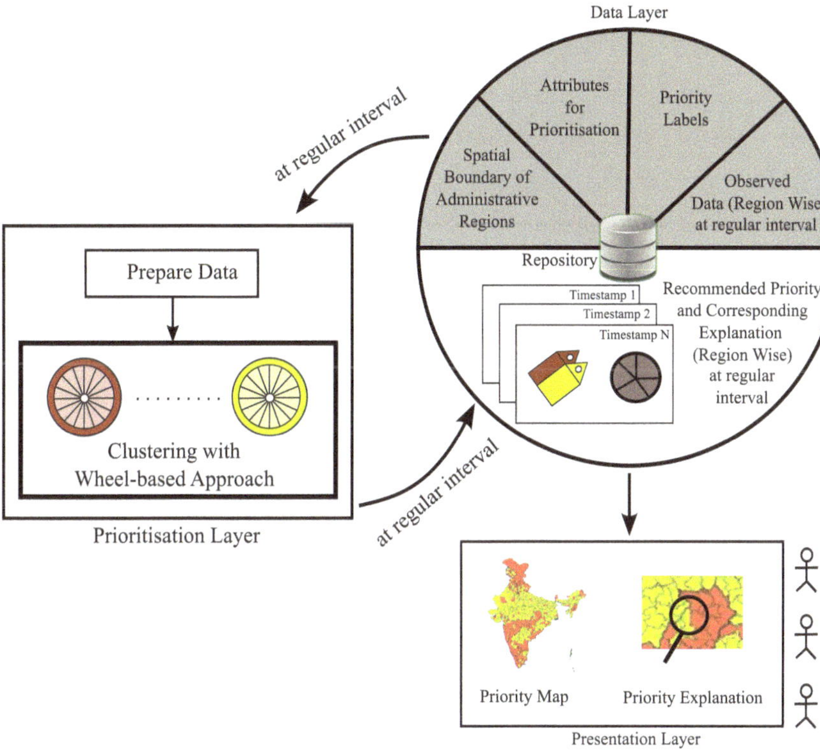


**Fig. 13.1** Proposed prioritisation platform

### *13.2.1 Data Layer*

The proposed prioritisation platform incorporates this layer to manage two types of data. The first type enables customisation of the platform for multipurpose usage and it requires pre-configuration. The second type is presentable data, either collected from diverse sources based on the defined configuration or generated after processing the collected data. The proposed platform employs a generic schema to organise these two types of data in structured format (refer Table 13.1).

The relation region here is responsible for storing the spatial information pertaining to the relevant administrative areas. This schema can handle hierarchical relationships, with the option to specify a parent-region-id for sub-areas within larger regions. On the other hand, the attribute relation contains information associated with various parameters to be used for prioritisation. The data-endpoint-url here defines the web-service that can be utilised for getting the corresponding value at a certain time. Additionally, the proposed platform mandates the inclusion of a response variable for prioritisation, which is indicated by the is-the-response property. The aim of the proposed platform is to assign a predefined priority label to each region. The

**Table 13.1** Schema for managing information at the data layer

| Configurable data | | |
|---|---|---|
| Region | Attribute | Label |
| (r1) region-id | (a1) attribute-id | (l1) label-id |
| (r2) name | (a2) attribute-name | (l2) label-name |
| (r3) parent-region-id | (a3) data-endpoint-url | (l3) rank |
| (r4) centroid | (a4) is-the-response | (l4) colour |
| (r4) centroid | | |
| (r5) boundary | | |
| *Collected and processed data* | | |
| Observation | Priority | Explanation |
| (o1) region-id | (p1) region-id | (e1) region-id |
| (o2) obs-timestamp | (p2) priority-date | (e2) priority-date |
| (o3) attribute-id | (p3) priority-label | (e3) attribute-id |
| (o4) observed-value | | (e4) contribution |

[keys] region: r1; attribute: a1; label: l1; observation: o1, o2, o3; priority: p1, p2; explanation: e1, e2, e3
[dependencies] r3 → r1, o1 → r1, o3 → a1, p1 → r1, p3 → l1, e1 → r1, e3 → a1

label relation within this schema maintains a comprehensive list of all priority labels along with their corresponding ranks and colours for visualisation.

The platform expects the above three relations to be configured properly so that it can start fetching the observed data from the endpoints and process them for priority assignment at regular intervals. The observation and priority store these two pieces of information respectively. Besides, explanation maintains the contribution of each attribute while determining the priority of each region. Furthermore, it is worth noting that the platform requires three additional configurations: (i) the interval at which attribute values should be fetched from endpoints, (ii) the interval at which region priorities should be measured, and (iii) the level of the region hierarchy that should be considered when measuring priorities.

## *13.2.2 Prioritisation Layer*

As mentioned earlier, the proposed approach requires a response variable for prioritisation. The number of volunteers to be deployed per thousand population, for example, could be one such response variable in case of flood relief. It would be a quite straightforward distribution of resources if only one parameter determines the response. However, multiple factors can stimulate the response in reality. Factors like reachability, availability of food, water, medicine, affected number of children, seniors, or their combinations, for instance, can influence the flood relief operation.

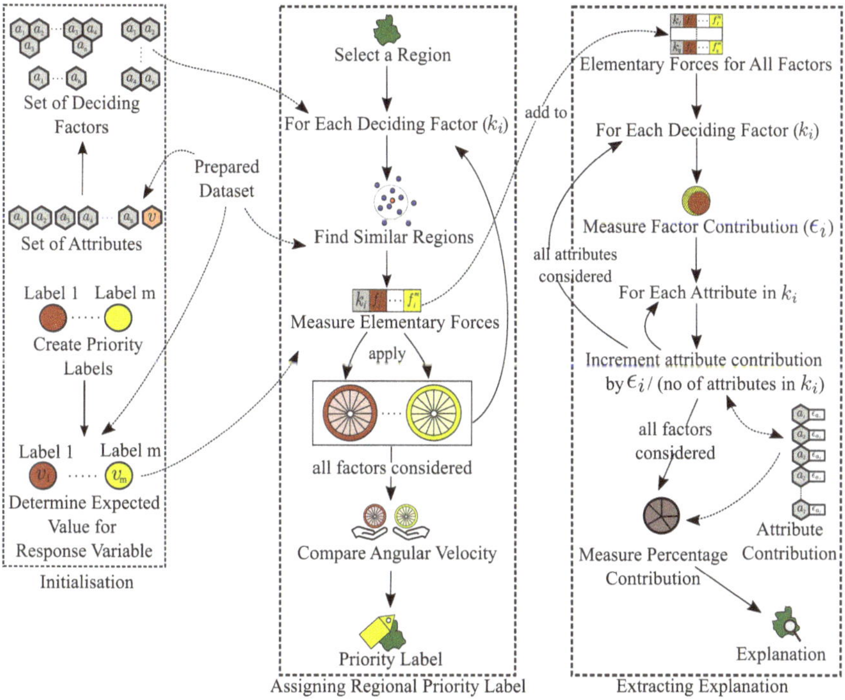


**Fig. 13.2** Clustering methodology adopted for assigning priority labels and extracting corresponding explanations

In order to handle this multivariate prioritisation, we have first assigned an estimated priority label against each region based on the distribution till the previous phase, assuming that the previous distribution aligns with the priorities at that time. Such an approach can lead to a cold-start problem during the first phase of distribution. However, it can be managed by estimating the response variable based on a single parameter and considering the initial availability of resources. This estimation has helped us in determining the expected value of the response variable for each priority zone. We have, thereafter, tried to measure the closeness of a particular region towards each priority zone based on different factors independently, and finally determine the actual priority label of the region. As presented in Fig. 13.2, this overall process starts with the initialisation of data, followed by determining the priorities and finally extracting explanations.

#### 13.2.2.1 Initialisation

In this initial step, the proposed platform processes the collected data at regular intervals and prepares them in a tabular format for prioritisation. Each row in this tabular data represents a region whereas the first $n$ columns specify $n$ attributes ($a_1$,

$a_2, a_3, ..., a_n$) and the last column indicates the response variable ($v$). As presented in Fig. 13.2, the wheel-based clustering process starts with the initialisation of the prepared dataset. It is important to mention here that the wheel-based approach is capable of handling these $n$ attributes and their different combinations as deciding factors. The set of deciding factors is therefore chosen first during the initialisation step. Besides, we have also initialised the expected value of the response variable in different priority zones. The value of the response variable in higher-priority zones is expected to be greater than that of lower-priority zones. We have therefore arranged the prepared dataset in descending order of response variable and divided it into $m$ almost equal parts, each representing one of the $m$ zones. The average value of the response variable in $j$ th part has helped us to determine the expected value of the response variable in $j$ th zone ($v_j$) here.

#### 13.2.2.2 Priority Modelling

In order to determine the priority of a region ($R$), the proposed platform first filters out the neighbouring regions concerning to each deciding factor considered here. The neighbours are not those regions which share a boundary with $R$. The platform rather considers regions with similar trends as neighbours to ensure a balanced distribution of resources across all regions. Concerning the factor $\{a_1, a_2\}$, for example, we have considered all those regions as neighbours of $R$ where the closer value of both $a_1$ and $a_2$ are observed. The margin ($\lambda$) of the wheel-based approach has helped us in this regard. Let the observed value of $a_1$ is $l_{a_1}$ and $a_2$ is $l_{a_2}$ in $R$. In this case, the platform considers all regions as neighbours where the value of $a_1$ and $a_2$ lies within $[l_{a_1} - \lambda\sigma_{a_1}, l_{a_1} + \lambda\sigma_{a_1}]$ and $[l_{a_2} - \lambda\sigma_{a_2}, l_{a_2} + \lambda\sigma_{a_2}]$ respectively. The $\sigma_{a_1}$ and $\sigma_{a_2}$ are the corresponding standard deviation here. Besides, the platform considers the value of $\lambda$ as 0.5 here.

The proposed platform measures $m$ elementary forces corresponding to each deciding factor in the next step and applies them on appropriate wheels representing various priority labels. Say $k_i$ is one such deciding factor and $\overline{v_i}$ is the average value of the response variable in the neighbouring regions of $R$ corresponding to $k_i$. If the elementary force on the wheel representing $j$ th priority label is $f_i^j$, then it can be measured using Eq. 13.1 as follows:

$$f_i^j = v_{max} - \left|\overline{v_i} - v_j\right| \tag{13.1}$$

Here $v_{max}$ is the upper limit of the response variable that can be determined either as per the availability of the resources or based on various policies. Once the elementary forces corresponding to all deciding factors are measured, the proposed platform aggregates them to determine the resultant forces for two wheels. Say, $F^j$ is the resultant force applied on the $j$ th wheel, and it is the aggregate of all elementary forces (refer Eq. 13.2).

$$F^j = \sum_{\forall i} f_i^j \tag{13.2}$$

The application of these resultant forces makes the wheels rotate with different angular velocities. The label corresponding to the wheel with highest angular velocity finally determines the priority label of the selected region.

#### 13.2.2.3 Explanation

The proposed platform considers all elementary forces to extract the explanation of assigned priority here. As presented in Fig. 13.2, it first measures the contribution of each factor in the final decision for this purpose. Let the contribution of factor $k_i$ is $\in_i$, and the corresponding elementary forces in wheels with highest and second-highest angular velocity are $\dot{f}_i$ and $\ddot{f}_i$ respectively. We have measured the $\in_i$ as per Eq. 13.3.

$$\in_i = \dot{f}_i - \ddot{f}_i \tag{13.3}$$

On availability of $\in_i$, the proposed platform decides the contribution of each attribute in $k_i$. It assumes that all constituting attributes are equally important and therefore divides the contribution uniformly among them. The aggregation of all such contributions in multiple deciding factors helped us in determining the overall contribution of each attribute here. It finally measures the percentage contribution of them to explain assigned priority.

### *13.2.3 Presentation Layer*

In this layer, the proposed platform uses a web-based dashboard to present the responsible recommendations to end-users. The dashboard incorporates three main features: (i) spatio-temporal priority maps, (ii) detailed tabular views, and (iii) explanations of the priority assigned to each region. We have provided sample illustrations of these features in the case study presented in Sect. 13.3. The spatio-temporal priority map delineates region boundaries based on the definitions stored in the region relation of the data layer and applies appropriate colour to these regions. This map enables policymakers to visualise the geospatial spread of the disaster and observe how the pattern evolves over time. Additionally, the tabular view presents more detailed information that includes the observed value of various attributes, the response variable, and the assigned priority label of each region. This detailed tabular view provides comprehensive data for further analysis and decision-making.

Besides, the dashboard explains each individual recommendation, thereby facilitating the policymakers to understand why a particular priority label is assigned to a region. Notably, the explanation includes the contribution of each attribute in the priority recommendation process. In order to foster more trust, the explanation also

**Table 13.2** List of attributes in the prepared dataset

| Attribute | Description |
|---|---|
| Confirmed per million | Confirmed cases in the region/population in million |
| Active per million | Active cases in the region/population in million |
| Deceased per million | Deceased patients in the region/population in million |
| Current growth | {Confirmed−(recovered + deceased)} cases in last two weeks/14 |
| Population density | Population/area in sq. km. of the region |
| Vaccination ratio | Number of vaccine doses administered/population of the region |

provides expected values of the considered attributes based on the overall trend. The availability of these expected values empowers policymakers to validate the explanations and instil greater confidence in the recommendations provided by the proposed platform.

## 13.3 Vaccine Mobilisation: A Case Study

In order to validate the functionalities, we have carried out a case study of Covid-19 vaccine distribution in India using the proposed prioritisation platform. All districts of Indian states are considered as the administrative regions of interest for this purpose. Besides, we have used the following five attributes for prioritisation: (i) confirmed per million, (ii) active per million, (iii) deceased per million, (iv) current growth, and (v) population density. The first three attributes specify the impact of the COVID-19 disease whereas the current growth indicates the recent trend of disease spread in a region. The population density is considered here to accommodate the regional susceptibility. In addition, the vaccination ratio, the average number of vaccination doses administered per person, is considered as the response variable here. Please refer to Table 13.2 for details about these attributes. It is also important to mention here that we have tried to divide all these regions into two zones: (i) high or red priority zone, and (ii) low or yellow priority zone.

In this case study, we have relied on the Indian Census 2011[1] to acquire the population of various districts. Notably, a few of districts of India underwent further division after 2011. We have checked various Government of India websites (under.gov.in domain) to obtain the population of these newly created districts. Besides, the caseloads and vaccination statistics used for prioritisation are taken from Covid19India[2] provided data-services. Importantly, these data services use different codification patterns for uniquely identifying the districts. We have used the Google Geocoding API to map the data to various regions considered in this case study. This API provides the coordinate of the data-service specified location. The spatial

[1] https://censusindia.gov.in/.

[2] http://covid19india.org.

intersection of this coordinate with the spatial boundary has helped us in identifying the appropriate district here. Once this information is available, we have kept it in a separate external repository and created multiple web-services to provide data in the desired format for subsequent usage.

Besides, we have configured the platform to gather data from the aforementioned web-services daily and updated the priority labels of all districts every fifteen days. We have carried out these operations on data of six months starting from 1st February 2021 to 15th July 2021. Significantly, $v_{max}$ is 2 in this case as two doses are mandated in India for each individual at that time. In addition, the wheel-based clustering approach considers upto third level factors for measuring elementary forces. It is worth emphasising that we have conducted a single trial in the wheel-based clustering, treating all factors as equally important. However, the wheel-based approach allows multiple trials by selecting the best factors randomly. In such cases, it is crucial to measure the factor importance and assign appropriate weights, especially without any training knowledge.

In this paper, Fig. 13.3 presents the priority map for all those twelve sets of recommendation and Fig. 13.4 provides detailed information about the same. Figure 13.5 shows sample explanations for two neighbouring districts (West and South Tripura) in the state of Tripura. As depicted in Fig. 13.5, the red priority is recommended for West Tripura primarily due to higher active cases and deceased rate. These two attributes together contribute more than 70% to this recommendation. In contrast, the platform assigns yellow priority to South Tripura. The lower rate of deceased and confirmed cases are the primary reasons behind this and these contributed almost 68% to the decision. The expected value of the attributes, based on the national trend, are also displayed while explaining a priority recommendation for further validation.

## 13.4 Analysis and Discussion

During any relief operations with limited resources, it is a common practice to identify regions that experience a higher severity of impact while having a smaller proportion of the affected population. This approach can be particularly relevant in the case of a pandemic like Covid-19, where prioritised vaccination efforts in these regions can help in controlling the further spread of the disease and also lowering the hospitalisation rates. Moreover, the lower share of the national population in these regions can additionally reduce the initial requirement of vaccine doses to a significant extent. Such a strategy can indirectly reduce the pressure on vaccine production units as well. However, the question is *whether the proposed platform is capable of identifying those regions*?

In order to find an answer, we have analysed the population and number of confirmed, active, and deceased cases in the red and yellow zones. Figure 13.6 presents the observed population distribution and the proportion of confirmed, active, and deceased cases in red and yellow zones. It exhibits all such proportions separately on twelve occasions when the framework updates the recommendations. The first

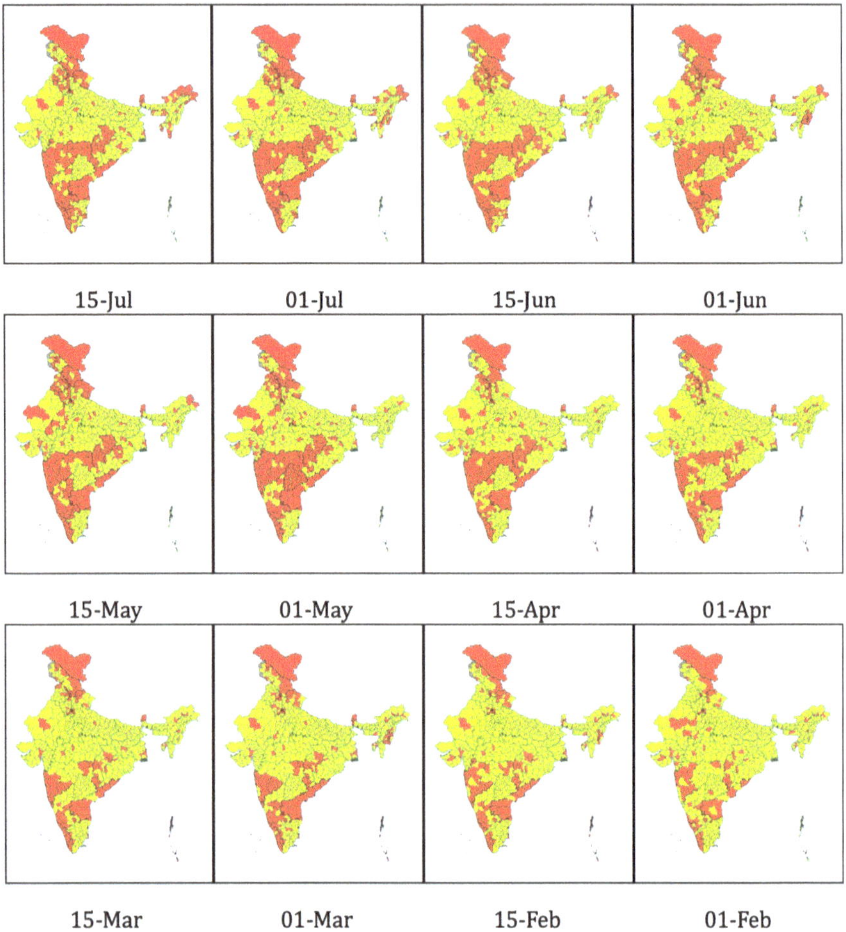


**Fig. 13.3** Priority maps of presentation layer

bar graph, showing the population distribution, clearly indicates that the population of red zone are significantly less than that of the yellow zone. The second, third and fourth bar graphs present the distribution of confirmed, active, and deceased cases respectively. It is quite evident that the confirmed, active and deceased cases are, in contrast, considerably higher in red zone and a similar trend is observed on almost all occasions. We have observed that the population of red zone is 40% whereas the confirmed, active or deceased cases is 70% during May 2021. Importantly, India was struggling with the peak of the second wave during that time.

Besides, we have also analysed the *correctness of the explanation* using (i) the contribution of an attribute in priority decision and (ii) the difference between expected and observed value of the corresponding attribute. We have presented sixty scatter plots in Fig. 13.7 for this purpose. Each row here represents one of

| # | State | District | Vaccination Ratio | Recommendation | Explanation |
|---|---|---|---|---|---|
| 1 | MAHARASHTRA | Mumbai | 1.4162 | Red | |
| 2 | LADAKH | Leh | 0.837 | Red | |
| 3 | LAKSHADWEEP | Lakshadweep | 0.758 | Red | |
| 4 | HARYANA | Gurugram | 0.7059 | Red | |
| 5 | TRIPURA | West Tripura | 0.6027 | Red | |
| 6 | LADAKH | Kargil | 0.5796 | Red | |
| 7 | PUDUCHERRY | Mahe | 0.5739 | Red | |
| 8 | WEST BENGAL | Kolkata | 0.5714 | Red | |
| 9 | TRIPURA | South Tripura | 0.563 | Yellow | |
| 10 | UTTAR PRADESH | Gautam Buddha Nagar | 0.5489 | Red | |

**Fig. 13.4** Detailed priority recommendations in presentation layer

the twelve different priority recommendation dates whereas each of the five columns corresponds to an attribute considered for prioritisation. The points on these plots represent different regions, with the colour indicating their recommended priority label. The horizontal axis of each plot indicates the difference between observed and expected attribute values. The vertical axis corresponds to the percentage contribution of the concerned attribute in deciding the priority label. The dashed vertical line in each plot is the reference level where no difference is noticed. Significantly, all regions plotted at the right side of this dashed vertical line observe a higher value than expectation and vice versa. The dashed horizontal line indicates the anticipated level of the contribution. Importantly, there are five input attributes considered here and all of them are supposed to contribute uniformly in the ideal cases. We have therefore considered 20% as the anticipated contribution of all attributes here.

In case of first four attributes, the majority of the red regions in Fig. 13.7 are at the right side of the dashed vertical line. We can therefore infer that the observed value of these attributes in red regions are generally greater than the expectation derived based on the national trend. Not only that, a significant proportion of the red regions are plotted above the dashed horizontal line; thus indicating a higher than anticipated contribution to the priority decision. The yellow regions, in contrast, are either towards the left or very close to the dashed vertical line. It suggests that the observed values in yellow regions are generally less than or very close to the national trend. However, the current growth attribute does not follow the same behaviour here. The pattern is similar to some extent during the month of April and May 2021 when India was going through the increasing trend of the second wave. However, the concentration of red and yellow regions overlapped significantly on other occasions. The more dominating nature of the other four attributes could be the reason behind this exceptional behaviour.

Red High priority should be given to West Tripura

Explanation

| # | Attribute | Expected Value (national trend for validation) | Observed Value | % Contribution in Decision |
|---|---|---|---|---|
| 1 | Active Per Million | 693 | 1996 | 50.52 |
| 2 | Deceased Per Million | 309 | 357 | 21.62 |
| 3 | Total Cases Per Million | 24137 | 26694 | 12.34 |
| 4 | Current Growth | -55 | -85 | 11.51 |
| 5 | Population Density | 368 | 974 | 4.00 |

(a) West Tripura

Yellow Less priority may be given to South Tripura

Explanation

| # | Attribute | Expected Value (national trend for validation) | Observed Value | % Contribution in Decision |
|---|---|---|---|---|
| 1 | Deceased Per Million | 309 | 83 | 38.57 |
| 2 | Total Cases Per Million | 24137 | 15637 | 30.35 |
| 3 | Current Growth | -27 | -28 | 20.49 |
| 4 | Population Density | 368 | 297 | 10.08 |
| 5 | Active Per Million | 693 | 1231 | 0.50 |

(b) South Tripura

**Fig. 13.5** Sample explanations of two neighbouring districts in Tripura

The proposed platform can also be used for identifying the variability of resource mobilisation across multiple regions and discovering the possible scope for further improvement. In case of vaccine distribution, the identification of such variability may help in detecting regions where the vaccination ratio can be increased at the cost of some other regions with lower severity. However, it is essential to know whether there is any scope available at all for such improvement. In order to analyse this, we have sorted the regions in descending order of the vaccination ratio and graphically represent them in Fig. 13.8. There are twelve plots here representing

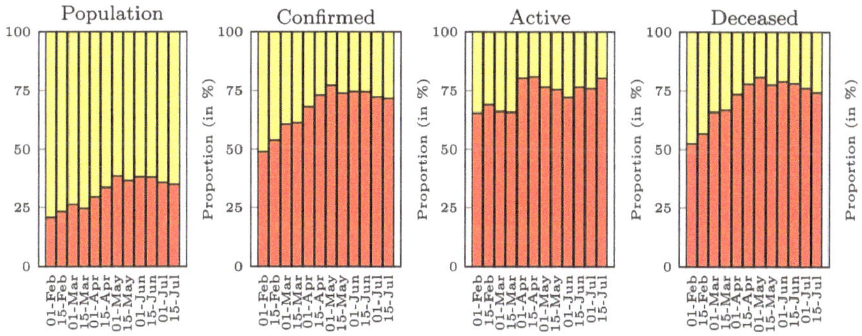


**Fig. 13.6** Proportion of population and COVID-19 cases in red and yellow zones

different recommendation dates. Each smallest square in these plots depicts a region whereas the colour of these smallest squares indicates the assigned priority label.

The proposed platform can also be used for identifying the variability of resource mobilisation across multiple regions and discovering the possible scope for further improvement. In case of vaccine distribution, the identification of such variability may help in detecting regions where the vaccination ratio can be increased at the cost of some other regions with lower severity. However, it is essential to know whether there is any scope available at all for such improvement. In order to analyse this, we have sorted the regions in descending order of the vaccination ratio and graphically represent them in Fig. 13.8. There are twelve plots here representing different recommendation dates. Each smallest square in these plots depicts a region whereas the colour of these smallest squares indicates the assigned priority label.

In ideal cases, the vaccination ratio of all red regions should be higher than the yellow labelled regions. They should therefore appear prior to yellow regions in all plots presented in Fig. 13.8. Although the majority of red regions appear towards the top of these plots, some yellow regions also coexist among them. Not only that, the vaccination ratio of some red regions turns out to be very low as they appear towards the bottom of these plots. We may use this matrix plot to identify the regions that require urgent attention.

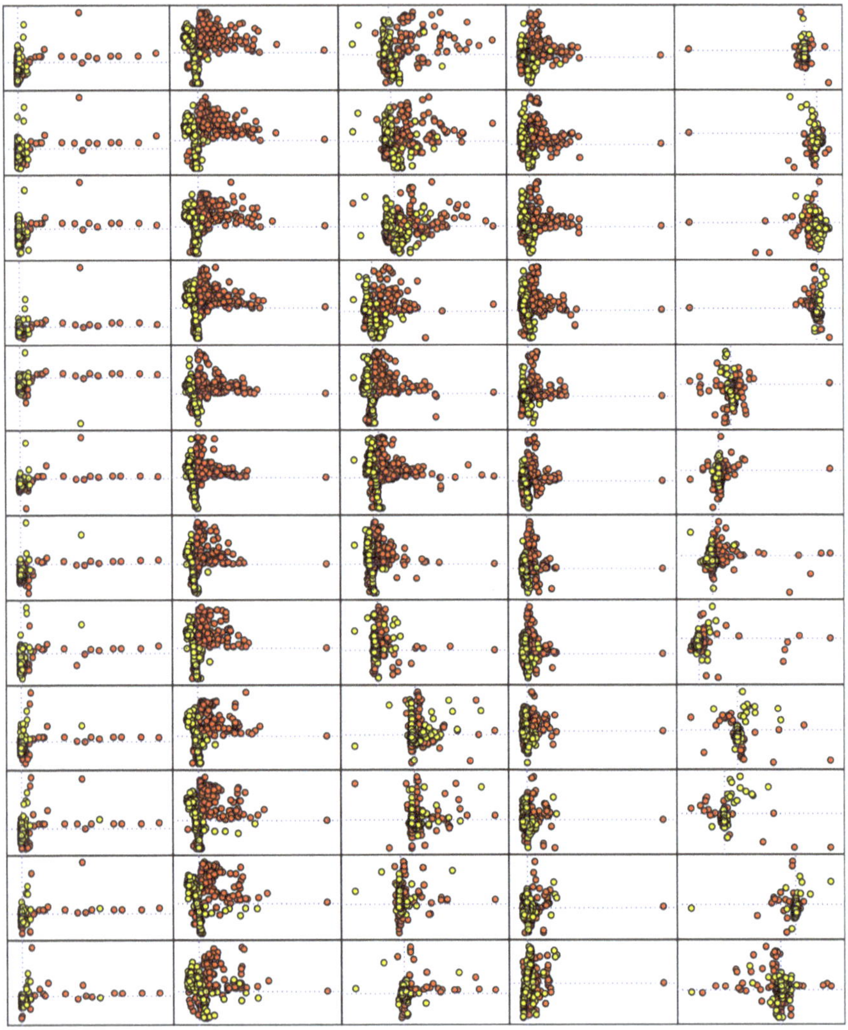

**Fig. 13.7** Scatter plots of attribute contribution (vertical axis) with respect to the difference between observed and expected attribute value (horizontal axis). Columns (left to right): (1) population density, (2) confirmed per million (3) active per million (4) deceased per million, (5) current growth. Rows (top to bottom): (1) 15-Jul-2021, (2) 01-Jul-2021, (3) 15-Jun-2021, (4) 01-Jun-2021, (5) 15-May-2021, (6) 01-May-2021, (7) 15-Apr-2021, (8) 01-Apr-2021, (9) 15-Mar-2021, (10) 01-Mar-2021, (11) 15-Feb-2021, (12) 01-Feb-2021

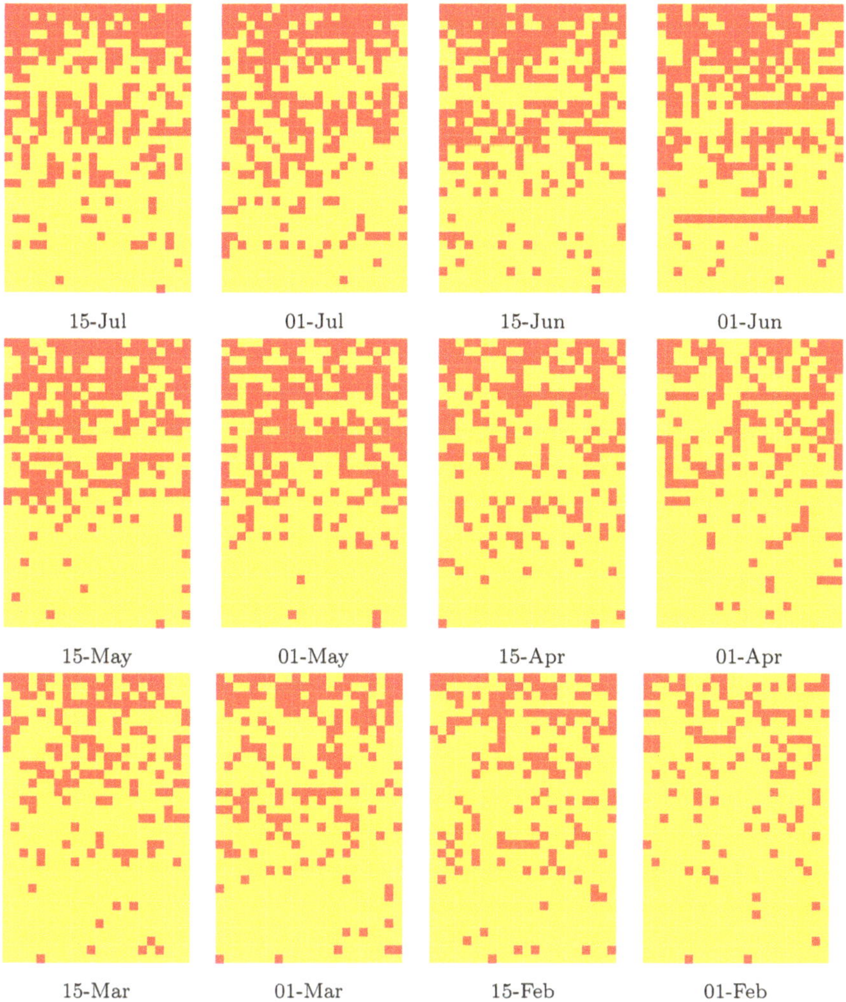


**Fig. 13.8** Recommended priority in descending order of vaccination ratio

## 13.5 Conclusion

In summary, we have proposed an explainable spatio-temporal platform in this paper to assist the mobilisation of high-demand and constrained resources, especially during emergencies like a pandemic, flood, cyclone or any widespread disaster. The proposed multi-purpose platform collects data from configured sources, processes and stores them in a local repository. On the availability of required data, it recommends the priority of various administrative regions at a scheduled interval. Besides, the platform explains each individual recommendation as well. It has a user-friendly dashboard which presents the regional priorities and explanations of each individual priority recommendation. Significantly, we have found that the proposed platform can be used for prioritising affected regions at regular intervals by considering multiple factors explaining the priorities and thereby assisting the policy-makers in making more informed decisions identifying the scope for further improvement in mobilising limited resources across multiple regions.

In this chapter, we have also presented a case study of Covid-19 vaccine distribution across various districts of India using this platform and analysed its' efficacy from various aspects. However, it is important to further analyse the efficacy of the proposed platform in other types of widespread disaster. Besides, we have limited the scope of this work to prioritising regions based on multiple factors and explaining the reasons only. An interesting future extension may involve recommending the allocation of limited resources to various regions of interest, indicating the percentage of resources to be allocated to each region. Besides, comparison of different explainable clustering methods could be another interesting future scope work that may be carried out on this prioritization framework. It is important to mention here that the developed framework is customizable and some other explainable model can always replace the wheel-based approach.

## References

1. Chaturvedi M (2021) Year ender 2021: Major natural disasters this year in India. https://www.indiatoday.in/information/story/year-ender-2021-major-natural-disasters-this-year-in-india-1893678-2021-12-29
2. Ghaffarian S, Taghikhah F (2023) Explainable artificial intelligence in disaster risk management: achievements and prospective futures. https://ssrn.com/abstract=4421109
3. Hayashi H, Asahara A, Sugaya N, Ogawa Y, Tomita H (2015) Spatio-temporal join technique for disaster estimation in large-scale natural disaster. In: Proceedings of the 6th ACM SIGSPATIAL International Workshop on GeoStreaming, pp 49–58
4. Sun W, Bocchini P, Davison BD (2020) Applications of artificial intelligence for disaster management. Nat Hazards 103(3):2631–2689
5. Benevolenza MA, DeRigne LA (2019) The impact of climate change and natural disasters on vulnerable populations: a systematic review of literature. J Hum Behav Soc Environ 29(2):266–281
6. Mercier-Laurent E (2020) The future of AI or AI for the future. In: Unimagined futures–ICT opportunities and challenges. Springer, pp 20–37
7. Rabta B, Wankmüller C, Reiner G (2018) A drone fleet model for last-mile distribution in disaster relief operations. Int J Disaster Risk Reduct 28:107–112
8. Sarma D, Das A, Dutta P, Bera UK (2020) A cost minimization resource allocation model for disaster relief operations with an information crowdsourcing-based MCDM approach. IEEE Trans Eng Manag 69(5):2454–2474
9. Ranjbar HR, Ardalan AA, Dehghani H, Saradjian MR (2018) Using high-resolution satellite imagery to provide a relief priority map after earthquake. Nat Hazards 90(3):1087–1113
10. Zhang Z, Yin D, Virrantaus K, Ye X, Wang S (2021) Modeling human activity dynamics: an object-class oriented space–time composite model based on social media and urban infrastructure data. Comput Urban Sci 1:1–13
11. Espejo-Dıaz JA, Guerrero WJ (2021) A multiagent approach to solving the dynamic postdisaster relief distribution problem. Oper Manag Res 14:177–193

12. Fikar C, Gronalt M, Hirsch P (2016) A decision support system for coordinated disaster relief distribution. Expert Syst Appl 57:104–116
13. Lin A, Wu H, Liang G, Cardenas-Tristan A, Wu X, Zhao C, Li D (2020) A big data-driven dynamic estimation model of relief supplies demand in urban flood disaster. Int J Disaster Risk Reduct 49:101682
14. Garrido RA, Aguirre I (2020) Emergency logistics for disaster management under spatio-temporal demand correlation: the earthquakes case. J Ind Manag Optim 16(5):2369–2387
15. Samany NN, Liu H, Aghataher R, Bayat M (2022) Ten GIS-based solutions for managing and controlling COVID-19 pandemic outbreak. SN Comput Sci 3(4):269
16. Guzzi PH, Tradigo G, Veltri P (2020) Spatio-temporal resource mapping for intensive care units at regional level for COVID-19 emergency in Italy. Int J Environ Res Public Health 17(10):3344
17. Grauer J, Löwen H, Liebchen B (2020) Strategic spatiotemporal vaccine distribution increases the survival rate in an infectious disease like Covid-19. Sci Rep 10(1):21594
18. Bao H, Zhou X, Zhang Y, Li Y, Xie Y (2020) Covid-gan: Estimating human mobility responses to covid-19 pandemic through spatio-temporal conditional generative adversarial networks. In: Proceedings of the 28th international conference on advances in geographic information systems, pp 273–282
19. Adadi A, Berrada M (2018) Peeking inside the blackbox: a survey on explainable artificial intelligence (XAI). IEEE Access 6:52138–52160
20. Tjoa E, Guan C (2020) A survey on explainable artificial intelligence (XAI): toward medical XAI. IEEE Trans Neural Netw Learn Syst
21. Hall P, Gill N (2019) An introduction to machine learning interpretability. O'Reilly Media, Incorporated
22. Khan A, Ghosh SK (2021) ExPrt: an explainable spatio-temporal framework for prioritizing the COVID-19 vaccination in India. Available at SSRN 3909772
23. Khan A, Ghosh SK (2023) Machine assistance for credit approval? Random wheel can recommend and explain. Expert Syst Appl 215:119231
24. Khan A, Ghosh SK, Ghosh D, Chattopadhyay S (2021) Random wheel: an algorithm for early classification of student performance with confidence. Eng Appl Artif Intell 102

# Correction to: XAI-Driven Analysis of Emerging Hegemonic Trends in Great Power Development Strategies Incorporating Fuzzy Logic

**Mădălin Ciprian Enescu, Cosmin-George Nicolăescu, Clementina Elena Niță, and Cristina Claudia Bizon**

## Correction to:
## Chapter 8 in: N. Bizon and B. Appasani (eds.), *Explainable Artificial Intelligence for Trustworthy Decisions in Smart Applications*, https://doi.org/10.1007/978-3-031-97007-8_8

In the original version of Chapter 8, the author name has been updated from Clementina Elena Nicolăescu to Clementina Elena Niță. The chapter has been updated with the change.

The updated version of this chapter can be found at
https://doi.org/10.1007/978-3-031-97007-8_8

 
N. Bizon and B. Appasani (eds.), *Explainable Artificial Intelligence for Trustworthy Decisions in Smart Applications*, https://doi.org/10.1007/978-3-031-97007-8_14

FSC
www.fsc.org
MIX
Papier | Fördert
gute Waldnutzung
FSC® C083411

Kolibri 360 GmbH
Ferdinand-Jühlke-Straße 7
99095 Erfurt, Deutschland
produktsicherheit@kolibri360.de